Finite Element Analysis of Composite Materials

Ever J. Barbero

Dedicado a mis padres, Sonia Eulalia y Ever Francisco.

Contents

Preface

Finite Element Analysis of Composite Materials deals with the analysis of structures made of composite materials, also called *composites*. The analysis of composites treated in this textbook includes the analysis of the material itself, at the micro-level, and the analysis of structures made of composite materials. This textbook evolved from the class notes of *MAE 646 Advanced Mechanics of Composite Materials* that I teach as a graduate course at West Virginia University. Although this is also a textbook on advanced mechanics of composite materials, the use of the finite element method is essential for the solution of the complex boundary value problems encountered in the advanced analysis of composites, and thus the title of the book.

There are a number of good textbooks on advanced mechanics of composite materials, but none carries the theory to a practical level by actually solving problems, as it is done in this textbook. Some books devoted exclusively to finite element analysis include some examples about modeling composites but fall quite short of dealing with the actual analysis and design issues of composite materials and composite structures. This textbook includes an explanation of the concepts involved in the detailed analysis of composites, a sound explanation of the mechanics needed to translate those concepts into a mathematical representation of the physical reality, and a detailed explanation of the solution of the resulting boundary value problems by using commercial Finite Element Analysis software such as ANSYS^{TM}. Furthermore, this textbook includes more than fifty fully developed examples interspersed with the theory, as well as more than seventy-five exercises at the end of chapters, and more than fifty separate pieces of ANSYS code used to explain in detail the solution of example problems. The reader will be able to reproduce the examples and complete the exercises. When a finite element analysis is called for, the reader will be able to do it with commercially or otherwise available software. A Web site is set up with links to download the necessary software unless it is easily available from Finite Element Analysis software vendors. ANSYS and MATLAB^{TM} code is explained in the examples, and the code can be downloaded from the Web site as well. Furthermore, the reader will be able to extend the capabilities of ANSYS by use of ANSYS Parametric Design Language (APDL), user material subroutines, and programmable postprocessing, as demonstrated in the examples included in this textbook.

Chapters 1 through 8 can be covered in a one-semester graduate course. Chapter 2 (Introduction to the Finite Element Method) contains a brief introduction intended for those readers who have not had a formal course or prior knowledge about the finite element method. Chapter 4 (Buckling) is not referenced in the remainder of the textbook and thus it could be omitted in favor of more exhaustive coverage of content in later chapters. Chapters 7 (Viscoelasticity) and 8 (Damage Mechanics) are placed consecutively to emphasize hereditary phenomena. However, Chapter 7 can be skipped if more emphasis on damage and/or delaminations is desired in a one-semester course. A complete continuum damage model with coupled damage-plasticity effects is presented in Chapter 9, immediately following the more fundamental treatment of damage in Chapter 8. Chapter 9 could be omitted for the sake of time, especially if the instructor desires to cover Chapter 10 (Delaminations) as part of a one-semester course.

The *inductive method* is applied as much as possible in this textbook. That is, topics are introduced with examples of increasing complexity, until sufficient physical understanding is reached to introduce the general theory without difficulty. This method will sometimes require that, at earlier stages of the presentation, certain facts, models, and relationships be accepted as fact, until they are completely proven later on. For example, in Chapter 7, viscoelastic models are introduced early to aid the reader in gaining an appreciation for the response of viscoelastic materials. This is done simultaneously with a cursory introduction to the superposition principle and the Laplace transform, which are formally introduced only later in the chapter. For those readers accustomed to the *deductive method*, this may seem odd, but many years of teaching have convinced me that students acquire and retain knowledge more efficiently in this way.

It is assumed that the reader is familiar with basic mechanics of composites as covered in introductory level textbooks such as my previous textbook *Introduction to Composite Material Design*. Furthermore, it is assumed that the reader masters a body of knowledge that is commonly acquired as part of a bachelor of science degree in any of the following disciplines: Aerospace, Mechanical, Civil, or similar. References to books and to other sections in this textbook, as well as footnotes are used to assist the reader in refreshing those concepts and to clarify the notation used. Prior knowledge of continuum mechanics, tensor analysis, and the finite element method would enhance the learning experience but are not necessary for studying with this textbook. The finite element method is used as a tool to solve practical problems. For the most part, ANSYS is used throughout the book. Finite element programming is limited to programming material models, post-processing algorithms, and so on. Basic knowledge of MATLAB is useful but not essential.

Only three software packages are used throughout the book. ANSYS is needed for finite element solution of numerous examples and suggested problems. MAT-LAB is needed for both symbolic and numerical solution of numerous examples and suggested problems. Additionally, BMI3^{TM}, which is available free of charge on the book's Web site, is used in Chapter 4. Several other programs are also mentioned, such as ABAQUS^{TM}and LS-DYNA^{TM} but they are not used in the examples. All the code printed in the examples is available on the book's Web site http://www.mae.wvu.edu/barbero/feacm/.

Composite materials are now ubiquitous in the marketplace, including extensive applications in aerospace, automotive, civil infrastructure, sporting goods, and so on. Their design is especially challenging because, unlike conventional materials such as metals, the composite material itself is designed concurrently with the composite structure. Preliminary design of composites is based on the assumption of a state of plane stress in the laminate. Furthermore, rough approximations are made about the geometry of the part, as well as the loading and support conditions. In this way, relatively simple analysis methods exist and computations can be carried out simply using algebra. However, preliminary analysis methods have a number of shortcomings that are remedied with advanced mechanics and finite element analysis, as explained in this textbook. Recent advances in commercial finite element analysis packages, with user friendly pre- and post-processing, as well as powerful user-programmable features, have made detailed analysis of composites quite accessible to the designer. This textbook bridges the gap between powerful finite element tools and practical problems in structural analysis of composites. I expect that many graduate students, practicing engineers, and instructors will find this to be a useful and practical textbook on finite element analysis of composite materials based on sound understanding of advanced mechanics of composite materials.

Ever J. Barbero

Acknowledgments

First and foremost, I wish to thank Joan Andreu Mayugo Majo, Domenico Bruno, and Fabrizio Greco for writing Chapters 9 and 10. Chapter 6 is largely based on the work of Raimondo Luciano and Elio Sacco, to whom I am indebted. I wish to thank Tom Damiani and Joan Andreu Mayugo Majo, who taught the course in 2004 and 2006, respectively, both making many corrections and additions to the course notes on which this textbook is based. Acknowledgement is due to those who reviewed the manuscript including Enrique Barbero, Guillermo Creus, Luis Godoy, Paolo Lonetti, Severino Marques, Pizhong Qiao, Timothy Norman, Sonia Sanchez, and Eduardo Sosa. Furthermore, recognition is due to those who helped me compile the solutions manual, including Hermann Alcazar, John Sandro Rivas, and Rajiv Dastane, among others. Also, I wish to thank my colleagues and students for their valuable suggestions and contributions to this textbook. Finally, my gratitude goes to my wife, Ana Maria, and to my children, Margaret and Daniel, who gave up many opportunities to bond with their dad so that I might write this book.

List of Symbols

Symbols Related to Mechanics of Orthotropic Materials

$\boldsymbol{\epsilon}$	Strain tensor
ε_{ij}	Strain components in tensor notation
ϵ_α	Strain components in contracted notation
ϵ_α^e	Elastic strain
ϵ_α^p	Plastic strain
λ	Lame constant
ν	Poisson's ratio
ν_{12}	Inplane Poisson's ratio
ν_{23}, ν_{13}	Interlaminar Poisson's ratios
ν_{xy}	Apparent laminate Poisson ratio x-y
$\boldsymbol{\sigma}$	Stress tensor
σ_{ij}	Stress components in tensor notation
σ_α	Stress components in contracted notation
$[a]$	Transformation matrix for vectors
e_i	Unit vector components in global coordinates
e_i'	Unit vector components in materials coordinates
f_i, f_{ij}	Tsai-Wu coefficients
k	Bulk modulus
l, m, n	Direction cosines
$\widetilde{u}(\varepsilon_{ij})$	Strain energy per unit volume
u_i	Displacement vector components
x_i	Global directions or axes
x_i'	Materials directions or axes
$\mathbf{C}$	Stiffness tensor
C_{ijkl}	Stiffness in index notation
$C_{\alpha,\beta}$	Stiffness in contracted notation
E	Young's modulus
E_1	Longitudinal modulus
E_2	Transverse modulus
E_2	Transverse-thickness modulus
E_x	Apparent laminate modulus in the global x-direction
$G = \mu$	Shear modulus
G_{12}	Inplane shear modulus

G_{23}, G_{13}	Interlaminar shear moduli
G_{xy}	Apparent laminate shear modulus x-y
I_{ij}	Second-order identity tensor
I_{ijkl}	Fourth-order identity tensor
Q'_{ij}	Lamina stiffness components in material coordinates
$[R]$	Reuter matrix
S	Compliance tensor
S_{ijkl}	Compliance in index notation
$S_{\alpha,\beta}$	Compliance in contracted notation
$[T]$	Coordinate transformation matrix for stress
$[\overline{T}]$	Coordinate transformation matrix for strain

Symbols Related to Finite Element Analysis

$\underline{\underline{\partial}}$	Strain-displacement equations in matrix form
$\underline{\epsilon}$	Six-element array of strain components
$\theta_x, \theta_y, \theta_z$	Rotation angles following the right-hand rule (Figure 2.9)
$\underline{\sigma}$	Six-element array of stress components
ϕ_x, ϕ_y	Rotation angles used in plate and shell theory
$\underline{a}$	Nodal displacement array
u_j^e	Unknown parameters in the discretization
$\underline{\underline{B}}$	Strain-displacement matrix
$\underline{\underline{C}}$	Stiffness matrix
$\underline{\underline{K}}$	Assembled global stiffness matrix
$\underline{\underline{K}}^e$	Element stiffness matrix
$\underline{N}$	Interpolation function array
N_j^e	Interpolation functions in the discretization
$\underline{P}^e$	Element force array
$\underline{P}$	Assembled global force array

Symbols Related to Elasticity and Strength of Laminates

γ_{xy}^0	Inplane shear strain
γ_{4u}	Ultimate interlaminar shear strain in the 2-3 plane
γ_{5u}	Ultimate interlaminar shear strain in the 1-3 plane
γ_{6u}	Ultimate inplane shear strain
$\epsilon_x^0, \epsilon_y^0$	Inplane strains
ϵ_{1t}	Ultimate longitudinal tensile strain
ϵ_{2t}	Ultimate transverse tensile strain
ϵ_{3t}	Ultimate transverse-thickness tensile strain
ϵ_{1c}	Ultimate longitudinal compressive strain
ϵ_{2c}	Ultimate transverse compressive strain
ϵ_{3c}	Ultimate transverse-thickness compressive strain
κ_x, κ_y	Bending curvatures

κ_{xy}	Twisting curvature
ϕ_x, ϕ_y	Rotations of the middle surface of the shell (Figure 2.9)
c_4, c_5, c_6	Tsai-Wu coupling coefficients
t_k	Lamina thickness
u_0, v_0, w_0	Displacements of the middle surface of the shell
z	Distance from the middle surface of the shell
A_{ij}	Components of the extensional stiffness matrix $[A]$
B_{ij}	Components of the bending-extension coupling matrix $[B]$
D_{ij}	Components of the bending stiffness matrix $[D]$
$[E_0]$	Extensional stiffness matrix $[A]$, in ANSYS notation
$[E_1]$	Bending-extension matrix $[B]$, in ANSYS notation
$[E_2]$	Bending stiffness matrix $[D]$, in ANSYS notation
F_{1t}	Longitudinal tensile strength
F_{2t}	Transverse tensile strength
F_{3t}	Transverse-thickness tensile strength
F_{1c}	Longitudinal compressive strength
F_{2c}	Transverse compressive strength
F_{3c}	Transverse-thickness compressive strength
F_4	Interlaminar shear strength in the 2-3 plane
F_5	Interlaminar shear strength in the 1-3 plane
F_6	Inplane shear strength
H_{ij}	Components of the interlaminar shear matrix $[H]$
I_F	Failure index
M_x, M_y, M_xy	Moments per unit length (Figure 3.3)
$\widehat{M_n}$	Applied bending moment per unit length
N_x, N_y, N_xy	Inplane forces per unit length (Figure 3.3)
$\widehat{N_n}$	Applied inplane force per unit length, normal to the edge
$\widehat{N_{ns}}$	Applied inplane shear force per unit length, tangential
$\left(\overline{Q}_{ij}\right)_k$	Lamina stiffness components in global coordinates, layer k
V_x, V_y	Shear forces per unit length (Figure 3.3)

Symbols Related to Buckling

λ, λ_i	Eigenvalues
s	Perturbation parameter
Λ	Load multiplier
$\Lambda^{(cr)}$	Bifurcation multiplier or critical load multiplier
$\Lambda^{(1)}$	Slope of the post-critical path
$\Lambda^{(2)}$	Curvature of the post-critical path
v	Eigenvectors (buckling modes)
$[K]$	Stiffness matrix
$[K_s]$	Stress stiffness matrix
P_{CR}	Critical load

Symbols Related to Free Edge Stresses

$\eta_{xy,x}, \eta_{xy,y}$	Coefficients of mutual influence
$\eta_{x,xy}, \eta_{y,xy}$	Alternate coefficients of mutual influence
F_{yz}	Interlaminar shear force y-z
F_{xz}	Interlaminar shear force x-z
M_z	Interlaminar moment

Symbols Related to Micromechanics

$\bar{\epsilon}_\alpha$	Average engineering strain components
$\bar{\varepsilon}_{ij}$	Average tensor strain components
$\epsilon_\alpha^0, \varepsilon_{ij}^0$	Far-field applied strain components
$\bar{\sigma}_\alpha$	Average stress components
$\mathbf{A}^i$	Strain concentration tensor, i-th phase, contracted notation
$2a_1, 2a_2, 2a_3$	Dimensions of the RVE
A_{ijkl}	Components of the strain concentration tensor
$\mathbf{B}^i$	Stress concentration tensor, i-th phase, contracted notation
B_{ijkl}	Components of the stress concentration tensor
I	6×6 identity matrix
P_{ijkl}	Eshelby tensor
V_f	Fiber volume fraction
V_m	Matrix volume fraction

Symbols Related to Viscoelasticity

$\dot{\varepsilon}$	Stress rate
η	Viscosity
θ	Age or aging time
$\dot{\sigma}$	Stress rate
τ	Time constant of the material or system
Γ	Gamma function
s	Laplace variable
t	Time
$C_{\alpha,\beta}(t)$	Stiffness tensor in the time domain
$C_{\alpha,\beta}(s)$	Stiffness tensor in the Laplace domain
$\widehat{C}_{\alpha,\beta}(s)$	Stiffness tensor in the Carson domain
$D(t)$	Compliance
$D_0, (D_i)_0$	Initial compliance values
$D_c(t)$	Creep component of the total compliance $D(t)$
D', D''	Storage and loss compliances
$E_0, (E_i)_0$	Initial moduli
E_∞	Equilibrium modulus
E, E_0, E_1, E_2	Parameters in the viscoelastic models (Figure 7.1)

$E(t)$	Relaxation
E', E''	Storage and loss moduli
$F[]$	Fourier transform
$(G_{ij})_0$	Initial shear moduli
$H(t - t_0)$	Heaviside step function
$H(\theta)$	Relaxation spectrum
$L[]$	Laplace transform
$L[]^{-1}$	Inverse Laplace transform

Symbols Related to Damage

α_{cr}	Critical misalignment angle at longitudinal compression failure
$\gamma(\delta)$	Damage hardening function
γ_0	Damage threshold
δ_{ij}	Kronecker delta
δ	Damage hardening variable
ε	Effective strain
$\bar{\varepsilon}$	Undamaged strain
ε^p	Plastic strain
$\dot{\gamma}$	Heat dissipation rate per unit volume
$\dot{\gamma}_s$	Internal entropy production rate
$\dot{\lambda}, \dot{\lambda}^d$	Damage multiplier
$\dot{\lambda}^p$	Yield multiplier
σ	Effective stress
$\bar{\sigma}$	Undamaged stress
φ, φ^*	Strain energy density, and complementary SED
χ	Gibbs energy density
ψ	Helmholtz free energy density
Λ	Standard deviation of fiber misalignment
$\boldsymbol{\Omega} = \Omega_{ij}$	Integrity tensor
d_i	Eigenvalues of the damage tensor
f^d	Damage flow surface
f^p	Yield flow surface
$f(x), F(x)$	Probability density, and its cumulative probability
g^d	Damage surface
g^p	Yield surface
m	Weibull modulus
p	Yield hardening variable
$u(\varepsilon_{ij})$	Internal energy density
A_{ijkl}	Tension-compression damage constitutive tensor
B_{ijkl}	Shear damage constitutive tensor
B_a	Dimensionless number (8.57)
$\overline{C}_{\alpha,\beta}$	Stiffness matrix in the undamaged configuration
$\mathbf{C}^{ed}$	Tangent stiffness tensor

D_{ij}	Damage tensor
D_{1t}^{cr}	Critical damage at longitudinal tensile failure
D_{1c}^{cr}	Critical damage at longitudinal compression failure
D_{2t}^{cr}	Critical damage at transverse tensile failure
$E(D)$	Effective modulus
$\overline{E}$	Undamaged (virgin) modulus
$G_c = 2\gamma_c$	Surface energy
J_{ijkl}	Normal damage constitutive tensor
M_{ijkl}	Damage effect tensor
$R(p)$	Yield hardening function
R_0	Yield threshold
Y_{ij}	Thermodynamic force tensor

Symbols Related to Delaminations

α, β, γ	Mixed mode fracture propagation parameters
ℓ	Delamination length for 2D delaminations
ψ_{xi}, ψ_{yi}	Rotation of normals to the middle surface of the plate
Ω	Volume of the body
Ω_D	Delaminated region
Π_e	Potential energy, elastic
Π^r	Potential energy, total
$\dot{\Gamma}$	Dissipation rate
Λ	Interface strain energy density per unit area
$\partial\Omega$	Boundary of the body
d	One-dimensional damage state variable
k_{xy}, k_z	Displacement continuity parameters
$[A_i], [B_i], [D_i]$	Laminate stiffness sub-matrices
$G(\ell)$	Energy release rate (ERR), total, in 2D
G	Energy release rate (ERR), total, in 3D
G_I, G_{II}, G_{III}	Energy release rate (ERR) of modes I, II, and III
G_c	Critical energy release rate (ERR), total, in 3D
G_I^c	Critical energy release rate mode I
$[H_i]$	Laminate interlaminar shear stiffness matrix
K_I, K_{II}, K_{III}	Stress intensity factors (SIF) of modes I, II, and III
N_i, M_i, T_i	Stress resultants
U	Internal energy
W	Work done by the body on its surroundings

List of Examples

Chapter 1

Mechanics of Orthotropic Materials

This chapter provides the foundation for the rest of the book. Basic concepts of mechanics, tailored for composite materials, are presented, including coordinate transformations, constitutive equations, and so on. Continuum mechanics is used to describe deformation and stress in an orthotropic material. The basic equations are reviewed in Sections 1.2 to 1.9. Tensor operations are reviewed in Section 1.10 because they are used in the rest of the chapter. Coordinate transformations are required to express quantities such as stress, strain, and stiffness in material axes, global axes, and so on. They are reviewed in Sections 1.10 to 1.11.

This chapter is heavily referenced in the rest of the book, and thus readers who are already versed in continuum mechanics may choose to come back to review this material as needed.

1.1 Material Coordinate System

A single lamina of fiber reinforced composite behaves as an orthotropic material. That is, the material has three mutually perpendicular planes of symmetry. The intersection of these three planes defines three axes that coincide with the fiber direction (x'_1), the thickness coordinate (x'_3), and a third direction $x'_2 = x'_3 \times x'_1$ perpendicular to the other two[1][1].

1.2 Displacements

Under the action of forces, every point in a body may translate and rotate as a rigid body as well as deform to occupy a new region. The displacements u_i of any point P in the body (Figure 1.1) are defined in terms of the three components of the vector u_i (in a rectangular Cartesian coordinate system) as $u_i = (u_1, u_2, u_3)$. An alternate

---

[1]$\times$ denotes vector cross product.

notation for displacements is $u_i = (u, v, w)$. Displacement is a vector or first-order tensor quantity

$$\mathbf{u} = u_i = (u_1, u_2, u_3) \; ; \; i = 1...3 \tag{1.1}$$

where boldface (e.g., $\mathbf{u}$) indicates a *tensor* writen in tensor notation, in this case a vector (or first-order tensor). In this book, all tensors are boldfaced (e.g., $\boldsymbol{\sigma}$), but their components are not (e.g., σ_{ij}). The order of the tensor (i.e., first, second, fourth, etc.) must be inferred from context, or as in (1.1), by looking at the number of subscripts of the same entity written in index notation (e.g., u_i).

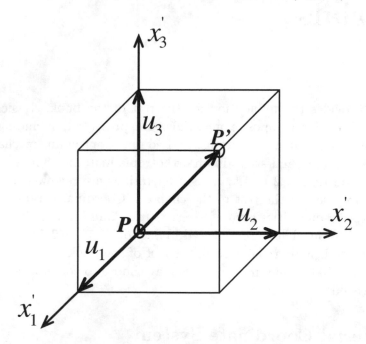

Fig. 1.1: Notation for displacement components.

1.3   Strain

For full geometric non-linear analysis, the components of the Lagrangian strain tensor are [2]

$$L_{ij} = \frac{1}{2}(u_{i,j} + u_{j,i} + u_{r,i}u_{r,j}) \tag{1.2}$$

where

$$u_{i,j} = \frac{\partial u_i}{\partial x_j} \tag{1.3}$$

If the gradients of the displacements are so small that products of partial derivatives of u_i are negligible compared with linear (first-order) derivative terms, then the (infinitesimal) strain tensor ε_{ij} is given by [2]

$$\boldsymbol{\varepsilon} = \varepsilon_{ij} = \frac{1}{2}(u_{i,j} + u_{j,i}) \tag{1.4}$$

Again, boldface indicates a tensor, the order of which is implied from the context. For example ε is a one-dimensional strain and $\boldsymbol{\varepsilon}$ is the second-order tensor of strain. Index notation (e.g., $= \varepsilon_{ij}$) is used most of the time and the tensor character of variables (scalar, vector, second order, and so on) is easily understood from context.

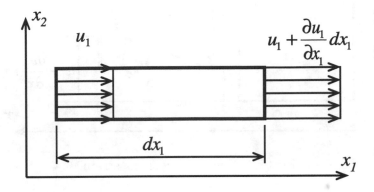

Fig. 1.2: Normal strain.

From the definition (1.4), strain is a second-order, symmetric tensor (i.e., $\varepsilon_{ij} = \varepsilon_{ji}$). In expanded form the strains are defined by

$$\varepsilon_{11} = \frac{\partial u_1}{\partial x_1} = \epsilon_1 \; ; \quad 2\varepsilon_{12} = 2\varepsilon_{21} = \left(\frac{\partial u_1}{\partial x_2} + \frac{\partial u_2}{\partial x_1}\right) = \gamma_6 = \epsilon_6$$

$$\varepsilon_{22} = \frac{\partial u_2}{\partial x_2} = \epsilon_2 \; ; \quad 2\varepsilon_{13} = 2\varepsilon_{31} = \left(\frac{\partial u_1}{\partial x_3} + \frac{\partial u_3}{\partial x_1}\right) = \gamma_5 = \epsilon_5$$

$$\varepsilon_{33} = \frac{\partial u_3}{\partial x_3} = \epsilon_3 \; ; \quad 2\varepsilon_{23} = 2\varepsilon_{32} = \left(\frac{\partial u_2}{\partial x_3} + \frac{\partial u_3}{\partial x_2}\right) = \gamma_4 = \epsilon_4 \tag{1.5}$$

where ϵ_α with $\alpha = 1..6$ are defined in Section 1.5. The normal components of strain ($i = j$) represent the change in length per unit length (Figure 1.2). The shear components of strain ($i \neq j$) represent one-half the change in an original right angle (Figure 1.3). The engineering shear strain $\gamma_\alpha = 2\varepsilon_{ij}$, for $i \neq j$ is often used instead of the tensor shear strain because the shear modulus G is defined by $\tau = G\gamma$ in strength of materials [3]. The strain tensor, being of second order, can be displayed as a matrix

$$[\varepsilon] = \begin{bmatrix} \varepsilon_{11} & \varepsilon_{12} & \varepsilon_{13} \\ \varepsilon_{12} & \varepsilon_{22} & \varepsilon_{23} \\ \varepsilon_{13} & \varepsilon_{23} & \varepsilon_{33} \end{bmatrix} = \begin{bmatrix} \epsilon_1 & \epsilon_6/2 & \epsilon_5/2 \\ \epsilon_6/2 & \epsilon_2 & \epsilon_4/2 \\ \epsilon_5/2 & \epsilon_4/2 & \epsilon_3 \end{bmatrix} \tag{1.6}$$

where [ ] is used to denote matrices.

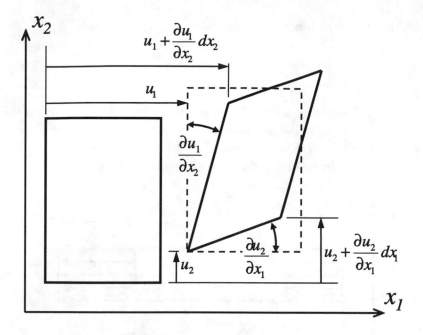

Fig. 1.3: Engineering shear strain.

1.4   Stress

The stress vector associated to a plane passing through a point is the force per unit area acting on the plane passing through the point. A second-order tensor, called stress tensor, completely describes the state of stress at a point. The stress tensor can be expressed in terms of the components acting on three mutually perpendicular planes aligned with the orthogonal coordinate directions as indicated in Figure 1.4. The tensor notation for stress is σ_{ij} with $(i, j = 1, 2, 3)$, where the first subscript corresponds to the direction of the normal to the plane of interest and the second subscript corresponds to the direction of the stress. Tensile normal stresses $(i = j)$ are defined to be positive when the normal to the plane and the stress component directions are either both positive or both negative. All components of stress depicted in Figure 1.4 have a positive sense. Force and moment equilibrium of the element in Figure 1.4 requires that the stress tensor be symmetric (i.e., $\sigma_{ij} = \sigma_{ji}$) [3]. The stress tensor, being of second order, can be displayed as a matrix

$$[\sigma] = \begin{bmatrix} \sigma_{11} & \sigma_{12} & \sigma_{13} \\ \sigma_{12} & \sigma_{22} & \sigma_{23} \\ \sigma_{13} & \sigma_{23} & \sigma_{33} \end{bmatrix} = \begin{bmatrix} \sigma_1 & \sigma_6 & \sigma_5 \\ \sigma_6 & \sigma_2 & \sigma_4 \\ \sigma_5 & \sigma_4 & \sigma_3 \end{bmatrix} \qquad (1.7)$$

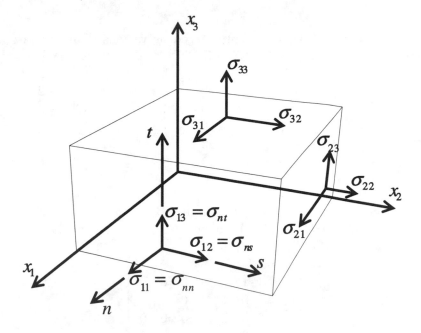

Fig. 1.4: Stress components.

1.5 Contracted Notation

Since the stress is symmetric, it can be written in contracted notation as

$$\sigma_\alpha = \sigma_{ij} = \sigma_{ji} \qquad (1.8)$$

with the contraction rule defined as follows

$$\begin{aligned} \alpha &= i && if \quad i = j \\ \alpha &= 9 - i - j && if \quad i \neq j \end{aligned} \qquad (1.9)$$

resulting in the contracted version of stress components shown in (1.7). The same applies to the strain tensor, resulting in the contracted version of strain shown in (1.6). Note that the six components of stress σ_α with $\alpha = 1 \ldots 6$ can be arranged into a column array, denoted by curly brackets { } as in (1.10), but $\{\sigma\}$ is not a vector, but just a convenient way to arrange the six unique components of a symmetric second-order tensor.

1.5.1 Alternate Contracted Notation

Some FEA software packages use different contracted notations, as shown in Table 1.1. For example, to transform stresses or strains from standard notation to ABAQUS notation, a transformation matrix can be used as follows

$$\{\sigma_A\} = [T]\{\sigma\} \qquad (1.10)$$

Table 1.1: Contracted notation convention used by various FEA software packages

Standard	ABAQUS	LS-DYNA	ANSYS
$11 \longrightarrow 1$	$11 \longrightarrow 1$	$11 \longrightarrow 1$	$11 \longrightarrow 1$
$22 \longrightarrow 2$	$22 \longrightarrow 2$	$22 \longrightarrow 2$	$22 \longrightarrow 2$
$33 \longrightarrow 3$	$33 \longrightarrow 3$	$33 \longrightarrow 3$	$33 \longrightarrow 3$
$23 \longrightarrow 4$	$12 \longrightarrow 4$	$12 \longrightarrow 4$	$12 \longrightarrow 4$
$13 \longrightarrow 5$	$13 \longrightarrow 5$	$23 \longrightarrow 5$	$23 \longrightarrow 5$
$12 \longrightarrow 6$	$23 \longrightarrow 6$	$13 \longrightarrow 6$	$13 \longrightarrow 6$

where the subscript $()_A$ denotes a quantity in ABAQUS notation. Also note that $\{\ \}$ denotes a column array, in this case of six elements, and $[\ ]$ denotes a matrix, in this case the 6×6 rotation matrix given by

$$[T] = \begin{bmatrix} 1 & 0 & 0 & 0 & 0 & 0 \\ 0 & 1 & 0 & 0 & 0 & 0 \\ 0 & 0 & 1 & 0 & 0 & 0 \\ 0 & 0 & 0 & 0 & 0 & 1 \\ 0 & 0 & 0 & 0 & 1 & 0 \\ 0 & 0 & 0 & 1 & 0 & 0 \end{bmatrix} \tag{1.11}$$

The stiffness matrix transforms as follows

$$[C_A] = [T]^T[C][T] \tag{1.12}$$

For LS-DYNA and ANSYS, the transformation matrix is

$$[T] = \begin{bmatrix} 1 & 0 & 0 & 0 & 0 & 0 \\ 0 & 1 & 0 & 0 & 0 & 0 \\ 0 & 0 & 1 & 0 & 0 & 0 \\ 0 & 0 & 0 & 0 & 0 & 1 \\ 0 & 0 & 0 & 1 & 0 & 0 \\ 0 & 0 & 0 & 0 & 1 & 0 \end{bmatrix} \tag{1.13}$$

1.6   Equilibrium and Virtual Work

The three equations of equilibrium at every point in a body are written in tensor notation as

$$\sigma_{ij,j} + f_i = 0 \tag{1.14}$$

where f_i is the body force per unit volume and $(\ )_{,j} = \dfrac{\partial}{\partial x_j}$. When body forces are negligible, the expanded form of the equilibrium equations, written in terms of a global x-y-z coordinate system, is

$$\frac{\partial \sigma_{xx}}{\partial x} + \frac{\partial \sigma_{xy}}{\partial y} + \frac{\partial \sigma_{xz}}{\partial z} = 0$$

$$\frac{\partial \sigma_{xy}}{\partial x} + \frac{\partial \sigma_{yy}}{\partial y} + \frac{\partial \sigma_{yz}}{\partial z} = 0$$

$$\frac{\partial \sigma_{xz}}{\partial x} + \frac{\partial \sigma_{yz}}{\partial y} + \frac{\partial \sigma_{zz}}{\partial z} = 0 \tag{1.15}$$

The principle of virtual work (PVW) provides an alternative to the equations of equilibrium[4]. Since the PVW is an integral expression, it is more convenient than (1.14) for finite element formulation. The PVW reads

$$\int_V \sigma_{ij} \delta \epsilon_{ij} dV - \int_S t_i \delta u_i dS - \int_V f_i \delta u_i dV = 0 \tag{1.16}$$

where t_i are the surface tractions per unit area acting on the surface S. The negative sign means that work is done by external forces (t_i, f_i) on the body. The forces and the displacements follow the same sign convention; that is, a component is positive when it points in the positive direction of the respective axis. The first term in (1.16) is the virtual work performed by the internal stresses and it is positive following the same sign convention.

Example 1.1 *Find the displacement function $u(x)$ for a slender rod of cross-sectional area A, length L, modulus E and density ρ, hanging from the top end and subjected to its own weight alone. Use a coordinate x pointing downward with origin at the top end.*

Solution to Example 1.1 *We assume a quadratic displacement function*

$$u(x) = C_0 + C_1 x + C_2 x^2$$

Using the boundary condition (B.C.) at the top yields $C_0 = 0$. The PVW (1.16) simplifies because the only non-zero strain is ϵ_x and there is no surface tractions. Using the Hooke's law

$$\int_0^L E\epsilon_x \delta \epsilon_x A dx - \int_0^L \rho g \delta u A dx = 0$$

From the assumed displacement

$$\delta u = x\delta C_1 + x^2 \delta C_2$$

$$\epsilon_x = \frac{du}{dx} = C_1 + 2xC_2$$

$$\delta \epsilon_x = \delta C_1 + 2x\delta C_2$$

Substituting

$$EA \int_o^L (C_1 + 2xC_2)(\delta C_1 + 2x\delta C_2)dx - \rho g A \int_0^L (x\delta C_1 + x^2 \delta C_2)dx = 0$$

Integrating and collecting terms in δC_1 and δC_2 separately

$$(EC_2 L^2 + EC_1 L - \frac{\rho g L^2}{2})\delta C_1 + (\frac{4}{3}EC_2 L^3 + EC_1 L^2 - \frac{\rho g L^3}{3})\delta C_2 = 0$$

Since δC_1 and δC_2 have arbitrary (virtual) values, two equations in two unknowns are obtained, one inside each parenthesis. Solving them we get

$$C_1 = \frac{L\rho g}{E} \; ; \; C_2 = -\frac{\rho g}{2E}$$

Substituting back into $u(x)$

$$u(x) = \frac{\rho g}{2E}(2L - x)x$$

which coincides with the exact solution from strength of materials.

1.7   Boundary Conditions

1.7.1   Traction Boundary Conditions

The solution of problems in solid mechanics requires that boundary conditions be specified. The boundary conditions may be specified in terms of components of displacement, stress, or a combination of both. For any point on an arbitrary surface, the traction T_i is defined as the vector consisting of the three components of stress acting on the surface at the point of interest. As indicated in Figure 1.4 the traction vector consists of one component of normal stress, σ_{nn}, and two components of shear stress, σ_{nt} and σ_{ns}. The traction vector can be written using Cauchy's law

$$T_i = \sigma_{ji}n_j = \sum_j^3 \sigma_{ji}n_j \qquad (1.17)$$

where n_j is the unit normal to the surface at the point under consideration.[2] For a plane perpendicular to the x_1 axis $n_i = (1,0,0)$ and the components of the traction are $T_1 = \sigma_{11}$, $T_2 = \sigma_{12}$, and $T_3 = \sigma_{13}$.

1.7.2   Free Surface Boundary Conditions

The condition that a surface be free of stress is equivalent to all components of traction being zero, i.e., $T_n = \sigma_{nn} = 0$, $T_t = \sigma_{nt} = 0$, and $T_s = \sigma_{ns} = 0$. It is possible that only selected components of the traction be zero while others are non-zero. For example, pure pressure loading corresponds to non-zero normal stress and zero shear stresses.

---

[2]Einstein's summation convention can be introduced with (1.17) as an example. Any pair of repeated indices implies a summation over all the values of the index in question. Furthermore, each pair of repeated indices represents a *contraction*. That is, the order of resulting tensor, in this case order one for T_i, is two less than the sum of the orders of the tensors involved in the operation. The resulting tensor keeps only the *free* indices that are not involved in the contraction–in this case only i remains.

1.8 Continuity Conditions

1.8.1 Traction Continuity

Equilibrium (action and reaction) requires that the traction components T_i must be continuous across any surface. Mathematically this is stated as $T_i^+ = T_i^-$ or using (1.17), $\sigma_{ji}^+ n_j^+ = \sigma_{ji}^- n_j^-$. In terms of individual stress components, $\sigma_{nn}^+ = \sigma_{nn}^-$, $\sigma_{nt}^+ = \sigma_{nt}^-$, and $\sigma_{ns}^+ = \sigma_{ns}^-$ (Figure 1.5). Thus, the normal and shear components of stress acting on a surface must be continuous across that surface. There are no continuity requirements on the other three components of stress. That is, it is possible that $\sigma_{tt}^+ \neq \sigma_{tt}^-$, $\sigma_{ss}^+ \neq \sigma_{ss}^-$, and $\sigma_{ts}^+ \neq \sigma_{ts}^-$. Lack of continuity of the two normal and one shear components of stress is very common because the material properties are discontinuous across layer boundaries.

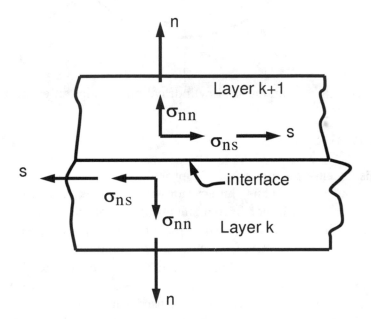

Fig. 1.5: Traction continuity across an interface.

1.8.2 Displacement Continuity

Certain conditions on displacements must be satisfied along any surface in a perfectly bonded continuum. Consider for example buckling of a cylinder under external pressure (Figure 1.6). The displacements associated with the material from either side of the line A-A must be identical $u_i^+ = u_i^-$. The continuity conditions must be satisfied at every point in a perfectly bonded continuum. However, continuity is not required in the presence of de-bonding or sliding between regions or phases of a material. For the example shown, continuity of slope must be satisfied also $(\dfrac{\partial w^+}{\partial \theta} = \dfrac{\partial w^-}{\partial \theta})$, where w is the radial displacement.

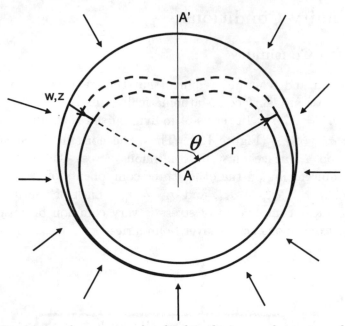

Fig. 1.6: Buckling of an encased cylindrical pipe under external pressure.

1.9 Compatibility

The strain displacement equations (1.5) provide six equations for only three unknown displacements u_i. Thus, integration of equations (1.5) to determine the
unknown displacements will not have a single-valued solution unless the strains ε_{ij}
satisfy certain conditions. Arbitrary specification of the ε_{ij} could result in discontinuities in the material, including gaps and/or overlapping regions.

The necessary conditions for single-valued displacements are the *compatibility
conditions*. Although these six equations are available [2], they are not used here
because the displacement method, which is used throughout this book, does not
require them. That is, in solving problems, the form of displacements u_i is always
assumed a priori. Then, the strains are computed with (1.5), and the stress with
(1.45). Finally, equilibrium is enforced by using the PVW (1.16).

1.10 Coordinate Transformations

The coordinates of point P in the prime coordinate system can be found from its
coordinates in the unprimed system. From Figure 1.7, the coordinates of point P
are

$$
\begin{aligned}
x_1' &= x_1 \cos\theta + x_2 \sin\theta \\
x_2' &= -x_1 \sin\theta + x_2 \cos\theta \\
x_3' &= x_3
\end{aligned}
\tag{1.18}
$$

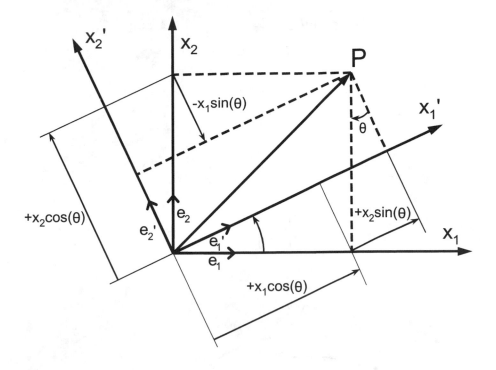

Fig. 1.7: Coordinate transformation.

or

$$x'_i = a_{ij}x_j = l_i x_1 + m_i x_2 + n_i x_3 \qquad (1.19)$$

or in matrix notation

$$\{x'\} = [a]\{x\} \qquad (1.20)$$

where a_{ij} are the components of the unit vectors of the primed system e'_i on the unprimed system x_j, by rows [2]

$$a_{ij} = \cos(x'_i, x_j) = \begin{array}{c|c|c|c} & x_1 & x_2 & x_3 \\ \hline e'_1 & l_1 & m_1 & n_1 \\ \hline e'_2 & l_2 & m_2 & n_2 \\ \hline e'_3 & l_3 & m_3 & n_3 \end{array} \qquad (1.21)$$

If primed coordinates denote the material coordinates and unprimed denote the global coordinates, then (1.19) transforms vectors from global to local coordinates. The inverse transformation simply uses the transpose matrix

$$\{x\} = [a]^T \{x'\} \qquad (1.22)$$

Example 1.2 *A composite layer has fiber orientation $\theta = 30°$. Construct the [a] matrix by calculating the direction cosines of the material system, i.e., the components of the unit vectors of the material system (x'_i) on the global system (x_j).*

Solution to Example 1.2 *From Figure 1.7 and (1.19) we have*

$$l_1 = \cos\theta = \frac{\sqrt{3}}{2}$$

$$m_1 = \sin\theta = \frac{1}{2}$$

$$n_1 = 0$$

$$l_2 = -\sin\theta = -\frac{1}{2}$$

$$m_2 = \cos\theta = \frac{\sqrt{3}}{2}$$

$$n_2 = 0$$

$$l_3 = 0$$

$$m_3 = 0$$

$$n_3 = 1$$

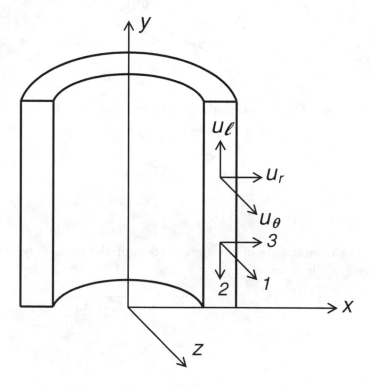

Fig. 1.8: Coordinate transformation for axial-symmetric analysis.

Example 1.3 *A fiber reinforced composite tube is wound in the hoop direction (1-direction). Formulas for the stiffness values (E_1, E_2, etc.) are given in that system. However, when analyzing the cross-section of this material with generalized plane strain elements (CAX4 in ABAQUS), the model must be constructed in the global system (X-Y-Z). It is therefore necessary to provide the stiffness values in the global system as E_x, E_y, etc. Construct the transformation matrix $[a]^T$ to go from material coordinates (1-2-3) to global coordinates (X-Y-Z) in Figure 1.8.*

Solution to Example 1.3 *First, construct* [a] *using the definition (1.21). Taking each unit vector (1-2-3) at a time we construct the matrix* [a] *by rows. The i-th row contains the components of (i=1,2,3) along (X-Y-Z).*

[a]	X	Y	Z
1	0	0	1
2	0	−1	0
3	1	0	0

The required transformation is just the transpose of the matrix above.

1.10.1 Stress Transformation

A second-order tensor σ_{pq} can be thought as the (un-contracted) outer product[3] of two vectors V_p and V_q

$$\sigma_{pq} = V_p \otimes V_q \tag{1.23}$$

each of which transforms as (1.19)

$$\sigma'_{ij} = a_{ip}V_p \otimes a_{jq}V_q \tag{1.24}$$

So

$$\sigma'_{ij} = a_{ip}a_{jq}\sigma_{pq} \tag{1.25}$$

For example, expand σ'_{11} in contracted notation

$$\sigma'_1 = a_{11}^2\sigma_1 + a_{12}^2\sigma_2 + a_{13}^2\sigma_3 + 2a_{11}a_{12}\sigma_6 + 2a_{11}a_{13}\sigma_5 + 2a_{12}a_{13}\sigma_4 \tag{1.26}$$

Expanding σ'_{12} in contracted notation yields

$$\sigma'_6 = a_{11}a_{21}\sigma_1 + a_{12}a_{22}\sigma_2 + a_{13}a_{23}\sigma_3 + (a_{11}a_{22} + a_{12}a_{21})\sigma_6 \tag{1.27}$$
$$+ (a_{11}a_{23} + a_{13}a_{21})\sigma_5 + (a_{12}a_{23} + a_{13}a_{22})\sigma_4$$

The following algorithm is used to obtain a 6×6 coordinate transformation matrix [T] such that (1.25) is rewritten in contracted notation as

$$\sigma'_\alpha = T_{\alpha\beta}\sigma_\beta \tag{1.28}$$

If $\alpha \leq 3$ and $\beta \leq 3$ then $i = j$ and $p = q$, so

$$T_{\alpha\beta} = a_{ip}a_{ip} = a_{ip}^2 \quad \text{no sum on } i \tag{1.29}$$

If $\alpha \leq 3$ and $\beta > 3$ then $i = j$ but $p \neq q$, and taking into account that switching p by q yields the same value of $\beta = 9 - p - q$ as per (1.9) we have

$$T_{\alpha\beta} = a_{ip}a_{iq} + a_{iq}a_{ip} = 2a_{ip}a_{iq} \quad \text{no sum on } i \tag{1.30}$$

---

[3]The outer product preserves all indices of the entities involved, thus creating a tensor of order equal to the sum of the order of the entities involved.

If $\alpha > 3$, then $i \neq j$, but we want only one stress, say σ_{ij}, not σ_{ji} because they are numerically equal. In fact $\sigma_\alpha = \sigma_{ij} = \sigma_{ji}$ with $\alpha = 9 - i - j$. If in addition $\beta \leq 3$ then $p = q$ and we get

$$T_{\alpha\beta} = a_{ip}a_{jp} \quad \text{no sum on } i \tag{1.31}$$

When $\alpha > 3$ and $\beta > 3$, $i \neq j$ and $p \neq q$ so we get

$$T_{\alpha\beta} = a_{ip}a_{jq} + a_{iq}a_{jp} \quad \text{no sum on } i \tag{1.32}$$

which completes the derivation of $T_{\alpha\beta}$. Expanding (1.29-1.32) and using (1.21) we get

$$[T] = \begin{bmatrix} l_1^2 & m_1^2 & n_1^2 & 2m_1n_1 & 2l_1n_1 & 2l_1m_1 \\ l_2^2 & m_2^2 & n_2^2 & 2m_2n_2 & 2l_2n_2 & 2l_2m_2 \\ l_3^2 & m_3^2 & n_3^2 & 2m_3n_3 & 2l_3n_3 & 2l_3m_3 \\ l_2l_3 & m_2m_3 & n_2n_3 & m_2n_3 + n_2m_3 & l_2n_3 + n_2l_3 & l_2m_3 + m_2l_3 \\ l_1l_3 & m_1m_3 & n_1n_3 & m_1n_3 + n_1m_3 & l_1n_3 + n_1l_3 & l_1m_3 + m_1l_3 \\ l_1l_2 & m_1m_2 & n_1n_2 & m_1n_2 + n_1m_2 & l_1n_2 + n_1l_2 & l_1m_2 + m_1l_2 \end{bmatrix} \tag{1.33}$$

A MATLAB program that can be used to generate (1.33) is shown next (also available in [5]).

```
% Derivation of the transformation matrix [T]
clear all;
syms T alpha R
syms a a11 a12 a13 a21 a22 a23 a31 a32 a33
a = [a11,a12,a13;
     a21,a22,a23;
     a31,a32,a33];
% it can be done in terms of l,m,n's as well
syms a l1 m1 n1 l2 m2 n2 l3 m3 n3
a = [l1,m1,n1;l2,m2,n2;l3,m3,n3]
T(1:6,1:6) = 0;
for i=1:1:3
for j=1:1:3
 if i==j; alpha = j; else alpha = 9-i-j; end
 for p=1:1:3
 for q=1:1:3
  if p==q beta = p; else beta = 9-p-q; end
  T(alpha,beta) = 0;
  if alpha<=3 & beta<= 3; T(alpha,beta)=a(i,p)*a(i,p); end
  if alpha> 3 & beta<= 3; T(alpha,beta)=a(i,p)*a(j,p); end
  if alpha<=3 & beta>3; T(alpha,beta)=a(i,q)*a(i,p)+a(i,p)*a(i,q);end
  if alpha>3 & beta>3; T(alpha,beta)=a(i,p)*a(j,q)+a(i,q)*a(j,p);end
 end
 end
end
end
T
R = eye(6,6); R(4,4)=2; R(5,5)=2; R(6,6)=2; % Reuter matrix
Tbar = R*T*R^(-1)
```

1.10.2   Strain Transformation

The tensor components of strain ε_{ij} transform in the same way as the stress components

$$\varepsilon'_{ij} = a_{ip}a_{jq}\varepsilon_{pq} \tag{1.34}$$

or

$$\varepsilon'_\alpha = T_{\alpha\beta}\varepsilon_\beta \tag{1.35}$$

with $T_{\alpha\beta}$ given by (1.33). However, the three engineering shear strains $\gamma_{xz}, \gamma_{yz}, \gamma_{xy}$ are normally used instead of tensor shear strains $\varepsilon_{xz}, \varepsilon_{yz}, \varepsilon_{xy}$. The engineering strains (ϵ instead of ε) are defined in (1.5). They can be obtained from the tensor components by the following relationship

$$\epsilon_\delta = R_{\delta\gamma}\varepsilon_\gamma \tag{1.36}$$

with the Reuter matrix given by

$$[R] = \begin{bmatrix} 1 & 0 & 0 & 0 & 0 & 0 \\ 0 & 1 & 0 & 0 & 0 & 0 \\ 0 & 0 & 1 & 0 & 0 & 0 \\ 0 & 0 & 0 & 2 & 0 & 0 \\ 0 & 0 & 0 & 0 & 2 & 0 \\ 0 & 0 & 0 & 0 & 0 & 2 \end{bmatrix} \tag{1.37}$$

Then, the coordinate transformation of engineering strain results from (1.35) and (1.36) as

$$\epsilon'_\alpha = \overline{T}_{\alpha\beta}\epsilon_\beta \tag{1.38}$$

with

$$[\overline{T}] = [R][T][R]^{-1} \tag{1.39}$$

used only to transform engineering strains. Explicitly we have

$$[\overline{T}] = \begin{bmatrix} l_1^2 & m_1^2 & n_1^2 & m_1n_1 & l_1n_1 & l_1m_1 \\ l_2^2 & m_2^2 & n_2^2 & m_2n_2 & l_2n_2 & l_2m_2 \\ l_3^2 & m_3^2 & n_3^2 & m_3n_3 & l_3n_3 & l_3m_3 \\ 2l_2l_3 & 2m_2m_3 & 2n_2n_3 & m_2n_3 + n_2m_3 & l_2n_3 + n_2l_3 & l_2m_3 + m_2l_3 \\ 2l_1l_3 & 2m_1m_3 & 2n_1n_3 & m_1n_3 + n_1m_3 & l_1n_3 + n_1l_3 & l_1m_3 + m_1l_3 \\ 2l_1l_2 & 2m_1m_2 & 2n_1n_2 & m_1n_2 + n_1m_2 & l_1n_2 + n_1l_2 & l_1m_2 + m_1l_2 \end{bmatrix} \tag{1.40}$$

1.11   Transformation of Constitutive Equations

The constitutive equations that relate stress σ to strain ε are defined using tensor strains (ε, not ϵ), as

$$\sigma' = \mathbf{C}' : \varepsilon'$$
$$\sigma'_{ij} = C'_{ijkl}\varepsilon'_{kl} \tag{1.41}$$

where both *tensor* and *index* notations have been used.[4]

For simplicity consider an orthotropic material (Section 1.12.3). Then, it is possible to write σ'_{11}, and σ'_{12} as

$$
\begin{aligned}
\sigma'_{11} &= C'_{1111}\varepsilon'_{11} + C'_{1122}\varepsilon'_{22} + C'_{1133}\varepsilon'_{33} \\
\sigma'_{12} &= C'_{1212}\varepsilon'_{12} + C'_{1221}\varepsilon'_{21} = 2C'_{1212}\varepsilon'_{12}
\end{aligned}
\tag{1.42}
$$

Rewriting (1.42) in contracted notation, it is clear that in contracted notation all the shear strains appear twice, as follows

$$
\begin{aligned}
\sigma'_1 &= C'_{11}\varepsilon'_1 + C'_{12}\varepsilon'_2 + C'_{13}\varepsilon'_3 \\
\sigma'_6 &= 2C'_{66}\varepsilon'_6
\end{aligned}
\tag{1.43}
$$

The factor 2 in front of the tensor shear strains is caused by two facts, the minor symmetry of the tensors C and ε (see (1.5,1.58-1.59) and the contraction of the last two indices of C_{ijkl} with the strain ε_{kl} in (1.42). Therefore, *any double contraction of tensors with minor symmetry needs to be corrected by a Reuter matrix (1.37) when written in the contracted notation*. Next, (1.41) can be written as

$$
\sigma'_\alpha = C'_{\alpha\beta} R_{\beta\delta}\varepsilon'_\delta
\tag{1.44}
$$

Note that the Reuter matrix in (1.44) can be combined with the tensor strains using (1.36), to write

$$
\sigma'_\alpha = C'_{\alpha\beta}\epsilon'_\beta
\tag{1.45}
$$

in terms of engineering strains. To obtain the stiffness matrix $[C]$ in the global coordinate system, introduce (1.28) and (1.38) into (1.45) so that

$$
T_{\alpha\delta}\sigma_\delta = C'_{\alpha\beta}\overline{T}_{\beta\gamma}\epsilon_\gamma
\tag{1.46}
$$

It can be shown that

$$
[T]^{-1} = [\overline{T}]^T
\tag{1.47}
$$

Therefore

$$
\{\sigma\} = [C]\{\epsilon\}
\tag{1.48}
$$

Finally,

$$
[C] = [\overline{T}]^T[C'][\overline{T}]
\tag{1.49}
$$

and

$$
[C'] = [\overline{T}]^{-T}[C][\overline{T}]^{-1} = [T][C][T]^T
\tag{1.50}
$$

The compliance matrix is the inverse of the stiffness matrix, not the inverse of the fourth order tensor C_{ijkl}. Therefore,

$$
[S'] = [C']^{-1}
\tag{1.51}
$$

---

[4]A double contraction involves contraction of two indices, in this case k and l, and it is denoted by : in tensor notation. Also note the use of boldface to indicate tensors in tensor notation.

Taking into account (1.47) and (1.49), the compliance matrix transforms as

$$[S] = [T]^T[S'][T] \tag{1.52}$$

$$[S'] = [T]^{-T}[S][T]^{-1} = [\overline{T}][S][\overline{T}]^T \tag{1.53}$$

For an orthotropic material, the compliance matrix $[S']$ is defined in the material coordinate system as

$$[S'] = \begin{bmatrix} \dfrac{1}{E_1} & \dfrac{-\nu_{21}}{E_2} & \dfrac{-\nu_{31}}{E_3} & 0 & 0 & 0 \\ \dfrac{-\nu_{12}}{E_1} & \dfrac{1}{E_2} & \dfrac{-\nu_{32}}{E_3} & 0 & 0 & 0 \\ \dfrac{-\nu_{13}}{E_1} & \dfrac{-\nu_{23}}{E_2} & \dfrac{1}{E_3} & 0 & 0 & 0 \\ 0 & 0 & 0 & \dfrac{1}{G_{23}} & 0 & 0 \\ 0 & 0 & 0 & 0 & \dfrac{1}{G_{13}} & 0 \\ 0 & 0 & 0 & 0 & 0 & \dfrac{1}{G_{12}} \end{bmatrix} \tag{1.54}$$

where E_i, G_{ij}, and ν_{ij}, are the elastic moduli, shear moduli, and Poisson's ratios, respectively. Furthermore, the subscripts indicate the material axes, i.e.,

$$\nu_{ij} = \nu_{x'_i x'_j} \text{ and } E_{ii} = E_{x'_i} \tag{1.55}$$

Since $[S']$ is symmetric, the following must be satisfied

$$\frac{\nu_{ij}}{E_{ii}} = \frac{\nu_{ji}}{E_{jj}} \text{ , } i,j = 1..3 \tag{1.56}$$

Furthermore, Poisson's ratios are defined so that the lateral strain is given by

$$\nu_j = -\nu_{ij}\epsilon_i \tag{1.57}$$

In ANSYS, ν_{xy} corresponds to PRXY, PRXZ, and PRYZ, not to NUXY and so on.

After computing S_{ij}, the components of stress are obtained by using (1.45) or (1.48). This formulation predicts realistic behavior for finite displacement and rotations as long as the strains are small. This formulation is expensive to use since it needs 18 state variables, 12 components of the strain displacement matrix computed in the initial configuration, and 6 direction cosines.

1.12 3D Constitutive Equations

Hooke's law in three dimensions (3D) takes the form of (1.41). The 3D stiffness tensor C_{ijkl} is a fourth-order tensor with 81 components. For anisotropic materials only 21 components are independent, the remaining 60 components can be written in terms of the other 21. The one dimensional case (1D), studied in strength of materials, is recovered when all the stress components are zero except $\sigma_{11}(\sigma_{ij} = 0$ if $i \neq 1, j \neq 1)$. Only for the 1D case, $\sigma_{11} = \sigma, \epsilon_{11} = \epsilon, C_{1111} = E$. All the derivations in this Section (1.12) are carried out in material coordinates *but for simplicity the prime symbol* (') *is omitted, in this section only.*

In (1.41), exchanging the dummy indexes i by j, and k by l we have

$$\sigma_{ji} = C_{jilk}\epsilon_{lk} \tag{1.58}$$

The fact that the stress and strain tensors are symmetric can be written as: $\sigma_{ij} = \sigma_{ji}$ and $\epsilon_{ij} = \epsilon_{ji}$. Then, it follows that

$$C_{ijkl} = C_{jikl} = C_{ijlk} = C_{jilk} \tag{1.59}$$

which effectively reduces the number of independent components from 81 to 36. For example, $C_{1213} = C_{2131}$ and so on. Then, the 36 independent components can be written as a 6×6 matrix.

By postulating the existence of a function called strain energy density $\tilde{u}$, it is possible to demonstrate that the number of independent constants reduces to 21 (see [2] and (8.86)). Expanding the strain energy density in a Taylor power series

$$\tilde{u} = \rho u(\varepsilon_{ij}) = \tilde{u}_0 + \beta_{ij}\varepsilon_{ij} + \frac{1}{2}\alpha_{ijkl}\varepsilon_{ij}\varepsilon_{kl} + \ldots \tag{1.60}$$

from which the stress tensor is defined as (see (8.86))

$$\sigma_{ij} = \rho\frac{\partial u}{\partial \varepsilon_{ij}} = \frac{\partial \tilde{u}}{\partial \varepsilon_{ij}} = \frac{1}{2}\left(\alpha_{ijkl}\varepsilon_{kl} + \alpha_{ijkl}\varepsilon_{ij}\right) \tag{1.61}$$

Ignoring the higher-order terms and assuming the reference state to be free of residual stress when the strains vanish, $\beta_{ij} = 0$. Since the subscripts are dummy indexes, they can be renamed

$$\sigma_{ij} = \frac{1}{2}\left(\alpha_{ijkl} + \alpha_{klij}\right)\varepsilon_{kl} = C_{ijkl}\varepsilon_{kl} \tag{1.62}$$

where it is clear that the stiffness tensor

$$C_{ijkl} = C_{klij} = \frac{1}{2}\left(\alpha_{ijkl} + \alpha_{klij}\right) \tag{1.63}$$

has major symmetry. Therefore the 6×6 stiffness matrix is symmetric.

Using contracted notation, the generalized Hooke's law becomes

$$\begin{Bmatrix} \sigma_1 \\ \sigma_2 \\ \sigma_3 \\ \sigma_4 \\ \sigma_5 \\ \sigma_6 \end{Bmatrix} = \begin{bmatrix} C_{11} & C_{12} & C_{13} & C_{14} & C_{15} & C_{16} \\ C_{12} & C_{22} & C_{23} & C_{24} & C_{25} & C_{26} \\ C_{13} & C_{23} & C_{33} & C_{34} & C_{35} & C_{36} \\ C_{14} & C_{24} & C_{34} & C_{44} & C_{45} & C_{46} \\ C_{15} & C_{25} & C_{35} & C_{45} & C_{55} & C_{56} \\ C_{16} & C_{26} & C_{36} & C_{46} & C_{56} & C_{66} \end{bmatrix} \begin{Bmatrix} \epsilon_1 \\ \epsilon_2 \\ \epsilon_3 \\ \gamma_4 \\ \gamma_5 \\ \gamma_6 \end{Bmatrix} \tag{1.64}$$

Once again, the 1D case is covered when $\sigma_p = 0$ if $p \neq 1$. Then, $\sigma_1 = \sigma, \epsilon_1 = \epsilon, C_{11} = E$.

1.12.1 Anisotropic Material

Equation (1.64) represents a fully anisotropic material. Such a material has properties that change with the orientation. For example, the material body depicted in Figure 1.9 deforms differently in the directions P, T, and Q, even if the forces applied along the directions P, T, and Q are equal. The number of constants required to describe anisotropic materials is 21.

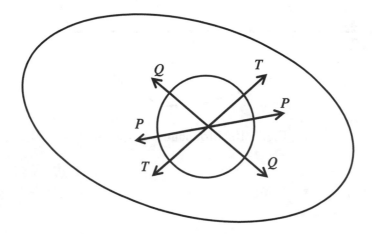

Fig. 1.9: Anisotropic material.

The inverse of the stiffness matrix is the compliance matrix $[S] = [C]^{-1}$. The constitutive equation (3D Hooke's law) is written in terms of compliances as follows

$$\left\{ \begin{array}{c} \epsilon_1 \\ \epsilon_2 \\ \epsilon_3 \\ \gamma_4 \\ \gamma_5 \\ \gamma_6 \end{array} \right\} = \left[\begin{array}{cccccc} S_{11} & S_{12} & S_{13} & S_{14} & S_{15} & S_{16} \\ S_{12} & S_{22} & S_{23} & S_{24} & S_{25} & S_{26} \\ S_{13} & S_{23} & S_{33} & S_{34} & S_{35} & S_{36} \\ S_{14} & S_{24} & S_{34} & S_{44} & S_{45} & S_{46} \\ S_{15} & S_{25} & S_{35} & S_{45} & S_{55} & S_{56} \\ S_{16} & S_{26} & S_{36} & S_{46} & S_{56} & S_{66} \end{array} \right] \left\{ \begin{array}{c} \sigma_1 \\ \sigma_2 \\ \sigma_3 \\ \sigma_4 \\ \sigma_5 \\ \sigma_6 \end{array} \right\} \qquad (1.65)$$

The [S] matrix is also symmetric and it has 21 independent constants. For the 1-D case, $\sigma = 0$ if $p \neq 1$. Then, $\sigma_1 = \sigma, \epsilon_1 = \epsilon, S_{11} = 1/E$.

1.12.2 Monoclinic Material

If a material has one plane of symmetry (Figure 1.10) it is called monoclinic and 13 constants are required to describe it. One plane of symmetry means that the properties are the same at symmetric points (z and -z as in Figure 1.10).

When the material is symmetric about the 1-2 plane, the material properties are identical upon reflection with respect to the 1-2 plane. For such reflection the

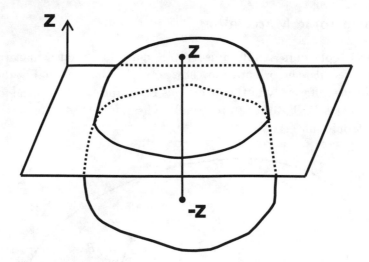

Fig. 1.10: Monoclinic material.

a-matrix (1.21) is

$$
\begin{array}{c}
\\
e_1'' \\
e_2'' \\
e_3''
\end{array}
\begin{array}{ccc}
x_1 & x_2 & x_3 \\
\end{array}
\begin{bmatrix}
1 & 0 & 0 \\
0 & 1 & 0 \\
0 & 0 & -1
\end{bmatrix}
=
\begin{bmatrix}
l_1 & m_1 & n_1 \\
l_2 & m_2 & n_2 \\
l_3 & m_3 & n_3
\end{bmatrix}
\tag{1.66}
$$

where $''$ has been used to avoid confusion with the material coordinate system that is denoted without $()'$ in this Section but with $()'$ elsewhere in this book. From (1.39) we get

$$
[T] =
\begin{bmatrix}
1 & 0 & 0 & 0 & 0 & 0 \\
0 & 1 & 0 & 0 & 0 & 0 \\
0 & 0 & 1 & 0 & 0 & 0 \\
0 & 0 & 0 & -1 & 0 & 0 \\
0 & 0 & 0 & 0 & -1 & 0 \\
0 & 0 & 0 & 0 & 0 & 1
\end{bmatrix}
\tag{1.67}
$$

The effect of $[T]$ is to multiply rows and columns 4 and 5 in $[C]$ by -1. The diagonal terms C_{44} and C_{55} remain positive because they are multiplied twice. Therefore, $C_{i4}'' = -C_{i4}$ *with* $i \neq 4,5$, $C_{i5}'' = -C_{i5}$ *with* $i \neq 4,5$, with everything else unchanged. Since the material properties in a monoclinic material cannot change by a reflection, it must be $C_{4i} = C_{i4} = 0$ *with* $i \neq 4,5$, $C_{5i} = C_{i5} = 0$ *with* $i \neq 4,5$. That is, 3D

Hooke's law reduces to

$$
\left\{
\begin{array}{c}
\sigma_1 \\
\sigma_2 \\
\sigma_3 \\
\sigma_4 \\
\sigma_5 \\
\sigma_6
\end{array}
\right\}
=
\left[
\begin{array}{cccccc}
C_{11} & C_{12} & C_{13} & 0 & 0 & C_{16} \\
C_{12} & C_{22} & C_{23} & 0 & 0 & C_{26} \\
C_{13} & C_{23} & C_{33} & 0 & 0 & C_{36} \\
0 & 0 & 0 & C_{44} & C_{45} & 0 \\
0 & 0 & 0 & C_{45} & C_{55} & 0 \\
C_{16} & C_{26} & C_{36} & 0 & 0 & C_{66}
\end{array}
\right]
\left\{
\begin{array}{c}
\epsilon_1 \\
\epsilon_2 \\
\epsilon_3 \\
\gamma_4 \\
\gamma_5 \\
\gamma_6
\end{array}
\right\}
\tag{1.68}
$$

and in terms of the compliances to

$$
\left\{
\begin{array}{c}
\epsilon_1 \\
\epsilon_2 \\
\epsilon_3 \\
\gamma_4 \\
\gamma_5 \\
\gamma_6
\end{array}
\right\}
=
\left[
\begin{array}{cccccc}
S_{11} & S_{12} & S_{13} & 0 & 0 & S_{16} \\
S_{12} & S_{22} & S_{23} & 0 & 0 & S_{26} \\
S_{13} & S_{23} & S_{33} & 0 & 0 & S_{36} \\
0 & 0 & 0 & S_{44} & S_{45} & 0 \\
0 & 0 & 0 & S_{45} & S_{55} & 0 \\
S_{16} & S_{26} & S_{36} & 0 & 0 & S_{66}
\end{array}
\right]
\left\{
\begin{array}{c}
\sigma_1 \\
\sigma_2 \\
\sigma_3 \\
\sigma_4 \\
\sigma_5 \\
\sigma_6
\end{array}
\right\}
\tag{1.69}
$$

1.12.3 Orthotropic Material

An orthotropic material has three planes of symmetry that coincide with the co-ordinate planes. It can be shown that if two orthogonal planes of symmetry exist, there is always a third orthogonal plane of symmetry. Nine constants are required to describe this type of material.

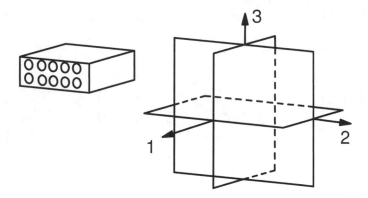

Fig. 1.11: Orthotropic material.

The symmetry planes can be Cartesian, as depicted in Figure 1.11, or they may correspond to any other coordinate representation (cylindrical, spherical, etc.). For example, the trunk of a tree has cylindrical orthotropy because of the growth rings. However, most practical materials exhibit Cartesian orthotropy. A unidirectional fiber reinforced composite may be considered to be orthotropic. One plane of symmetry is perpendicular to the fiber direction, and the other two are orthogonal to the fiber direction and among themselves.

In addition to the reflection about the 1-2 plane discussed in Section 1.12.2, a second reflection about the 1-3 plane should not affect the properties of the orthotropic materials. In this case the a-matrix is

$$[a] = \begin{bmatrix} 1 & 0 & 0 \\ 0 & -1 & 0 \\ 0 & 0 & 1 \end{bmatrix} \tag{1.70}$$

The $\overline{T}$-matrix from (1.39) is

$$[T] = \begin{bmatrix} 1 & 0 & 0 & 0 & 0 & 0 \\ 0 & 1 & 0 & 0 & 0 & 0 \\ 0 & 0 & 1 & 0 & 0 & 0 \\ 0 & 0 & 0 & -1 & 0 & 0 \\ 0 & 0 & 0 & 0 & 1 & 0 \\ 0 & 0 & 0 & 0 & 0 & -1 \end{bmatrix} \tag{1.71}$$

This will make $C_{i6} = -C_{i6}$, $i \neq 4, 6$ and $C_{i4} = -C_{i4}$, $i \neq 4, 6$. Since the material has symmetry about the 1-3 plane, this means that $C_{i6} = C_{6i} = 0$, $i \neq 6$. In this case, 3D Hooke's law reduces to

$$\begin{Bmatrix} \sigma_1 \\ \sigma_2 \\ \sigma_3 \\ \sigma_4 \\ \sigma_5 \\ \sigma_6 \end{Bmatrix} = \begin{bmatrix} C_{11} & C_{12} & C_{13} & 0 & 0 & 0 \\ C_{12} & C_{22} & C_{23} & 0 & 0 & 0 \\ C_{13} & C_{23} & C_{33} & 0 & 0 & 0 \\ 0 & 0 & 0 & C_{44} & 0 & 0 \\ 0 & 0 & 0 & 0 & C_{55} & 0 \\ 0 & 0 & 0 & 0 & 0 & C_{66} \end{bmatrix} \begin{Bmatrix} \epsilon_1 \\ \epsilon_2 \\ \epsilon_3 \\ \gamma_4 \\ \gamma_5 \\ \gamma_6 \end{Bmatrix} \tag{1.72}$$

and in terms of the compliances to

$$\begin{Bmatrix} \epsilon_1 \\ \epsilon_2 \\ \epsilon_3 \\ \gamma_4 \\ \gamma_5 \\ \gamma_6 \end{Bmatrix} = \begin{bmatrix} S_{11} & S_{12} & S_{13} & 0 & 0 & 0 \\ S_{12} & S_{22} & S_{23} & 0 & 0 & 0 \\ S_{13} & S_{23} & S_{33} & 0 & 0 & 0 \\ 0 & 0 & 0 & S_{44} & 0 & 0 \\ 0 & 0 & 0 & 0 & S_{55} & 0 \\ 0 & 0 & 0 & 0 & 0 & S_{66} \end{bmatrix} \begin{Bmatrix} \sigma_1 \\ \sigma_2 \\ \sigma_3 \\ \sigma_4 \\ \sigma_5 \\ \sigma_6 \end{Bmatrix} \tag{1.73}$$

Note that if the material has two planes of symmetry, it automatically has three because applying the procedure once more for a third plane (the 2-3 plane) will not change (1.72-1.73).

1.12.4   Transversely Isotropic Material

A transversely isotropic material has one axis of symmetry. For example, the fiber direction of a unidirectional fiber reinforced composite can be considered an axis of symmetry if the fibers are randomly distributed in the cross section (Figure 1.12). In this case, any plane containing the fiber direction is a plane of symmetry. A

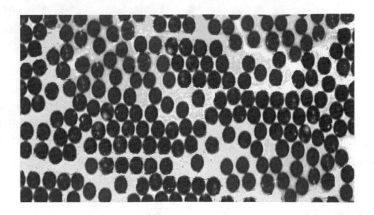

Fig. 1.12: Randomly distributed E-glass fibers with 200X magnification.

transversely isotropic material is described by five constants. When the axis of symmetry is the fiber direction (1-direction), 3D Hooke's law reduces to

$$
\begin{Bmatrix} \sigma_1 \\ \sigma_2 \\ \sigma_3 \\ \sigma_4 \\ \sigma_5 \\ \sigma_6 \end{Bmatrix} = \begin{bmatrix} C_{11} & C_{12} & C_{12} & 0 & 0 & 0 \\ C_{12} & C_{22} & C_{23} & 0 & 0 & 0 \\ C_{12} & C_{23} & C_{22} & 0 & 0 & 0 \\ 0 & 0 & 0 & (C_{22}-C_{23})/2 & 0 & 0 \\ 0 & 0 & 0 & 0 & C_{66} & 0 \\ 0 & 0 & 0 & 0 & 0 & C_{66} \end{bmatrix} \begin{Bmatrix} \epsilon_1 \\ \epsilon_2 \\ \epsilon_3 \\ \gamma_4 \\ \gamma_5 \\ \gamma_6 \end{Bmatrix} \tag{1.74}
$$

and in terms of the compliances to

$$
\begin{Bmatrix} \epsilon_1 \\ \epsilon_2 \\ \epsilon_3 \\ \gamma_4 \\ \gamma_5 \\ \gamma_6 \end{Bmatrix} = \begin{bmatrix} S_{11} & S_{12} & S_{12} & 0 & 0 & 0 \\ S_{12} & S_{22} & S_{23} & 0 & 0 & 0 \\ S_{12} & S_{23} & S_{22} & 0 & 0 & 0 \\ 0 & 0 & 0 & 2(S_{22}-S_{23}) & 0 & 0 \\ 0 & 0 & 0 & 0 & S_{66} & 0 \\ 0 & 0 & 0 & 0 & 0 & S_{66} \end{bmatrix} \begin{Bmatrix} \sigma_1 \\ \sigma_2 \\ \sigma_3 \\ \sigma_4 \\ \sigma_5 \\ \sigma_6 \end{Bmatrix} \tag{1.75}
$$

Note the equations would be different if the axis of symmetry is not the 1-direction. In terms of engineering properties (Section 1.13), and taking into account that the directions 2 and 3 are indistinguishable, the following relations apply for a transversely isotropic material:

$$
E_2 = E_3
$$
$$
\nu_{12} = \nu_{13} \tag{1.76}
$$
$$
G_{12} = G_{13}
$$

In addition, any two perpendicular directions on the plane 2-3 can be taken as axes. In other words, the plane 2-3 is isotropic. Therefore, the following holds in the 2-3 plane

$$
G_{23} = \frac{E_2}{2(1+\nu_{23})} \tag{1.77}
$$

just as it holds for isotropic materials (see Problem 1.13).

1.12.5   Isotropic Material

The most common materials of industrial use are isotropic, like aluminium, steel, etc. Isotropic materials have an infinite number of planes of symmetry, meaning that the properties are independent of the orientation. Only two constants are needed to represent the elastic properties. These two properties can be the Young's modulus E and the Poisson's ratio ν, but several other pairs of constants are used whenever it is convenient. However, any pair of properties has to be related to any other pair. For example, you could describe isotropic materials by E and G, but the shear modulus of isotropic materials is related to E and ν by

$$G = \frac{E}{2(1+\nu)} \tag{1.78}$$

Also, the Lame constants are sometimes used for convenience, in this case the two constants are

$$\lambda = \frac{E}{(1+\nu)(1-2\nu)} \tag{1.79}$$

$$\mu = G$$

To form yet another pair, any of the above properties could be substituted by the bulk modulus k, as follows

$$k = \frac{E}{3(1-2\nu)} \tag{1.80}$$

which relates the hydrostatic pressure p to the volumetric strain as

$$p = k(\epsilon_1 + \epsilon_2 + \epsilon_3) \tag{1.81}$$

For isotropic materials, the 3D Hooke's law is written in terms of only two constants C_{11} and C_{12} as

$$\begin{Bmatrix} \sigma_1 \\ \sigma_2 \\ \sigma_3 \\ \sigma_4 \\ \sigma_5 \\ \sigma_6 \end{Bmatrix} = \begin{bmatrix} C_{11} & C_{12} & C_{12} & 0 & 0 & 0 \\ C_{12} & C_{11} & C_{12} & 0 & 0 & 0 \\ C_{12} & C_{12} & C_{11} & 0 & 0 & 0 \\ 0 & 0 & 0 & \frac{(C_{11}-C_{12})}{2} & 0 & 0 \\ 0 & 0 & 0 & 0 & \frac{(C_{11}-C_{12})}{2} & 0 \\ 0 & 0 & 0 & 0 & 0 & \frac{(C_{11}-C_{12})}{2} \end{bmatrix} \begin{Bmatrix} \epsilon_1 \\ \epsilon_2 \\ \epsilon_3 \\ \gamma_4 \\ \gamma_5 \\ \gamma_6 \end{Bmatrix} \tag{1.82}$$

In terms of compliances, once again, two constants are used, S_{11} and S_{12} as follows

$$\begin{Bmatrix} \epsilon_1 \\ \epsilon_2 \\ \epsilon_3 \\ \gamma_4 \\ \gamma_5 \\ \gamma_6 \end{Bmatrix} = \begin{bmatrix} S_{11} & S_{12} & S_{12} & 0 & 0 & 0 \\ S_{12} & S_{11} & S_{12} & 0 & 0 & 0 \\ S_{12} & S_{12} & S_{11} & 0 & 0 & 0 \\ 0 & 0 & 0 & 2s & 0 & 0 \\ 0 & 0 & 0 & 0 & 2s & 0 \\ 0 & 0 & 0 & 0 & 0 & 2s \end{bmatrix} \begin{Bmatrix} \sigma_1 \\ \sigma_2 \\ \sigma_3 \\ \sigma_4 \\ \sigma_5 \\ \sigma_6 \end{Bmatrix} \tag{1.83}$$

$$s = S_{11} - S_{12}$$

Not only are the various constants related in pairs, but also certain restrictions apply on the values that these constants may have for real materials. Since the Young and shear moduli must always be positive, the Poisson's ratio must be $\nu > -1$. Furthermore, since the bulk modulus must be positive, we have $\nu < \frac{1}{2}$. Finally, the Poisson's ratio of isotropic materials is constrained by $-1 < \nu < \frac{1}{2}$.

1.13 Engineering Constants

Please note from here forward ()' denotes the material coordinate system. Our next task is to write the components of the stiffness and compliance matrices in terms of engineering constants for orthotropic materials. For this purpose it is easier to work with the compliance matrix, which is defined as the inverse of the stiffness matrix. In material coordinates $[S'] = [C']^{-1}$. The compliance matrix is used to write the relationship between strains and stresses in (1.73) for an orthotropic material. Let's rewrite the first of (1.73), which corresponds to the strain in the 1-direction (fiber direction)

$$\epsilon_1 = S'_{11}\sigma_1 + S'_{12}\sigma_2 + S'_{13}\sigma_3 \qquad (1.84)$$

and let's perform a thought experiment. Note that $[S']$ is used to emphasize the fact that we are working in the material coordinate system. First, apply a tensile stress along the 1-direction (fiber direction) as in Figure 1.13, with all the other stresses equal to zero, and compute the strain produced in the 1-direction, which is

$$\epsilon'_1 = \frac{\sigma_1{}'}{E_1{}'} \qquad (1.85)$$

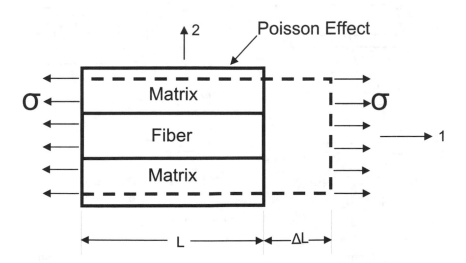

Fig. 1.13: Longitudinal loading.

Then, apply a stress in the 2-direction only, and compute the strain in the 1-

direction using the appropriate Poisson's ratio [1]

$$\epsilon_1' = -\nu_{21}\frac{\sigma_2'}{E_2} \tag{1.86}$$

Now, apply a stress in the 3-direction only, and compute the strain in the 1-direction using the appropriate Poisson's ratio,

$$\epsilon_1' = -\nu_{31}\frac{\sigma_3'}{E_3} \tag{1.87}$$

The total strain ϵ_1' is the sum of equations (1.85), (1.86) and (1.87)

$$\epsilon_1' = \frac{1}{E_1}\sigma_1' - \frac{\nu_{21}}{E_2}\sigma_2' - \frac{\nu_{31}}{E_3}\sigma_3' \tag{1.88}$$

Comparing (1.88) with (1.84) we conclude that

$$S_{11}' = \frac{1}{E_1}; S_{12}' = -\frac{\nu_{21}}{E_2}; S_{13}' = -\frac{\nu_{31}}{E_3} \tag{1.89}$$

Repeat the same procedure for the equations corresponding to ϵ_2' and ϵ_3' to obtain the coefficients in the second and third rows of the compliance matrix (1.73).

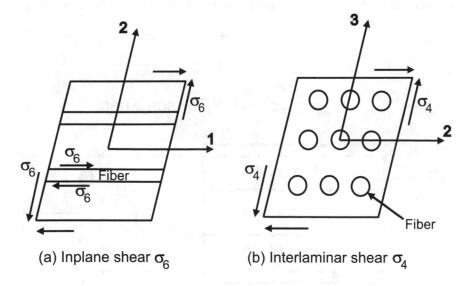

(a) Inplane shear σ_6 (b) Interlaminar shear σ_4

Fig. 1.14: Shear loading.

For the shear terms use the 4th, 5th and 6th rows of the compliance matrix (1.73). For example, from Figure 1.14 we write

$$\sigma_6' = \epsilon_6'G_{12} = 2\varepsilon_6'G_{12} \tag{1.90}$$

which compared to the 6th row of (1.73) leads to $S_{66} = 1/G_{12}$. Finally, it is possible to write

$$[S'] = \begin{bmatrix} \dfrac{1}{E_1} & -\dfrac{\nu_{21}}{E_2} & -\dfrac{\nu_{31}}{E_3} & 0 & 0 & 0 \\ -\dfrac{\nu_{12}}{E_1} & \dfrac{1}{E_2} & -\dfrac{\nu_{32}}{E_3} & 0 & 0 & 0 \\ -\dfrac{\nu_{13}}{E_1} & -\dfrac{\nu_{23}}{E_2} & \dfrac{1}{E_3} & 0 & 0 & 0 \\ 0 & 0 & 0 & \dfrac{1}{G_{23}} & 0 & 0 \\ 0 & 0 & 0 & 0 & \dfrac{1}{G_{13}} & 0 \\ 0 & 0 & 0 & 0 & 0 & \dfrac{1}{G_{12}} \end{bmatrix} \qquad (1.91)$$

in terms of twelve engineering constants. However only nine constants are independent because the matrix $[S']$ must be symmetric (see 1.94), so

$$[S'] = \begin{bmatrix} \dfrac{1}{E_1} & -\dfrac{\nu_{12}}{E_1} & -\dfrac{\nu_{13}}{E_1} & 0 & 0 & 0 \\ -\dfrac{\nu_{12}}{E_1} & \dfrac{1}{E_2} & -\dfrac{\nu_{23}}{E_2} & 0 & 0 & 0 \\ -\dfrac{\nu_{13}}{E_1} & -\dfrac{\nu_{23}}{E_2} & \dfrac{1}{E_3} & 0 & 0 & 0 \\ 0 & 0 & 0 & \dfrac{1}{G_{23}} & 0 & 0 \\ 0 & 0 & 0 & 0 & \dfrac{1}{G_{13}} & 0 \\ 0 & 0 & 0 & 0 & 0 & \dfrac{1}{G_{12}} \end{bmatrix} \qquad (1.92)$$

The stiffness matrix can be computed also in terms of engineering constants by inverting the above equation so that $[C'] = [S']^{-1}$, with components given in terms of engineering constants as

$$C'_{11} = \frac{1 - \nu_{23}\nu_{32}}{E_2 E_3 \Delta}$$

$$C'_{12} = \frac{\nu_{21} + \nu_{31}\nu_{23}}{E_2 E_3 \Delta} = \frac{\nu_{12} + \nu_{32}\nu_{13}}{E_1 E_3 \Delta}$$

$$C'_{13} = \frac{\nu_{31} + \nu_{21}\nu_{32}}{E_2 E_3 \Delta} = \frac{\nu_{13} + \nu_{12}\nu_{23}}{E_1 E_2 \Delta}$$

$$C'_{22} = \frac{1 - \nu_{13}\nu_{31}}{E_1 E_3 \Delta}$$

$$C'_{23} = \frac{\nu_{32} + \nu_{12}\nu_{31}}{E_1 E_3 \Delta} = \frac{\nu_{23} + \nu_{21}\nu_{13}}{E_1 E_2 \Delta}$$

$$C'_{33} = \frac{1 - \nu_{12}\nu_{21}}{E_1 E_2 \Delta}$$

$$C'_{44} = G_{23}$$

$$C'_{55} = G_{13}$$

$$C'_{66} = G_{12} \tag{1.93}$$

$$\Delta = \frac{1 - \nu_{12}\nu_{21} - \nu_{23}\nu_{32} - \nu_{31}\nu_{13} - 2\nu_{21}\nu_{32}\nu_{13}}{E_1 E_2 E_3}$$

So far both $[S']$ and $[C']$ are 6×6 matrices with 9 independent constants for the case of orthotropic materials. If the material is transversely isotropic $G_{13} = G_{12}, \nu_{13} = \nu_{12}, E_3 = E_2$.

1.13.1   Restrictions on Engineering Constants

It is important to note that because of the symmetry of the compliance matrix (1.91), the following restrictions on engineering constants apply

$$\frac{\nu_{ij}}{E_i} = \frac{\nu_{ji}}{E_j}; \quad i, j = 1..3; \quad i \neq j \tag{1.94}$$

Further restrictions on the values of the elastic constants can be derived from the fact that all diagonal terms in both the compliance and stiffness matrices must be positive. Since all the engineering elastic constants must be positive ($E_1, E_2, E_3, G_{12}, G_{23}, G_{31} > 0$), all the diagonal terms of the stiffness matrix (1.93) will be positive if the following two conditions are met. The first condition is that $(1 - \nu_{ij}\nu_{ji}) > 0$ for $i, j = 1..3$ and $i \neq j$, which leads to the following restriction on the values of the engineering constants

$$0 < \nu_{ij} < \sqrt{\frac{E_i}{E_j}}; \quad i, j = 1..3; \quad i \neq j \tag{1.95}$$

The second condition is that

$$\Delta = 1 - \nu_{12}\nu_{21} - \nu_{23}\nu_{32} - \nu_{31}\nu_{13} - 2\nu_{21}\nu_{32}\nu_{13} > 0 \tag{1.96}$$

These restrictions can be used to check experimental data. For example, consider an experimental program in which if E_1 and ν_{12} are measured in a longitudinal test (fibers in the direction of loading) by using two strain gages, one longitudinal and one transverse, and E_2 and ν_{21} are measured in the transverse tensile tests (fibers perpendicular to loading). For the test procedure to be valid, all the four data values, E_1, E_2, ν_{12} and ν_{21} must conform to (1.94-1.96) within the margin allowed by experimental errors.

Example 1.4 *Sonti et al. [6] performed a series of tests on pultruded glass-fiber reinforced composites. From tensile tests along the longitudinal axis, the average of eight tests gives $E_1 = 19.981\,GPa$ and $\nu_{12} = 0.274$. The average of eight tests in the transverse direction gives $E_2 = 11.389\,GPa$ and $\nu_{21} = 0.192$. Does this data fall within the constraints on elastic constants?*

Solution to Example 1.4 *First compute both sides of (1.94) for $i, j = 1, 2$ as*

$$\frac{E_1}{\nu_{12}} = \frac{19.981}{0.274} = 72.9\,GPa$$

$$\frac{E_2}{\nu_{21}} = \frac{11.389}{0.192} = 59.3\,GPa$$

The transverse data is 23% lower than expected. Either E_2 measured is too low or ν_{21} measured is 23% higher than what it should be. In any case a 23% difference deserves some scrutiny.

Next check (1.95)

$$abs(\nu_{12}) < \sqrt{\frac{E_1}{E_2}}$$

$$0.274 < 1.32$$

$$abs(\nu_{21}) < \sqrt{\frac{E_2}{E_1}}$$

$$0.192 < 0.75$$

Finally, there is insufficient data to evaluate the last of the restrictions on elastic constants from (1.96).

1.14   From 3D to Plane Stress Equations

Setting $\sigma_3 = 0$ in the compliance equations (1.73) of an orthotropic material implies that the third row and column of the compliance matrix are not used

$$\begin{Bmatrix} \epsilon_1' \\ \epsilon_2' \\ \epsilon_3' \\ \gamma_4' \\ \gamma_5' \\ \gamma_6' \end{Bmatrix} = \begin{bmatrix} S_{11}' & S_{12}' & S_{13}' & 0 & 0 & 0 \\ S_{12}' & S_{22}' & S_{23}' & 0 & 0 & 0 \\ S_{13}' & S_{23}' & S_{33}' & 0 & 0 & 0 \\ 0 & 0 & 0 & S_{44}' & 0 & 0 \\ 0 & 0 & 0 & 0 & S_{55}' & 0 \\ 0 & 0 & 0 & 0 & 0 & S_{66}' \end{bmatrix} \begin{Bmatrix} \sigma_1' \\ \sigma_2' \\ \sigma_3' = 0 \\ \sigma_4' \\ \sigma_5' \\ \sigma_6' \end{Bmatrix} \qquad (1.97)$$

so, the first two equations plus the last one can be written separately of the remaining, in terms of a 3×3 reduced compliance matrix $[S]$ and using $\gamma = 2\epsilon$, we have

$$
\left\{ \begin{array}{c} \epsilon'_1 \\ \epsilon'_2 \\ \gamma'_6 \end{array} \right\} = \left[\begin{array}{ccc} S'_{11} & S'_{12} & 0 \\ S'_{12} & S'_{22} & 0 \\ 0 & 0 & S'_{66} \end{array} \right] \left\{ \begin{array}{c} \sigma'_1 \\ \sigma'_2 \\ \sigma'_6 \end{array} \right\} \tag{1.98}
$$

The third equation is seldom used

$$
\epsilon'_3 = S'_{13}\sigma'_1 + S'_{23}\sigma'_2 \tag{1.99}
$$

and the remaining two equations can be written separately as

$$
\left\{ \begin{array}{c} \gamma'_4 \\ \gamma'_5 \end{array} \right\} = \left[\begin{array}{cc} S'_{44} & 0 \\ 0 & S'_{55} \end{array} \right] \left\{ \begin{array}{c} \sigma'_4 \\ \sigma'_5 \end{array} \right\} \tag{1.100}
$$

To compute stress components from strains, (1.98) can be inverted to get $\{\sigma\} = [Q]\{\epsilon\}$ or

$$
\left\{ \begin{array}{c} \sigma'_1 \\ \sigma'_2 \\ \sigma'_6 \end{array} \right\} = \left[\begin{array}{ccc} Q'_{11} & Q'_{12} & 0 \\ Q'_{12} & Q'_{22} & 0 \\ 0 & 0 & Q'_{66} \end{array} \right] \left\{ \begin{array}{c} \epsilon'_1 \\ \epsilon'_2 \\ \gamma'_6 \end{array} \right\} \tag{1.101}
$$

where the matrix $[Q'] = [S'_{3\times3}]^{-1}$ is the reduced stiffness matrix for plane stress. Note that while the components of the reduced compliance matrix $[S'_{3\times3}]$ are numerically identical to the corresponding entries in the 6×6 compliance matrix, the components of the reduced stiffness matrix $[Q']$ are not numerically equal to the corresponding entries on the 6×6 stiffness matrix $[C']$, thus the change in name. This is because the inverse of a 3×3 matrix produces different values than the inverse of a 6×6 matrix. The set of equations is completed by writing

$$
\left\{ \begin{array}{c} \sigma'_4 \\ \sigma'_5 \end{array} \right\} = \left[\begin{array}{cc} C'_{44} & 0 \\ 0 & C'_{55} \end{array} \right] \left\{ \begin{array}{c} \gamma'_4 \\ \gamma'_6 \end{array} \right\} \tag{1.102}
$$

where the coefficient C'_{44} and C'_{55} are numerically equal to the corresponding entries in the 6×6 stiffness matrix because the 2×2 matrix in (1.102) is diagonal.

Example 1.5 *Show that the change in the thickness $t\epsilon_3$ of a plate is negligible when compared to the in-plane elongations $a\epsilon_1$ and $b\epsilon_2$. Use the data from a composite plate with thickness $t = 0.635$ mm, and dimensions $a = 279$ mm and $b = 203$ mm. Take $E_1 = 19.981$ GPa, $E_2 = 11.389$ GPa, $\nu_{12} = 0.274$.*

Solution to Example 1.5 *Assuming that the 0.635 mm thick glass-reinforced Polyester plate is transversely isotropic, take $E_3 = E_2 = 11.389$ GPa, $\nu_{13} = \nu_{12} = 0.274$, $G_{31} = G_{12}$. Sonti et al. [6] report the average of eight torsion tests as $G_{12} = 3.789$ GPa. Lacking experimental data, assume $\nu_{23} \approx \nu_m = 0.3$, $G_{23} \approx G_m = 0.385$ GPa, with the properties of the Polyester matrix taken from [1]. The remaining properties in (1.92) can be obtained,* *using (1.94), as*

$$\nu_{21} = \nu_{12}\frac{E_2}{E_1} = 0.274 \left(\frac{11.389}{19.981}\right) = 0.156$$

$$\nu_{31} = \nu_{13}\frac{E_3}{E_1} = 0.274 \left(\frac{11.389}{19.981}\right) = 0.156$$

$$\nu_{32} = \nu_{23}\frac{E_3}{E_2} = 0.3 \left(\frac{11.389}{11.389}\right) = 0.3$$

Because transverse isotropy $G_{13} = G_{12} = 3.789$ *GPa. Now, assume a state of stress* $\sigma'_1 = \sigma'_2 = 0.1$ *GPa,* $\sigma'_4 = \sigma'_5 = \sigma'_6 = 0$ *and* $\sigma'_3 = 0$ *because of the assumption of plane stress. Using (1.98) we get*

$$\epsilon'_1 = S'_{11}\sigma'_1 + S'_{12}\sigma'_2 = \frac{0.1}{19.981} - \frac{0.1(0.156)}{11.389} = 3.635 \ 10^{-3}$$

$$\epsilon'_2 = S'_{12}\sigma'_1 + S'_{22}\sigma'_2 = -\frac{0.1(0.156)}{11.389} + \frac{0.1}{11.389} = 7.411 \ 10^{-3}$$

$$\epsilon'_3 = S'_{13}\sigma'_1 + S'_{23}\sigma'_2 = -\frac{0.274(0.1)}{19.981} - \frac{0.3(0.1)}{11.389} = -4.005 \ 10^{-3}$$

Finally

$$t\epsilon'_3 = -0.635(4.005 \ 10^{-3}) = -2.543 \ 10^{-3} \ mm$$

$$a\epsilon'_1 = 279(3.635 \ 10^{-3}) = 1.014 \ mm$$

$$b\epsilon'_2 = 203(7.411 \ 10^{-3}) = 1.504 \ mm$$

Since the elongation in the transverse direction is so small, it is neglected in the derivation of the plate equations in [1].

1.15   Apparent Laminate Properties

The global stiffness matrix $[C]$ of a symmetric laminate with N layers is built by adding the global matrices of the layers multiplied by the thickness ratio t_k/t of each layer, where t is the laminate thickness and t_k denotes the thickness of the k-th layer

$$[C] = \sum_{k=1}^{N} \frac{t_k}{t}[C_k] \tag{1.103}$$

Note that compliances cannot be added nor averaged. The laminate compliance is obtained inverting the 6×6 stiffness matrix, as

$$[S] = [C]^{-1} \tag{1.104}$$

A laminate is called balanced if the total thickness of layers oriented with respect to the global direction at $+\theta$ and $-\theta$ are the same. Such laminate has orthotropic stiffness $[C]$ and compliance $[S]$. In terms of the apparent engineering properties of the laminate, the compliance is

$$[S] = \begin{bmatrix} \dfrac{1}{E_x} & -\dfrac{\nu_{yx}}{E_x} & -\dfrac{\nu_{zx}}{E_x} & 0 & 0 & 0 \\[8pt] -\dfrac{\nu_{xy}}{E_y} & \dfrac{1}{E_y} & -\dfrac{\nu_{zy}}{E_y} & 0 & 0 & 0 \\[8pt] -\dfrac{\nu_{xz}}{E_z} & -\dfrac{\nu_{yz}}{E_z} & \dfrac{1}{E_z} & 0 & 0 & 0 \\[8pt] 0 & 0 & 0 & \dfrac{1}{G_{yz}} & 0 & 0 \\[8pt] 0 & 0 & 0 & 0 & \dfrac{1}{G_{xz}} & 0 \\[8pt] 0 & 0 & 0 & 0 & 0 & \dfrac{1}{G_{xy}} \end{bmatrix} \tag{1.105}$$

Since the compliance must be symmetric, it must satisfy (1.94) with $i,j = x,y,z$. Therefore, it is possible to compute the apparent engineering properties of a laminate in terms of the laminate compliance, as follows

$$\begin{aligned} E_x &= 1/S_{11} & \nu_{xy} &= -S_{21}/S_{11} \\ E_y &= 1/S_{22} & \nu_{xz} &= -S_{31}/S_{11} \\ E_z &= 1/S_{33} & \nu_{yz} &= -S_{32}/S_{22} \\ G_{yz} &= 1/S_{44} \\ G_{xz} &= 1/S_{55} \\ G_{xy} &= 1/S_{66} \end{aligned} \tag{1.106}$$

Example 1.6 *Compute the laminate properties of $[0/90/\pm 30]_S$ with $t_k = 1.5$ mm, $E_f = 241$ GPa, $\nu_f = 0.2$, $E_m = 3.12$ GPa, $\nu_m = 0.38$, fiber volume fraction $V_f = 0.6$, where f,m, denote fiber and matrix, respectively.*

Solution to Example 1.6 *First use periodic microstructure micromechanics (6.8) to obtain the layer properties (in MPa).*

$$\begin{aligned} E_1 &= 145,880 & G_{12} &= 4,386 & \nu_{12} &= \nu_{13} = 0.263 \\ E_2 &= 13,312 & G_{23} &= 4,528 & \nu_{23} &= 0.470 \end{aligned}$$

Then, compute the compliance matrix $[S']$ using (1.92), rotation matrix $[T]$ using (1.33), global compliance $[S]$ using (1.52) and global stiffness $[C] = [S]^{-1}$ for each layer. Then, average them using (1.103), invert it, and finally using (1.106) get

$$\begin{aligned} E_x &= 78,901 & G_{xy} &= 17,114 & \nu_{xy} &= 0.320 \\ E_y &= 47,604 & G_{yz} &= 4,475 & \nu_{yz} &= 0.364 \\ E_z &= 16,023 & G_{xz} &= 4,439 & \nu_{xz} &= 0.280 \end{aligned}$$

Suggested Problems

Problem 1.1 *Using the principle of virtual work (PVW), find a quadratic displacement function $u(x)$ in $0 < x < L$ of a tapered slender rod of length L, fixed at the origin and* loaded axially in tension at the free end. The cross section area changes lineally and the areas are $A_1 > A_2$ at the fixed and free ends, respectively. The material is homogeneous and isotropic with modulus E.

Problem 1.2 Using the principle of virtual work (PVW), find a quadratic rotation angle function $\theta(x)$ in $0 < x < L$ of a tapered slender shaft of circular cross section and length L, fixed at the origin and loaded by a torque T at the free end. The cross section area changes lineally and the areas are $A_1 > A_2$ at the fixed and free ends, respectively. The material is homogeneous and isotropic with shear modulus G.

Problem 1.3 Construct a rotation matrix $[a]$ resulting from three consecutive reflections about (a) the x-y plane, (b) the x-z plane, (c) the y-z plane. The resulting system does not follow the right-hand-rule.

Problem 1.4 Construct three rotation matrices $[a]$ for rotations $\theta = \pi$ about (a) the x-axis, (b) the y-axis, (c) the z-axis.

Problem 1.5 Write a computer program to evaluate the compliance and stiffness matrices in terms of engineering properties. Take the input from a file and the output to another file. Validate the program with your own examples. You may use material properties from [1, Table 1.1] and assume the material is transversely isotropic as per Section 1.12.4. Show all work in a report.

Problem 1.6 Write a computer program to transform the stiffness and compliance matrix from material coordinates C', S', to another coordinate system C, S, by a rotation $-\theta$ around the z-axis (Figure 1.7). The data C', S', θ, should be read from a file. The output C, S should be written to another file. Validate your program with your own examples. You may use material properties from [1, Table 1.1] and assume the material is transversely isotropic as per Section 1.12.4. Show all work in a report.

Problem 1.7 Verify numerically (1.93) against $[S]^{-1}$ for the material of your choice. You may use material properties from [1, Table 1.1] and assume the material is transversely isotropic as per Section 1.12.4.

Problem 1.8 The following data has been obtained experimentally for a composite based on a unidirectional carbon-epoxy prepreg (MR50 carbon fiber at 63% by volume in LTM25 Epoxy). Determine if the restrictions on elastic constants are satisfied.

$$E_1 = 156.403 \; GPa, \qquad E_2 = 7.786 \; GPa$$
$$\nu_{12} = 0.352, \qquad \nu_{21} = 0.016$$
$$G_{12} = 3.762 \; GPa$$
$$\sigma_{1t}^u = 1.826 \; GPa, \qquad \sigma_{1c}^u = 1.134 \; GPa$$
$$\sigma_{2t}^u = 19 \; MPa, \qquad \sigma_{2c}^u = 131 \; MPa$$
$$\sigma_6^u = 75 \; MPa$$
$$\epsilon_{1t}^u = 11,900 \; 10^{-6}, \qquad \epsilon_{1c}^u = 8,180 \; 10^{-6}$$
$$\epsilon_{2t}^u = 2,480 \; 10^{-6}, \qquad \epsilon_{2c}^u = 22,100 \; 10^{-6}$$
$$\gamma_{12}^u = 20,000 \; 10^{-6}$$

Problem 1.9 Explain contracted notation for stresses and strains.

Problem 1.10 *What is an orthotropic material and how many constants are needed to describe it?*

Problem 1.11 *What is a transversely isotropic material and how many constants are needed to describe it?*

Problem 1.12 *Use the three rotations matrices in Problem 1.4 to verify (1.47) numerically.*

Problem 1.13 *Prove (1.77) using (1.75) and (1.92).*

Problem 1.14 *Demonstrate that a material having two perpendicular planes of symmetry also has a third. Apply a reflection about the 2-3 plane to (1.72) using the procedure in Section 1.12.3.*

Problem 1.15 *What is a plane stress assumption?*

Problem 1.16 *Write a computer program to evaluate the laminate engineering properties for symmetric balanced laminates. All layers are of the same material. Input data consists of all the engineering constants for a transversely isotropic material, number of layers N, thickness and angle for all the layers t_k, θ_k with $k = 1...N$. Use Section 1.15, 1.12.4, and 1.13.*

References

[1] E. J. Barbero, Introduction to Composite Materials Design, Taylor & Francis, Philadelphia, PA, 1999.

[2] D. Frederick, T.-S. Chang, Continuum Mechanics, Scientific Publishers, Cambridge, 1972.

[3] F. P. Beer, E. R. Johnston Jr., J. T. DeWolf, Mechanics of Materials, 3rd Ed., McGraw Hill, Boston, MA, 2001.

[4] J. N. Reddy, Energy and Variational Methods in Applied Mechanics, John Wiley, New York, 1984.

[5] E. J. Barbero, Web resource: http://www.mae.wvu.edu/barbero/feacm/

[6] S. S. Sonti, E. J. Barbero, T. Winegardner, Mechanical Properties of Pultruded E-Glass/Vinyl Ester Composites, 50th Annual Conference, Composites Institute, Society of the Plastics Industry (February 1995) pp. 10-C/1-7.

Chapter 2

Introduction to the Finite Element Method

In this textbook, the finite element method (FEM) is used as a tool to solve practical problems. For the most part, commercial packages, mainly ANSYS, are used in the examples. Finite element programming is limited to programming material models and post processing algorithms. When commercial codes lack needed features, other codes are used, which are provided in [1]. Therefore, some understanding of the finite element method and associated technology are necessary. This chapter contains a brief introduction intended for those readers who have not had a formal course or prior knowledge about the finite element method.

2.1 Basic FEM procedure

Consider the axial deformation of a bar. The ordinary differential equation (ODE) describing the deformation of the bar is

$$-\frac{d}{dx}\left(EA\frac{du}{dx}\right) - f = 0 \quad ; \quad 0 \le x \le L \tag{2.1}$$

where E, A are the modulus and cross section area of the bar, respectively, and f is the distributed force. The boundary conditions for the case illustrated in Figure 2.1 are

$$u(0) = 0$$
$$\left[\left(EA\frac{du}{dx}\right)\right]_{x=L} = P \tag{2.2}$$

As it is customary in strength of materials textbooks, the real bar shown in Figure 2.1(a) is mathematically modeled as a line in Figure 2.1(b). The bar occupies the domain $[0, L]$ along the real axis x.

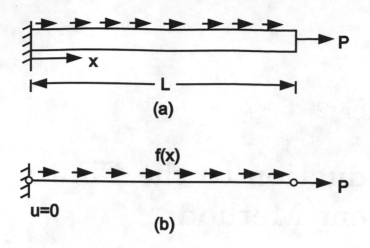

Fig. 2.1: Physical and mathematical (idealization) model.

2.1.1 Discretization

The next step is to divide the domain into discrete elements, as shown in Figure 2.2.

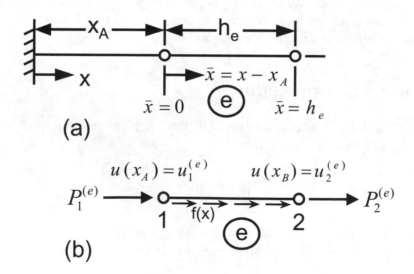

Fig. 2.2: Discretization into elements.

2.1.2 Element Equations

To derive the element equations, an integral form of the ordinary differential equation (ODE) is used, which is obtained by integrating the product of the ODE times a weight function v as follows

$$0 = \int_{x_A}^{x_B} v \left[-\frac{d}{dx} \left(EA \frac{du}{dx} \right) - f \right] dx. \tag{2.3}$$

This is called a weak form because the solution $u(x)$ does not have to satisfy the ODE (2.1) for all and every of the infinite values of x in $[0, L]$, in a strong sense. Instead, the solution $u(x)$ only has to satisfy the ODE in (2.3) in a weighted average sense. It is therefore easier to find a weak solution than a strong one. Although for the case of the bar, the strong (exact) solution is known, most problems of composite mechanics do not have exact solution. The governing equation is obtained by integrating (2.3) by parts as follows

$$0 = \int_{x_A}^{x_B} EA \frac{dv}{dx} \frac{du}{dx} dx - \int_{x_A}^{x_B} v f dx - \left[v \left(EA \frac{du}{dx} \right) \right]_{x_A}^{x_B} \tag{2.4}$$

where $v(x)$ is a weight function, which is usually set equal to the primary variable $u(x)$. From the boundary term, it is concluded that

-specifying $v(x)$ at x_A or x_B is an essential boundary condition

-specifying $\left(EA \frac{du}{dx} \right)$ at either end is the natural boundary condition

While $u(x)$ is the *primary variable*, $\left(EA \frac{du}{dx} \right) = EA\epsilon_x = A\sigma_x$ is the *secondary variable*. Let

$$u(x_A) = u_1^e$$
$$u(x_B) = u_2^e$$
$$-\left[\left(EA \frac{du}{dx} \right) \right]_{x_A} = P_1^e$$
$$\left[\left(EA \frac{du}{dx} \right) \right]_{x_B} = P_2^e \tag{2.5}$$

Then, the governing equation becomes

$$0 = \int_{x_A}^{x_B} \left(EA \frac{dv}{dx} \frac{du}{dx} - vf \right) dx - P_1^e v(x_A) - P_2^e v(x_B) = B(v, u) - l(v) \tag{2.6}$$

with

$$B(u, v) = \int_{x_A}^{x_B} EA \frac{dv}{dx} \frac{du}{dx} dx$$
$$l(v) = \int_{x_A}^{x_B} v f dx + P_1^e v(x_A) + P_2^e v(x_B) \tag{2.7}$$

2.1.3 Approximation over an Element

Now, the unknown $u(x)$ is approximated as a linear combination (series expansion) of known functions $N_i^e(x)$ and unknown coefficients a_j^e, as

$$u_e(x) = \sum_{j=1}^{n} a_j^e N_j^e(x)$$

where a_j^e are the coefficients to be found and $N_j^e(x)$ are the interpolation functions. For the weight function $v(x)$, the Ritz method can be used [2], in which $v(x) = N_j^e(x)$. Substituting in the governing equation (2.6) we get

$$0 = \sum_{j=1}^{n} \left(\int_{x_A}^{x_B} EA \frac{dN_i^e}{dx} \frac{dN_j^e}{dx} dx \right) a_j^e - \int_{x_A}^{x_B} N_i^e f dx - P_1^e N_i^e(x_A) - P_2^e N_i^e(x_B) \quad (2.8)$$

or

$$\sum_{j=1}^{n} \left(\int_{x_A}^{x_B} EA \frac{dN_i^e}{dx} \frac{dN_j^e}{dx} dx \right) a_j^e = \int_{x_A}^{x_B} N_i^e f dx + P_1^e N_i^e(x_A) + P_2^e N_i^e(x_B) \quad (2.9)$$

which can be written as

$$\sum_{j=1}^{n} K_{ij}^e a_j^e = F_i^e \quad (2.10)$$

or in matrix form

$$[K^e]\{a^e\} = \{F^e\} \quad (2.11)$$

where $[K^e]$ is the element stiffness matrix, $\{F^e\}$ is the element vector equivalent force and $\{a^e\}$ are the element unknown parameters.

2.1.4   Interpolation Functions

Although any complete set of linearly independent functions could be used as interpolation functions, it is convenient to choose the function in such a way that the unknown coefficients represent the nodal displacements, that is $a_i = u_i$. For a two-node element with $x_e \leq x \leq x_{e+1}$, the following linear interpolation functions (Figure 2.3) can be used

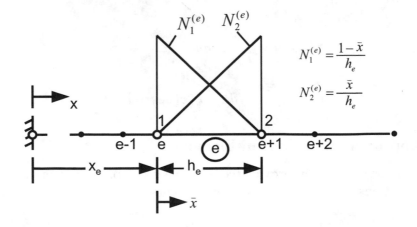

Fig. 2.3: Linear interpolation functions for a two-node element bar.

$$N_1^e = \frac{x_{e+1} - x}{x_{e+1} - x_e}$$

$$N_2^e = \frac{x - x_e}{x_{e+1} - x_e} \tag{2.12}$$

which satisfy the following conditions

$$N_i^e(x_j) = \begin{cases} 0 \ if \ i \neq j \\ 1 \ if \ i = j \end{cases} \tag{2.13}$$

$$\sum_{i=1}^{2} N_i^e(x) = 1 \tag{2.14}$$

Many other interpolation functions can be used, each one with some advantages and disadvantages. The interpolation functions are intimately related to the number of nodes of the element. Figure 2.4 illustrates the shape of the interpolation functions N_1 and N_5 (corresponding to nodes 1 and 5) in an eight-node shell element.

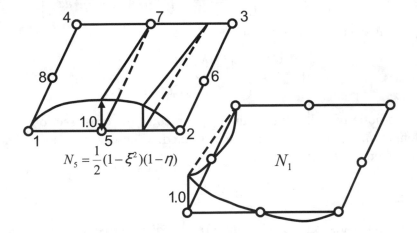

Fig. 2.4: Two-dimensional interpolation functions.

Broadly speaking, more nodes per element imply more accuracy and less need for a fine mesh, but also imply higher cost in terms of computer time. Figure 2.5 illustrates how the approximate solution converges to the exact one as the number of elements increases from 2 to 4 or as the number of nodes in the element increases from 2 for the linear element to 3 for the quadratic element.

2.1.5 Element Equations for a Specific Problem

With interpolation functions that satisfy the conditions in (2.13-2.14), it is possible to rewrite (2.11) as

$$[K^e]\{u^e\} = \{F^e\} \tag{2.15}$$

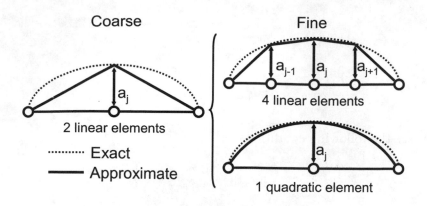

Fig. 2.5: Discretization error.

where $\{u^e\}$ are the nodal displacements, $[K^e]$ is the element stiffness matrix given by

$$[K^e] = \begin{bmatrix} \int_{x_A}^{x_B} EA \dfrac{dN_1^e}{dx}\dfrac{dN_1^e}{dx}dx & \int_{x_A}^{x_B} EA \dfrac{dN_1^e}{dx}\dfrac{dN_2^e}{dx}dx \\ \int_{x_A}^{x_B} EA \dfrac{dN_2^e}{dx}\dfrac{dN_1^e}{dx}dx & \int_{x_A}^{x_B} EA \dfrac{dN_2^e}{dx}\dfrac{dN_2^e}{dx}dx \end{bmatrix} \qquad (2.16)$$

and $\{F^e\}$ is the element force vector

$$\{F_i^e\} = \left\{ \begin{array}{c} \int_{x_A}^{x_B} N_1^e f dx + P_1^e \\ \int_{x_A}^{x_B} N_2^e f dx + P_2^e \end{array} \right\} \qquad (2.17)$$

For a two-node bar element number e, the constant cross-section area A_e, the element length h_e, and the modulus E are fixed. These values define the tensile-compression element stiffness as

$$k^e = \frac{EA_e}{h_e} \qquad (2.18)$$

The external loads on the element are the distributed force f_e, the force at end number 1, P_1^e, and the force at end number 2 , P_2^e. Using these values, the linear interpolation functions (2.12), as well as (2.16) and (2.17), the element matrix stiffness and the equivalent nodal forces become

$$[K^e] = \begin{bmatrix} k^e & -k^e \\ -k^e & k^e \end{bmatrix} = \frac{EA_e}{h_e} \begin{bmatrix} 1 & -1 \\ -1 & 1 \end{bmatrix} \qquad (2.19)$$

$$\{F^e\} = \frac{f_e h_e}{2} \left\{ \begin{array}{c} 1 \\ 1 \end{array} \right\} + \left\{ \begin{array}{c} P_1^e \\ P_2^e \end{array} \right\} \qquad (2.20)$$

2.1.6 Assembly of Element Equations

The element unknown parameters correspond to displacements at the element nodes. Since a node must have the same displacement on both adjacent elements, the value is unique. For example, using the connectivity of elements shown in Figure 2.6, unique labels are assigned to the displacements, using capital letters. While a superscript denotes an element number, a subscript indicates a nodal number, as follows

$$u_1^1 = U_1$$
$$u_2^1 = U_2 = u_1^2$$
$$u_1^2 = U_3 = u_1^3$$
$$u_2^3 = U_4 \tag{2.21}$$

Fig. 2.6: Connectivity between three two-node elements.

Now, the element equations can be assembled into the global system. First, the contribution of element #1 is

$$\begin{bmatrix} k^1 & -k^1 & 0 & 0 \\ -k^1 & k^1 & 0 & 0 \\ 0 & 0 & 0 & 0 \\ 0 & 0 & 0 & 0 \end{bmatrix} \begin{Bmatrix} U_1 \\ U_2 \\ U_3 \\ U_4 \end{Bmatrix} = \begin{Bmatrix} f_1 h_1/2 \\ f_1 h_1/2 \\ 0 \\ 0 \end{Bmatrix} + \begin{Bmatrix} P_1^1 \\ P_2^1 \\ 0 \\ 0 \end{Bmatrix} \tag{2.22}$$

Add contribution of element #2, as follows

$$\begin{bmatrix} k^1 & -k^1 & 0 & 0 \\ -k^1 & k^1 + k^2 & -k^2 & 0 \\ 0 & -k^2 & k^2 & 0 \\ 0 & 0 & 0 & 0 \end{bmatrix} \begin{Bmatrix} U_1 \\ U_2 \\ U_3 \\ U_4 \end{Bmatrix} = \begin{Bmatrix} f_1 h_1/2 \\ f_1 h_1/2 + f_2 h_2/2 \\ f_2 h_2/2 \\ 0 \end{Bmatrix} + \begin{Bmatrix} P_1^1 \\ P_2^1 + P_1^2 \\ P_2^2 \\ 0 \end{Bmatrix} \tag{2.23}$$

Finally, add element #3 to obtain the fully assembled system, as follows

$$\begin{bmatrix} k^1 & -k^1 & 0 & 0 \\ -k^1 & k^1 + k^2 & -k^2 & 0 \\ 0 & -k^2 & k^2 + k^3 & -k^3 \\ 0 & 0 & -k^3 & k^3 \end{bmatrix} \begin{Bmatrix} U_1 \\ U_2 \\ U_3 \\ U_4 \end{Bmatrix} = \begin{Bmatrix} f_1 h_1/2 \\ f_1 h_1/2 + f_2 h_2/2 \\ f_2 h_2/2 + f_3 h_3/2 \\ f_3 h_3/2 \end{Bmatrix} + \begin{Bmatrix} P_1^1 \\ P_2^1 + P_1^2 \\ P_2^2 + P_1^3 \\ P_3^2 \end{Bmatrix} \tag{2.24}$$

2.1.7 Boundary Conditions

By equilibrium (see Figure 2.2), the internal loads cancel whenever two elements share a node, or

$$P_2^1 + P_1^2 = 0$$
$$P_2^2 + P_1^3 = 0 \tag{2.25}$$

The remaining P_1^1 and P_3^2 are the forces at the end of the bar. If either end of the bar is fixed, then the displacement must be set to zero at that end. Say the end at $x = 0$ is fixed, then $U_1 = 0$. If the end at $x = L$ is free, then P_3^2 must be specified, since $U_4 \neq 0$. If it is not specified, then it is assumed that the force is zero.

2.1.8   Solution of the Equations

Since $U_1 = 0$, eliminating the first row and column of the stiffness matrix, a 3×3 system of algebraic equations is obtained, and solved for 3 unknowns: U_2, U_3, U_4. Once a solution for U_2 is found, the reaction P_1^1 is computed from the first equation of (2.24), as follows

$$-k^1 U_2 = \frac{f_1 h_1}{2} + P_1^1 \qquad (2.26)$$

2.1.9   Solution Inside the Elements

Now, the solution U_i at 4 points along the bar is available. Next, the solution at any location x can be computed by interpolating with the interpolation functions, as follows

$$U^e(x) = \sum_{j=1}^{2} U_j^e N_j^e(x) \qquad (2.27)$$

or

$$u(x) = \begin{cases} U_1 N_1^1(x) + U_2 N_2^1(x) & if \ \ 0 \leq x \leq h_1 \\ U_2 N_1^2(x) + U_3 N_2^2(x) & if \ \ h_1 \leq x \leq h_1 + h \\ U_3 N_1^3(x) + U_4 N_2^3(x) & if \ \ h_1 + h_2 \leq x \leq h_1 + h \end{cases} \qquad (2.28)$$

2.1.10   Derived Results

Strains

Strains are computed using (1.5) directly from the known displacements inside the element. For example,

$$\epsilon_x = \frac{du}{dx} = \sum_{j=1}^{2} U_j^e \frac{dN_j^e}{dx} \qquad (2.29)$$

Note that if $N_j^e(x)$ are linear functions, the strains are constant over the element. In general the quality of the strains is one order of magnitude poorer than the primary variable (displacements).

Stresses

Stress values are usually computed from strains through the constitutive equations. In this example, with one-dimensional stress-strain behavior

$$\sigma_x = E \, \epsilon_x \qquad (2.30)$$

Note that the quality of stresses is the same as that of the strains.

2.2   General FEM Procedure

The derivation of the element equations, assembly, and solution for any type of
elements is similar to that of the one-dimensional bar element described in Section
2.1, with the exception that the principle of virtual work (1.16) is used instead of
the governing equation (2.1). The PVW provides a weak form similar to that in
(2.4). Expanding (1.16) for full 3D state of deformation, the internal virtual work
is

$$\delta W_I = \int (\sigma_{xx}\delta\epsilon_{xx} + \sigma_{yy}\delta\epsilon_{yy} + \sigma_{zz}\delta\epsilon_{zz} + \sigma_{yz}\delta\gamma_{yz} + \sigma_{xz}\delta\gamma_{xz} + \sigma_{xy}\delta\gamma_{xy})\, dV$$

$$= \int_V \underline{\sigma}^T \delta\underline{\epsilon}\, dV \tag{2.31}$$

where

$$\underline{\sigma}^T = [\sigma_{xx}, \sigma_{yy}, \sigma_{zz}, \sigma_{yz}, \sigma_{xz}, \sigma_{xy}]$$

$$\delta\underline{\epsilon}^T = [\delta\epsilon_{xx}, \delta\epsilon_{yy}, \delta\epsilon_{zz}, \delta\gamma_{yz}, \delta\gamma_{xz}, \delta\gamma_{xy}] \tag{2.32}$$

Next, the external work is

$$\delta W_E = \int_V \underline{f}^T \delta\underline{u}\, dV + \int_S \underline{t}^T \delta\underline{u}\, dS \tag{2.33}$$

where the volume forces per unit volume and surface forces per unit area are

$$\underline{f}^T = [f_x, f_y, f_z]$$

$$\underline{t}^T = [t_x, t_y, t_z] \tag{2.34}$$

Here, underline (_) denotes a one-dimensional array, not necessarily a vector.
For example, $\underline{u}$ is a vector but $\underline{\sigma}$ are the six components of stress arranged in
a six-element array. The virtual strains are strains that would be produced by
virtual displacements $\delta\underline{u}\,(\underline{x})$. Therefore, virtual strains are computed from virtual
displacements using the strain-displacement equations (1.5). In matrix notation

$$\underline{\epsilon} = \underline{\underline{\partial}}\, \underline{u}$$

$$\delta\underline{\epsilon} = \underline{\underline{\partial}}\, \delta\underline{u} \tag{2.35}$$

where

$$\underline{\underline{\partial}} = \begin{bmatrix} \dfrac{\partial}{\partial x} & 0 & 0 & \dfrac{\partial}{\partial y} & 0 & \dfrac{\partial}{\partial z} \\[2mm] 0 & \dfrac{\partial}{\partial y} & 0 & \dfrac{\partial}{\partial x} & \dfrac{\partial}{\partial z} & 0 \\[2mm] 0 & 0 & \dfrac{\partial}{\partial z} & 0 & \dfrac{\partial}{\partial y} & \dfrac{\partial}{\partial x} \end{bmatrix} \tag{2.36}$$

Then, the PVW is written in matrix notation as

$$\int_V \underline{\sigma}^T \underline{\underline{\partial}} \, \underline{\delta u} \, dV = \int_V \underline{f}^T \underline{\delta u} \, dV + \int_S \underline{t}^T \underline{\delta u} \, dS \qquad (2.37)$$

The integrals over the volume V and surface S of the body can be broken element by element over m elements, as

$$\sum_{e=1}^m \left[\int_{V_e} \underline{\sigma}^T \underline{\underline{\partial}} \, \underline{\delta u} \, dV \right] = \sum_{e=1}^m \left[\int_{V_e} \underline{f}^T \underline{\delta u} \, dV + \int_{S_e} \underline{t}^T \underline{\delta u} \, dS \right] \qquad (2.38)$$

Whenever two elements share a surface, the contributions of the second integral cancel out, just as the internal loads canceled in Section 2.1.7. The stress components are given by the constitutive equations. For a linear material

$$\underline{\sigma} = \underline{\underline{C}} \, \underline{\epsilon} \qquad (2.39)$$

with $\underline{\underline{C}}$ given by (1.72). The internal virtual work over each element becomes

$$\delta W_I^e = \int_{V_e} \underline{\sigma}^T \underline{\delta \epsilon} \, dV = \int_{V_e} \underline{\epsilon}^T \underline{\underline{C}} \, \underline{\delta \epsilon} \, dV \qquad (2.40)$$

The expansion of the displacements can be written in matrix form as

$$\underline{u} = \underline{N} \, \underline{a} \qquad (2.41)$$

where $\underline{N}$ contains the element interpolation functions and $\underline{a}$ the nodal displacements of the element, just as in Section 2.1.4. Therefore, the strains are

$$\underline{\epsilon} = \underline{\underline{\partial}} \, \underline{u} = \underline{\underline{\partial}} \, \underline{N} \, \underline{a} = \underline{\underline{B}} \, \underline{a} \qquad (2.42)$$

where $\underline{\underline{B}} = \underline{\underline{\partial}} \, \underline{N}$ is the strain-displacement matrix. Now, the discretized form of the internal virtual work over an element can be computed as

$$\delta W_I^e = \int_{V_e} \underline{a}^T \underline{\underline{B}}^T \underline{\underline{C}} \, \underline{\underline{B}} \, \underline{\delta a} \, dV = \underline{a}^T \int_{V_e} \underline{\underline{B}}^T \underline{\underline{C}} \, \underline{\underline{B}} \, dV \underline{\delta a} = \underline{a}^T \underline{\underline{K}}^e \, \underline{\delta a} \qquad (2.43)$$

where the element stiffness matrix K^e is

$$\underline{\underline{K}}^e = \int_{V_e} \underline{\underline{B}}^T \underline{\underline{C}} \, \underline{\underline{B}} \, dV \qquad (2.44)$$

The external virtual work becomes

$$\delta W_E^e = \int_{V_e} \underline{f}^T \underline{\delta u} \, dV + \int_{S_e} \underline{t}^T \underline{\delta u} \, dS$$

$$= \left(\int_{V_e} \underline{f}^T \underline{N} \, dV + \int_{S_e} \underline{t}^T \underline{N} \, dS \right) \underline{\delta a} = (\underline{P}^e)^T \underline{\delta a} \qquad (2.45)$$

where the element force vector is

$$\underline{P}^e = \int_{V_e} \underline{N}^T \underline{f} \, dV + \int_{S_e} \underline{N}^T \underline{t} \, dS \tag{2.46}$$

The integrals over the element volume V_e and element surface S_e are usually evaluated numerically by the Gauss integration procedure. For the volume integral, such a procedure needs evaluation of the integrand at a few points inside the volume. Such points, which are called Gauss points, are important for two reasons. First, the constitutive matrix C is evaluated at those locations. Second, the most accurate values of strains (and stresses) are obtained at those locations too.

The assembly of the element equations δW_I^e and δW_E^e into the PVW for the whole body is done similarly to the process in Section 2.1.6. Obviously the process is more complicated than for bar elements. The details of such process and its computer programming are part of finite element technology, which is outside the scope of this textbook. Eventually all the element stiffness matrices K^e and element force vectors P^e are assembled into a global system for the whole body

$$\underline{\underline{K}} \, \underline{a} = \underline{P} \tag{2.47}$$

Next, boundary conditions are applied on the system (2.47) in a systematic way resembling the procedure in Section 2.1.6. Next, the algebraic system of equations (2.47) is solved to find the nodal displacement array $\underline{a}$ over the whole body. Since the nodal displacements results for every element can be found somewhere in $\underline{a}$, it is possible to go back to (2.35) and to (2.39) to compute the strains and stresses anywhere inside the elements.

Example 2.1 *Compute the element stiffness matrix and the equivalent force vector of a bar discretized with two-node (linear) element using equations (2.44) and (2.46). Also compare the result with the expression (2.19-2.20).*

Solution to Example 2.1 *Let A_e be the transverse area of the bar and h_e the element length, with $x_e = 0$ and $x_{e+1} = h_e$. Substituting these values in the linear interpolation functions from equation (2.12), the interpolation functions arrays are obtained as follows*

$$\underline{N}^T = \begin{bmatrix} N_1^e \\ N_2^e \end{bmatrix} = \begin{bmatrix} \dfrac{x_{e+1} - x}{x_{e+1} - x_e} \\ \dfrac{x - x_e}{x_{e+1} - x_e} \end{bmatrix} = \begin{bmatrix} 1 - x/h_e \\ x/h_e \end{bmatrix}$$

The strain-displacement array is obtained as

$$\underline{B}^T = \underline{\underline{\partial}} \, \underline{N}^T = \begin{bmatrix} \partial N_1^e / \partial x \\ \partial N_2^e / \partial x \end{bmatrix} = \begin{bmatrix} -1/h_e \\ 1/h_e \end{bmatrix}$$

The bar element has a one-dimensional strain-stress state with linear elastic behavior. Therefore

$$\underline{C} = E$$

Then, using equation (2.44) we can write

$$\underline{\underline{K}}^e = \int_{V_e} \underline{B}^T \underline{\underline{C}} \, \underline{B} \, dV = \int_0^{h_e} \begin{bmatrix} -1/h_e & 1/h_e \end{bmatrix} E \begin{bmatrix} -1/h_e \\ 1/h_e \end{bmatrix} A_e \, dx$$

The element stiffness matrix is obtained by integration

$$[K^e] = \frac{EA_e}{h_e} \begin{bmatrix} 1 & -1 \\ -1 & 1 \end{bmatrix}$$

To calculate the equivalent vector force, f_e is defined as the distributed force on element, P_1^e is the force at end $x = 0$, and P_2^e is the force at end $x = h_e$. Substituting into equation (2.46) we obtain

$$\underline{P}^e = \int_{V_e} \underline{N}^T \underline{f} \, dV + \int_{S_e} \underline{N}^T \underline{t} \, dS = \int_0^{h_e} \begin{bmatrix} 1 - x/h_e \\ x/h_e \end{bmatrix} f_e dx + \begin{bmatrix} P_1^e \\ P_2^e \end{bmatrix}$$

The element equivalent force vector is obtained by integration

$$\underline{P}^e = \frac{f_e h_e}{2} \begin{bmatrix} 1 \\ 1 \end{bmatrix} + \begin{bmatrix} P_1^e \\ P_2^e \end{bmatrix}$$

2.3   FE Analysis with CAE systems

Nowadays, a large number of the commercial programs exist with many finite element analysis capabilities for different engineering disciplines. They help solve a variety of problems from a simple linear static analysis to nonlinear transient analysis. A few of these codes, such as ANSYS^{TM} or ABAQUS^{TM}, have special capabilities to analyze composite materials and they accept user programmed element formulations and custom constitutive equations.

These types of computer aided engineering (CAE) systems are commonly organized into three different blocks: the pre-processor, the processor, and the post-processor. In the first block, commonly called pre-processor, the model is built defining material properties, element formulation, and geometry. Loads and boundary conditions may also be entered here, but they can also be entered during the solution phase. With this information, the processor can compute the stiffness matrix of the model, as well as the force vector. Next, the equilibrium equations are solved and the solution is obtained in the form of displacement values. In the last block, the post-processor, derived results, such as stress, strain, and failure ratios, are computed. The solution can be reviewed using graphic tools.

In the remainder of this chapter, a general description of the procedures and the specific steps for a basic finite element analysis (FEA) are presented using examples. However, a complete description of the capabilities and procedures of any commercial FEA code is outside the scope of this textbook. Refer to Appendix E for an introduction to the software interface.

2.3.1   Pre-process: Model Generation

In order to introduce all the information needed to compute the stiffness matrix, different aspects such as element type and material properties must be defined. Also, the geometry must be specified by entering the position of all nodes and their connectivity, element by element.

Element Type

Usually, FE programs have an *element library* that contains many different element types. The element type determines the element formulation used. For example, the degree of freedom set, the interpolation functions, whether the element is for 2D or 3D space, etc. Above all, the element type identifies the element category: bar tensile-compression, beam bending, solid, shell, laminate shell, etc. Each commercial code identifies element formulations with different labels. Identification labels and basic characteristics of a few element formulations in ANSYS and ABAQUS are shown in Table 2.1.

Table 2.1: Some linear structural elements available in ANSYS and ABAQUS

ANSYS	ABAQUS	nodes	DOF	Element Description
LINK1	T2D2	2	$u_X\ u_Y$	line bar/truss, 2D space
LINK3	T3D2	2	$u_X\ u_Y\ u_Z$	line bar/truss, 3D space
BEAM3	B21	2	$u_X\ u_Y$ $\theta_X\ \theta_Y$	line beam in 2D space
BEAM4	B31	2	$u_X\ u_Y\ u_Z$ $\theta_X\ \theta_Y\ \theta_Z$	line beam in 3D space
PLANE42	CPE4	4	$u_X\ u_Y$	solid 4-node quadrilateral in 2D space
PLANE82	CPE8	8	$u_X\ u_Y$	solid 8-node quadrilateral in 2D space
SOLID45	C3D8	8	$u_X\ u_Y\ u_Z$	solid 8-node hexahedra in 3D space
SHELL63	S4	4	$u_X\ u_Y\ u_Z$ $\theta_X\ \theta_Y\ \theta_Z$	shell 4-node quadrilateral in 3D space
SHELL91		4	$u_X\ u_Y\ u_Z$ $\theta_X\ \theta_Y\ \theta_Z$	layered shell 4-node quad. in 3D space
SHELL99		4	$u_X\ u_Y\ u_Z$ $\theta_X\ \theta_Y\ \theta_Z$	layered shell 4-node quad. in 3D space

Each element type has different options. For example, on a planar solid element, an option allows one to choose between plane strain and plane stress analysis.

Sometimes the elements need real constants. These are properties that depend on the element type, such as cross-sectional properties of a beam element, the ply sequence in a laminated shell element, and so on. For example, some real constants for a 2D beam element are: cross area, moment of inertia, and height. Not all element types require real constants. Although the real constants are associated to element type, different elements of the same category may have different real constant values.

Material Definition

The elements must be associated to a material. Depending on the analysis, material properties can be linear (linear elastic analysis) or nonlinear (e.g. damage mechanics analysis), isotropic or orthotropic, constant or temperature-dependent.

For structural analysis, elastic properties must be defined according to Section (1.12). Other mechanical properties, such as ultimate strength, density, and thermal dilatation coefficients are optional and their definition depends on the analysis characteristics.

2.3.2   Model Geometry

The model geometry is obtained specifying all nodes, their position, and the element connectivity. The connectivity information allows the program to assemble the element stiffness matrix and the element equivalent force vector to obtain the global equilibrium equations, as shown in Section 2.1.6.

There are two ways to generate the model. The first is to directly create a mesh. The second is to use solid modeling, then mesh the solid automatically to get the node and element distribution. Each method has its advantages and disadvantages.

Direct Mesh Generation

Direct meshing was the only method before solid modeling became widespread among commercial packages. It is still the only option with some older packages, although in those cases it is always possible to use a general purpose pre-processor to generate the mesh. In this case the pre-processor is likely to allow for solid modeling. In direct meshing, the user creates nodes, then connects the nodes into elements. Afterward, the user applies boundary conditions and loads directly on nodes and/or elements. Direct meshing is used in Example 2.2 and in Example 2.5.

Solid Modeling

In solid modeling, the user creates a geometric representation of the geometry using solid model constructs, such as volumes, areas, lines, and points (or keypoints). Boundary conditions, loads, and material properties can be assigned to parts of the solid model before meshing. The models are meshed by the same program just prior to the FEM solution. One additional advantage of solid modeling is that re-meshing can be done without loosing, or having to remove, the loads and boundary conditions. Solid modeling and automatic meshing are used in Example 2.3 and Example 2.4.

Example 2.2 *Using ANSYS, generate the curved beam shown in Figure 2.7. Because the thickness is small and constant, use planar solid elements with plane stress analysis. Use direct generation for generating the mesh geometry.*

Solution to Example 2.2 *The commands listed below, which are available on the Web site [1], define the model geometry by using* Direct Mesh Generation. *The characters after* (!)

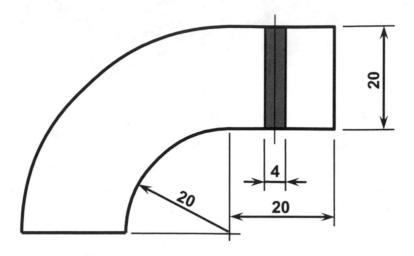

Fig. 2.7: Curved beam.

are comments. These commands can be typed one line at a time in the ANSYS command window (see Appendix E). Alternatively, in the ANSYS command window, read the text file by entering /input,file,ext, *where* file *is the name of the file, and* ext *is the file extension.*

```
/TITLE, Bending curved beam (Direct Generation)
/PREP7                  ! Start pre-processor module
ET,1,PLANE42            ! Element type #1: PLANE42
KEYOPT,1,3,3            ! Key option #3 = 3, plane stress
R,1,4                   ! Real constant #1: Th = 4 mm
MP,EX,1,195000          ! Material #1: E=195000 MPa
MP,NUXY,1,0.3           ! Material #1: Poisson coefficient 0.3
! Nodes and elements generation
CSYS,1                  ! Activate polar coordinate system
N,1,20,180             ! Define node #1: radius=20mm, angle=180
N,10,20,90             ! Define node #10: radius=20mm, angle=90
FILL,1,10              ! Fill nodes between node 1 and 10
NGEN,9,20,1,10,,2.5    ! Generate node rows increase radius 2.5 mm
CSYS,0                  ! Activate cartesian coordinate system
N,15,20,20             ! Define node #15: x=20mm, y=20mm
FILL,10,15             ! Fill nodes between node 10 and 15
NGEN,9,20,11,15,,,2.5   !Generate node rows increasing y 2.5 mm
E,1,2,22,21            ! Define element #1, joining nodes 1,2,22,21
EGEN,8,20,1            ! Generate a row of elements
EGEN,14,1,ALL          ! Generate the rest of elements
FINISH                  ! Exit pre-processor module
```

Example 2.3 *Using ANSYS, generate the same geometry defined in Example 2.2, but this time use Solid Modeling commands for generating the mesh geometry.*

Solution to Example 2.3 *The commands listed below generate the geometry using Solid Modeling [1].*

```
/TITLE, Bending curved beam (Solid Modeling)
/PREP7                  ! Start pre-processor module
ET,1,PLANE42            ! Element type #1: PLANE42
KEYOPT,1,3,3            ! Key option #3 = 3, plane stress
R,1,4                   ! Real constant #1: Th = 4 mm
MP,EX,1,195000          ! Material #1: E=195000 MPa
MP,NUXY,1,0.3           ! Material #1: Poisson coefficient 0.3
! Geometry generation
CYL4,0,0,40,90,20,180    ! Generate curved area
BLC4,0,20,20,20         ! Generate rectangular area
AGLUE,all               ! Glue both areas
LESIZE,all,,,8          ! Define divisions of elements by lines
LESIZE,1,,,10           ! Re-define divisions of some lines
LESIZE,3,,,10
LESIZE,9,,,5
LESIZE,10,,,5
MSHKEY,1                ! Force quadrilateral elements mesh
AMESH,all               ! Mesh all areas
FINISH                  ! Exit pre-processor module
```

Example 2.4 *Using ANSYS, generate a model for a dome (Figure 2.8) with different types of elements (shell and beam elements), using different real constants sets and two materials. Use solid modeling to generate the mesh geometry.*

Solution to Example 2.4 *The element types in ANSYS are defined by the ET command [3]. The element types can be defined by their library names (see Table 2.1) and given reference numbers to be used later. For example, the commands shown below define two element types, BEAM4 and SHELL63, and assign them type reference numbers 1 and 2 respectively.*

```
ET,1,BEAM4
ET,2,SHELL63
```

In ANSYS, the real constants sets are defined similarly to element types, using reference numbers for each set. For example, two sets are defined below for shell elements and one for 3D beam element as follows

```
R,1,6                   ! Define real const. #1 Th = 6 mm
R,2,4                   ! Define real const. #2 Th = 4 mm
R,3,100,833,833,10,10    ! Define real const. #3 10x10 sect.beam
                        ! A=100, Izz=Iyy=1/12*10**4, thz=thy=10
```

For material definition, MP can be used along with the appropriate property label; e.g. EX for Young's modulus, NUXY for Poisson ratio, etc. For isotropic material, only the X-direction properties need to be defined. The remaining properties in the other directions default to the X-direction values. Also a reference number is used for each material. For example, the following code defines two materials

```
MP,EX,1,200E3           ! Define material #1 Young's modulus 200000MPa
MP,NUXY,1,0.29          ! Define material #1 Poisson modulus 0.29
MP,EX,2,190E3           ! Define material #2 Young's modulus 190000MPa
MP,NUXY,2,0.27          ! Define material #2 Poisson modulus 0.27
```

The commands shown above define a database with a table of elements type, another table with real constant sets, and another with materials. The reference number of each table can be selected by using the commands TYPE, REAL and MAT, before defining the mesh.

The ANSYS command sequence for this example is listed below. These commands can be typed on the command window, or in a text file [1], then, on the command window enter /input,file,ext, *where* file *is the name of the file, and* ext *is the file extension (see Appendix E).*

```
/PREP7                 ! Start pre-processor module
! Define element types
ET,1,BEAM4             ! Define element type #1 BEAM4 - 3D
ET,2,SHELL63           ! Define element type #2 sHELL63 - 3D

! Define real constants
R,1,6                  ! Define real const. #1 Th = 6 mm
R,2,4                  ! Define real const. #2 Th = 4 mm
R,3,100,833,833,10,10   ! Define real const. #3 10x10 sect.beam

! Define materials
MP,EX,1,200E3          ! Define material #1 E=200000MPa
MP,NUXY,1,0.29         ! Define material #1 Poisson modulus
MP,EX,2,190E3          ! Define material #2 E=190000MPa
MP,NUXY,2,0.27         ! Define material #2 Poisson modulus

! Create geometry by solid modeling
SPH4, , ,500                      ! Define sphere radius 500 mm
BLOCK,-600,600,-600,600,-600,0   ! Define blocks for subtract ...
BLOCK, 300,600,-600,600,0,600    ! ...to sphere
BLOCK,-300,-600,-600,600,0,600
BLOCK,-600,600,300,600,0,600
BLOCK,-600,600,-300,-600,0,600
! Boolean operations to obtain the final volume, areas and lines
VADD,2,3,4,5,6         ! Add all blocks
VSBV,1,7               ! Subtract blocks to the sphere
WPAVE,0,0,200          ! Offset working plane z=+200 mm
VSBW,ALL               ! Divide Volume by working plane

! Mesh geometry
ESIZE,20               ! Define element size

TYPE,2                 ! Assign shell to next defined elements
REAL,1                 ! Assign Th = 6 mm to next defined elements
MAT,2                  ! Assign mater. #2 to next defined elements
AMESH,8,9              ! Mesh areas 8 and 9 (top surface dome).

REAL,2                 ! Assign Th = 4 mm to next defined elements
AMESH,12,15            ! Mesh areas 12,13,14 and 15 (side surface dome)

TYPE,1                 ! Assign beam to next defined elements
```

```
REAL,3              ! Assign cross section 10x10 squared
MAT,1               ! Assign mater. #1 to next defined elements
LMESH,1,2           ! Mesh lines 1 and 2 (columns)
LMESH,4,5           ! Mesh lines 4 and 5 (columns)

FINISH              ! Exit pre-processor module
```

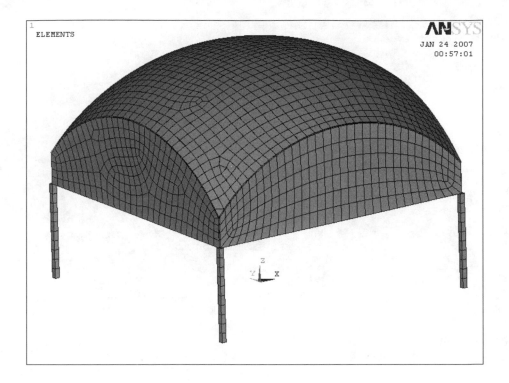

Fig. 2.8: Mesh obtained by the command sequence used to generate a dome.

2.3.3 Load States

Load states include boundary conditions and externally applied forcing functions. Load states in structural analysis are defined by forces, pressures, inertial forces (as gravity), and specified displacements, all applied to the model.

Specification of different kinds of loads for the FE model are explained in the following sections. The reactions obtained by fixing a nodal degree of freedom (displacements and rotations) are shown in Section 2.3.4. Next, the loads from forces (or moments) on nodes or forces distributed over the elements are discussed in Section 2.3.5.

2.3.4 Boundary Conditions

The boundary conditions are the known values of the degrees of freedom (DOF) on the boundary. In structural analysis, the DOF are displacements and rotations. With this information, the FE program knows which values of $\underline{a}$ in (2.47) are known or unknown.

Constrained Displacements and Rotations

In general, a node can have more than one DOF. For example, if the FE model uses beam elements in 2D space, there are three DOF: the horizontal displacement, the vertical displacement, and the rotation around an axis perpendicular to the plane. Constraining different sets of DOF results in different boundary conditions being applied. In the 2D beam element case, constraining only the horizontal and vertical displacements results in a simple support, but constraining all the DOF results in a clamped condition.

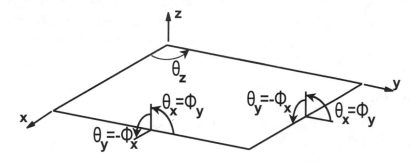

Fig. 2.9: Convention for rotations of a plate or shell

Symmetry Conditions

Symmetry conditions can be applied to reduce the size of the model without loss of accuracy. Four types of symmetry must exist concurrently: symmetry of geometry, boundary conditions, material, and loads. Under these conditions the solution will also be symmetric. For example, symmetry with respect to the $y - z$ plane means that the nodes on the symmetry plane have the following constraints

$$u_x = 0 \quad ; \quad \theta_y = 0 \quad ; \quad \theta_z = 0 \tag{2.48}$$

where u_x is the displacement along the x-direction, θ_y and θ_z are the rotations around the y and z axis, respectively (Figure 2.9). Note that the definition of rotations used in shell theory (ϕ_i, see Section 3.1) is different than the usual definition of rotations θ_i that follows the right-hand rule. All rotations in ANSYS are described using right-hand-rule rotations θ_i. Symmetry boundary conditions on nodes in the symmetry plane involve the restriction of DOF translations out-of-plane with respect to the symmetry plane and restriction of the DOF rotations in-plane with respect to the symmetry plane. Symmetry boundary conditions are used in Example 2.6.

Antisymmetry Conditions

Antisymmetry conditions are similar to the symmetry conditions. They can be applied when the model exhibits antisymmetry of loads but otherwise the model exhibits symmetry of geometry, symmetry of boundary conditions, and symmetry of material. Antisymmetry boundary conditions on nodes in the antisymmetry plane involve restriction of DOF translations in the antisymmetry plane and restriction of DOF rotations out-of-plane with respect to the antisymmetry plane.

Periodicity Conditions

When the material, load, boundary conditions, and geometry are periodic with period $(x, y, z) = (2a_i, 2b_i, 2c_i)$, only a portion of the structure needs to be modeled, with dimensions $(2a_i, 2b_i, 2c_i)$. The fact that the structure repeats itself periodically means that the solution will also be periodic. Periodicity conditions can be imposed by different means. One possibility involves using constrained equations (CE) between DOF (see Section 6.2) or using Lagrange multipliers.

2.3.5   Loads

Loads can be applied on nodes by means of concentrated forces and moments, as shown in Example 2.5. Also, loads can be distributed over the elements as: surface loads, body loads, inertia loads, or other coupled-field loads (for example, thermal strains). Surface loads are used in Example 2.6.

A surface load is a distributed load applied over a surface, for example a pressure. A body load is a volumetric load, for example expansion of material by temperature increase in structural analysis. Inertia loads are those attributable to the inertia of a body, such as gravitational acceleration, angular velocity, and acceleration.

A concentrated load applied on a node is directly added to the force vector. However, the element interpolation functions are used to compute the equivalent forces vector due to distributed loads.

2.3.6   Solution Procedure

In the solution phase of the analysis, the solver subroutine included in the finite element program solves the simultaneous set of equations that the finite element method generates. The primary solution is obtained by solving for the nodal degree of freedom values, i.e., displacements and rotations, in structural analysis. Then, derived results, i.e., stresses and strains are computed. The element solution is usually calculated at the integration points.

Several methods of solving the system of simultaneous equations are available. Some methods are better for larger models, others are faster for nonlinear analysis, others allow one to distribute the solution by parallel computation. Commercial finite element programs solve these equations in batch mode. The frontal direct solution method is commonly used because it is rather efficient for FEA. When the analysis is nonlinear, the equations must be solved repeatedly, which increases the computational time significantly.

2.3.7 Post-process: Analysis and Results Visualization

Once the solution has been calculated, the post-processor can be used to review and to analyze the results. Results can be reviewed graphically or by listing the values numerically. Since a model usually contains a considerable amount of results, it may be better to use graphical tools. Post-processors of commercial codes produce contour plots of stress and strain distributions, deformed shapes, etc. The software usually includes derived calculations such as error estimation, load case combinations, or path operations.

Examples 2.5 and 2.6 include commands to review the results by listing and by graphic output, respectively.

Example 2.5 *Use a commercial FE code to find the axial displacement at the axially loaded end of a bar clamped at the other end. The bar is made of steel $E = 200,000\ MPa$, diameter $d = 9\ mm$, length $L = 750\ mm$, and load $P = 100,000\ N$. Also find the stress and strain. Use three two-node (linear) link elements.*

Solution to Example 2.5 *The ANSYS command sequence for this example is listed below. You can either type these commands on the command window, or you can type them on a text file [1], then, on the command window enter* /input,file,ext, *where* file *is the name of the file, and* ext *is the file extension (see Appendix E).*

```
/PREP7                 ! Start pre-processor module
ET,1,LINK1             ! Define element type #1 LINK1 - 2D bar
R,1,63.6173            ! Define real constant A=63.6173 mm^2
MP,EX,1,200E3          ! Define elastic modulus E=200000MPa
N,1                    ! Define node 1, coordinates=0,0,0
NGEN,4,1,1,,,250       ! Generate 3 additional nodes
                       ! distance between adjacent nodes 250mm
E,1,2                  ! Generate element 1 by node 1 to 2
EGEN,3,1,1             ! Generate element 2,3
FINISH                 ! Exit pre-processor module

/SOLU                  ! Start Solution module
OUTPR,ALL,LAST,
D,1,all                ! Define b.c. in node 1, totally fixed
F,4,FX,100E3           ! Define horizontal force in node 4
SOLVE                  ! Solve the current load state
/STAT,SOLU             ! Provides a solution status summary
FINISH                 ! Exit solution module

/POST1                 ! Start Post-processor module
PRNSOL,U,X             ! Print in a list the horizonal disp.
PRESOL,ELEM            ! Print all line element results
```

```
PRRSOL,FX              ! Print horizontal reactions
FINISH                 ! Exit post-processor module
```

A convenient combination of units for this case is Newton, mm, and MPa. The analysis results can be easily verified by strength of material calculations, as follows

$$U_x = \frac{PL}{AE} = \frac{(750)(100000)}{(63.617)(200000)} = 5.894 \ mm$$

$$\sigma = \frac{P}{A} = \frac{100000}{63.617} = 1571.9 \ MPa$$

$$\epsilon = \frac{\sigma}{E} = 7.859 \cdot 10^{-3}$$

Example 2.6 *Use a commercial FE code to find the stress concentration factor of a rectangular notched strap. The dimensions and the load state are defined in Figure 2.10. Use eight-noded (quadratic) quadrilateral plane stress elements.*

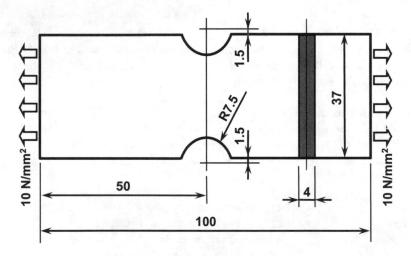

Fig. 2.10: Rectangular notched strap, axial load 10 N/mm².

Solution to Example 2.6 *The ANSYS command sequence for this example is listed below. These commands can be typed on the command window or in a text file [1], then, on the command window enter /input,file,ext, where file is the name of the file, and ext is the file extension (see Appendix E).*

```
/PREP7                 ! Start pre-processor module
ET,1,PLANE82           ! Define element type #1 PLANE82 8-node 2-D
KEYOPT,1,3,3           ! Key option #3 = 3, plane stress
R,1,4                  ! Define real constant th=2 mm
MP,EX,1,190E3          ! Define elastic modulus E=200000MPa
MP,PRXY,1,0.3          ! Define Poisson coefficient 0.3
BLC4,0,0,50,18.5       ! Define squared area 18.5x50 mm
CYL4,0,20,7.5          ! Define circular area radius 7,5 mm
ASBA,1,2               ! Subtract previous areas
```

```
ESIZE,1.5,0,          ! Define element size
MSHKEY,0              ! Free mesh
AMESH,all             ! Mesh
FINISH                ! Exit pre-processor module

/SOLU                 ! Start Solution module
DL, 1, ,SYMM          ! Define symmetry b. conditions in line 1
DL, 9, ,SYMM          ! Define symmetry b. conditions in line 9
SFL,2,PRES,-10        ! Apply pressure on line 2
SOLVE                 ! Solve the current load state
FINISH                ! Exit solution module

/POST1                ! Start Post-processor module
PLNSOL,S,X,2,1        ! Contour plot horizontal stress
PLNSOL,S,EQV,2,1      ! Contour plot Von Mises equivalent stress
PLVECT,S              ! Vector plot principal stress
FINISH                ! Exit post-processor module
```

The stress in the net area without stress concentration is

$$\sigma_o = \frac{P}{A} = \frac{10 \cdot 37 \cdot 4}{25 \cdot 4} = 14.8 \ MPa$$

The maximum horizontal stress close to the notch is 28 MPa obtained from the FE model. Therefore, the concentration factor is

$$k = \frac{\sigma_{max}}{\sigma_o} = 1.89$$

Suggested Problems

Problem 2.1 *Solve Example 2.5 explicitly as it is done in Section 2.1, using only two elements. Show all work.*

Problem 2.2 *Using the equations derived in Section 2.1, and the data of Example 2.5, compute the axial displacement at (a) $x = 700$ mm, (b) $x = 500$ mm.*

Problem 2.3 *Using the same procedure in Example 2.1 calculate the element stiffness matrix and the equivalent force vector of a three-node element bar with quadratic interpolation functions. The interpolation functions are*

$$N_1^e = \frac{x - x_2}{x_1 - x_2}\frac{x - x_3}{x_1 - x_3} \quad N_2^e = \frac{x - x_3}{x_2 - x_3}\frac{x - x_1}{x_2 - x_1} \quad N_3^e = \frac{x - x_1}{x_3 - x_1}\frac{x - x_2}{x_3 - x_2}$$

where x_1, x_2 and x_3 are the coordinate positions of node 1, 2 and 3 respectively. Use $x_1 = 0$, $x_2 = h/2$ and $x_3 = h$, where h is the element length. Show all work.

Problem 2.4 *Program a FE code using the element formulation obtained in Example 2.1 and the assembly procedure shown in Section 2.1.6. With this code, solve Example 2.5. Show all work in a report.*

Problem 2.5 *Program a FE code using the element formulation obtained in Problem 2.3 and the assembly procedure shown in Section 2.1.6. With this code, solve Example 2.5. Show all work in a report.*

References

[1] E. J. Barbero, Web resource: http://www.mae.wvu.edu/barbero/feacm/

[2] J. N. Reddy, Energy and Variational Methods in Applied Mechanics, John Wiley, New York, 1984.

[3] ANSYS, Structural Analysis Guide, ANSYS Inc., Cannonsburg, PA, 2005.

Chapter 3

Elasticity and Strength of Laminates

Most composite structures are built as assemblies of plates and shells. This is because the structure is more efficient when it carries membrane loads. Another important reason is that thick laminates are difficult to produce.

For example, consider a beam made of an homogeneous material with tensile and compressive strength σ_u subjected to bending moment M. Further, consider a solid beam of square cross section (Figure 3.1), equal width and depth $2c$, with area A, inertia I, and section modulus S given by

$$A = 4c^2$$
$$I = \frac{4}{3}c^4$$
$$S = \frac{I}{c} = \frac{4}{3}c^3 \tag{3.1}$$

When the stress on the surface of the beam reaches the failure stress σ_u, the bending moment per unit area is

$$m_u = \frac{M_u}{A} = \frac{S\sigma_u}{A} = \frac{1}{3}c\sigma_u \tag{3.2}$$

Now consider a square hollow tube (Figure 3.1) of dimensions $2c \times 2c$ and wall thickness t, with $2c >> t$, so that the following approximations are valid

$$A = 4(2c)t = 8ct$$
$$I = 2\left[\frac{t(2c)^3}{12} + c^2(2ct)\right] = \frac{16}{3}tc^3$$
$$S = \frac{I}{c} = \frac{16}{3}tc^2 \tag{3.3}$$

Then

$$m_u = \frac{M_u}{A} = \frac{S\sigma_u}{A} = \frac{\frac{16}{3}tc^2\sigma_u}{8ct} = \frac{2}{3}c\sigma_u \tag{3.4}$$

The failure moment per unit area m_u is twice as large for a hollow square tube with thin walls than for a solid section.

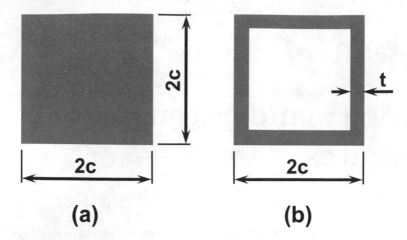

Fig. 3.1: Solid section (a) and hollow square tube (b).

Of course, the failure moment is limited by buckling of the thin walls (see Chapter 4). This is the reason buckling analysis is so important for composites. Most composite structures are designed under buckling constraints because the thicknesses are small and the material is very strong, so normally one does not encounter material failure as in metallic structures (e.g., yield stress) but structural failure such as buckling.

Plates are a particular case of shells, having no initial curvature. Therefore, only shells will be mentioned in the sequel. Shells are modeled as two-dimensional structures because two dimensions (length and width) are much larger than thickness. The thickness coordinate is eliminated from the governing equations so that the 3D problem simplifies to 2D. In the process, the thickness becomes a parameter that is known and supplied to the analysis model.

Modeling of laminated composites differs from modeling conventional materials in three aspects. First, the constitutive equations of each lamina are orthotropic (Section 1.12.3). Second, the constitutive equations of the element depend on the kinematic assumptions of the shell theory used and their implementation into the element. Finally, material symmetry is as important as geometric and load symmetry when trying to use symmetry conditions in the models.

3.1   Kinematic of Shells

Shell elements are based on various shell theories which in turn are based on kinematic assumptions. That is, there are some underlying assumptions about the likely type of deformation of the material through the thickness of the shell. These assumptions are needed to reduce the 3D governing equations to 2D. Such assumptions are more or less appropriate for various situations, as discussed next.

3.1.1 First-Order Shear Deformation Theory

The most popular composite shell theory is the first-order shear deformation theory (FSDT). It is based on the following assumptions:

1. A straight line drawn through the thickness of the shell in the undeformed configuration may rotate but it will remain straight when the shell deforms. The angles it forms (if any) with the normal to the undeformed mid-surface are denoted by ϕ_x and ϕ_y when measured in the $x - z$ and $y - z$ planes, respectively (Figures 2.9 and 3.2).

2. The thickness of the shell remains unchanged as the shell deforms.

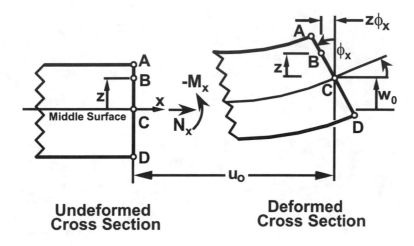

Undeformed Cross Section   **Deformed Cross Section**

Fig. 3.2: Assumed deformation in FSDT[1].

These assumptions are verified by experimental observation in most laminated shells when the following are true:

- The aspect ratio $r = a/t$, defined as the ratio between the shortest surface dimension a and the thickness t, is larger than 10.

- The stiffness of the laminae in shell coordinates (x, y, z) do not differ by more than two orders of magnitude. This restriction effectively rules out sandwich shells, where the core is much softer than the faces.

Based on the assumptions above, the displacement of a generic point B anywhere in the shell can be written in terms of the displacement and rotations at the mid-surface C as

---

[1]Reprinted from Introduction to Composite Materials Design, E. J. Barbero, Fig. 6.2, copyright (1999), with permission from Taylor & Francis.

$$u(x, y, z) = u_0(x, y) - z\phi_x(x, y)$$
$$v(x, y, z) = v_0(x, y) - z\phi_y(x, y)$$
$$w(x, y, z) = w_0(x, y)$$

(3.5)

The mid-surface variables on the right-hand side of (3.5) are functions of only two coordinates (x and y), thus the shell theory is 2D. On the left-hand side, the displacements are functions of three coordinates, and thus correspond to the 3D representation of the material. At the 3D level, we use the 3D constitutive equations (1.72) and the 3D strain-displacement equations (1.5), which now can be written in terms of 2D quantities as

$$\epsilon_x(x, y, z) = \frac{\partial u_0}{\partial x} - z\frac{\partial \phi_x}{\partial x} = \epsilon_x^0 + z\kappa_x$$

$$\epsilon_y(x, y, z) = \frac{\partial v_0}{\partial y} - z\frac{\partial \phi_y}{\partial y} = \epsilon_y^0 + z\kappa_y$$

$$\gamma_{xy}(x, y, z) = \frac{\partial u_0}{\partial y} + \frac{\partial v_0}{\partial x} - z\left(\frac{\partial \phi_x}{\partial y} + \frac{\partial \phi_y}{\partial x}\right) = \gamma_{xy}^0 + z\kappa_{xy}$$

$$\gamma_{yz}(x, y) = -\phi_y + \frac{\partial w_0}{\partial y}$$

$$\gamma_{xz}(x, y) = -\phi_x + \frac{\partial w_0}{\partial x}$$

$$\epsilon_z = 0$$

(3.6)

where

- The mid-surface strains ϵ_x^0, ϵ_y^0, γ_{xy}^0, also called membrane strains, represent stretching and in-plane shear of the mid-surface.

- The change in curvature κ_x, κ_y, κ_{xy}, which are close but not exactly the same as the geometric curvatures of the mid-surface. They are exactly that for the Kirchhoff theory discussed in Section 3.1.2.

- The interlaminar shear strains γ_{xz}, γ_{yz}, which are through-the-thickness shear deformations. These are small but not negligible for laminated composites because the interlaminar shear moduli G_{23}, G_{13} are small when compared with the in-plane modulus E_1. Metals are relatively stiff in shear ($G = E/2(1+\nu)$), and thus the interlaminar strains are negligible. In addition, the interlaminar shear strength of composites F_4, F_5 are relatively small when compared to the in-plane strength values F_{1t}, F_{1c}, thus making evaluation of interlaminar strains (and possibly stresses) a necessity. On the other hand, the shear strength of metals is comparable to their tensile strength, and since the interlaminar stress is always smaller than the in-plane stress, it is not necessary to check for interlaminar failure of metallic homogeneous shells. That is not the case for laminated metallic shells since the adhesive is not quite strong and it may fail by interlaminar shear.

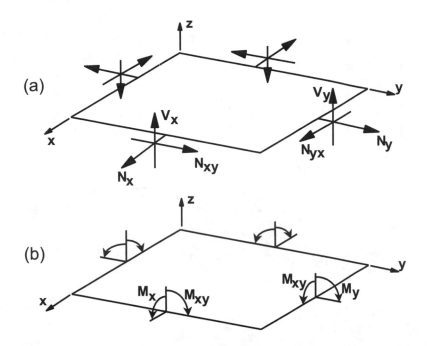

Fig. 3.3: Stress resultants acting on a plate or shell element: (a) forces per unit length, and (b) moments per unit length.[2]

While the 3D constitutive equations relate strains to stress, the laminate constitutive equations relate mid-surface strains and curvatures. The laminate constitutive equations are obtained by using the definition of stress resultants. While in 3D elasticity every material point is under stress, a shell is loaded by stress resultants (Figure 3.3), which are simply integrals of the stress components through the thickness of the shell, as follows

$$
\left\{ \begin{array}{c} N_x \\ N_y \\ N_{xy} \end{array} \right\} = \sum_{k=1}^{N} \int_{z_{k-1}}^{z_k} \left\{ \begin{array}{c} \sigma_x \\ \sigma_y \\ \sigma_{xy} \end{array} \right\}^k dz
$$

$$
\left\{ \begin{array}{c} V_y \\ V_x \end{array} \right\} = \sum_{k=1}^{N} \int_{z_{k-1}}^{z_k} \left\{ \begin{array}{c} \sigma_{yz} \\ \sigma_{xz} \end{array} \right\}^k dz
$$

$$
\left\{ \begin{array}{c} M_x \\ M_y \\ M_{xy} \end{array} \right\} = \sum_{k=1}^{N} \int_{z_{k-1}}^{z_k} \left\{ \begin{array}{c} \sigma_x \\ \sigma_y \\ \sigma_{xy} \end{array} \right\}^k z \, dz \qquad (3.7)
$$

[2]Reprinted from Introduction to Composite Materials Design, E. J. Barbero, Fig. 6.3, copyright (1999), with permission from Taylor & Francis.

where N is the number of layers, z_{k-1} and z_k are the coordinates at the bottom and top surfaces of the k-th layer, respectively. Replacing the plane stress version of the 3D constitutive equations in shell local coordinates (1.101-1.102) at each layer and performing the integration we get

$$\left\{\begin{array}{c} N_x \\ N_y \\ N_{xy} \\ M_x \\ M_y \\ M_{xy} \end{array}\right\} = \left[\begin{array}{cccccc} A_{11} & A_{12} & A_{16} & B_{11} & B_{12} & B_{16} \\ A_{12} & A_{22} & A_{26} & B_{12} & B_{22} & B_{26} \\ A_{16} & A_{26} & A_{66} & B_{16} & B_{26} & B_{66} \\ B_{11} & B_{12} & B_{16} & D_{11} & D_{12} & D_{16} \\ B_{12} & B_{22} & B_{26} & D_{12} & D_{22} & D_{26} \\ B_{16} & B_{26} & B_{66} & D_{16} & D_{26} & D_{66} \end{array}\right] \left\{\begin{array}{c} \epsilon_x^0 \\ \epsilon_y^0 \\ \gamma_{xy}^0 \\ \kappa_x \\ \kappa_y \\ \kappa_{xy} \end{array}\right\} \qquad (3.8)$$

$$\left\{\begin{array}{c} V_y \\ V_x \end{array}\right\} = \left[\begin{array}{cc} H_{44} & H_{45} \\ H_{45} & H_{55} \end{array}\right] \left\{\begin{array}{c} \gamma_{yz} \\ \gamma_{xz} \end{array}\right\}$$

where

$$A_{ij} = \sum_{k=1}^{N} \left(\overline{Q}_{ij}\right)_k t_k; \qquad i,j = 1,2,6$$

$$B_{ij} = \sum_{k=1}^{N} \left(\overline{Q}_{ij}\right)_k t_k \bar{z}_k; \qquad i,j = 1,2,6$$

$$D_{ij} = \sum_{k=1}^{N} \left(\overline{Q}_{ij}\right)_k \left(t_k \bar{z}_k^2 + \frac{t_k^3}{12}\right); \qquad i,j = 1,2,6$$

$$H_{ij} = \frac{5}{4} \sum_{k=1}^{N} \left(\overline{Q}_{ij}^*\right)_k \left[t_k - \frac{4}{t^2}\left(t_k \bar{z}_k^2 + \frac{t_k^3}{12}\right)\right]; \qquad i,j = 4,5 \qquad (3.9)$$

where $\left(\overline{Q}_{ij}\right)_k$ are the coefficients in global coordinates of the plane-stress stiffness matrix for layer number k, t_k is the thickness of layer k, and $\bar{z}_k$ is the coordinate of the middle surface of the kth layer. For an in-depth discussion of the meaning of various terms see [1]. In summary, the A_{ij} coefficients represent in-plane stiffness of the laminate, the D_{ij} coefficients represent bending stiffness, the B_{ij} represent bending-extension coupling, and the H_{ij} represent interlaminar shear stiffness. All these coefficients can be calculated by (3.9) and are implemented in widely available software packages such as CADEC [2].

When membrane and bending deformations are uncoupled (e.g., symmetric laminates), the governing equations of FSDT involve three variables for solving the bending problem (w^0, ϕ_x, ϕ_y) and two to solve the membrane problem (u^0, v^0). Bending-extension coupling means that all five variables will have to be found simultaneously, which is what FEA software codes do for every case, whether the problem is coupled or not.

The equilibrium equations of plates can be derived by using the PVW (see (1.16)). Furthermore, the governing equations can be derived by substituting the constitutive equations (3.8) into the equilibrium equations.

Simply Supported Boundary Conditions in Plates

Composite plates with coupling effects may have bending, shear, and membrane deformations coupled even if loaded by pure bending, pure shear, or pure in-plane loads (see p. 141 in [1]). While the term *simply supported* always means to restrict the transverse deflection $w(x, y)$, it does not uniquely define the boundary conditions on the in-plane displacements u_n and u_s, normal and tangent to the boundary, respectively. In the context of analytical solutions, it is customary to restrict either u_n or u_s. Therefore, the following possibilities exist

- SS-1: $w = u_s = \phi_s = 0; N_n = \widehat{N}_n; M_n = \widehat{M}_n$

- SS-2: $w = u_n = \phi_s = 0; N_{ns} = \widehat{N}_{ns}; M_n = \widehat{M}_n$

In type SS-1, a normal force and a moment are specified. In SS-2, a shear force and a moment are specified. If the laminate does not have bending-extension coupling, and the analysis is geometrically linear, transverse loads will not induce u_n. The naming convention for the rotations is the same as that used for moment resultants in Figure 3.3, where a subscript $()_n$ indicates the direction normal to the edge of the shell, and a subscript $()_s$ indicates the direction tangent to the edge (see also [4, Figure 6.2.1]). Furthermore $\widehat{()}$ represents a fixed known value that may or may not be zero.

3.1.2   Kirchhoff Theory

Historically, Kirchhoff theory was preferred because the governing equations can be written in terms of only one variable, the transverse deflection of the shell w_0. In the pre-information age, it was easier to obtain analytical solutions in terms of only one variable rather than in terms of the three variables needed in FSDT. This means that a wealth of closed form design equations and approximate solutions exist in engineering design manuals which are based on Kirchhoff theory [8]. Such simple design formulas can still be used for preliminary design of composite shells if we are careful and we understand their limitations. Metallic shells were and still are commonly modeled with Kirchhoff theory. The FSDT governing equations can be reduced to Kirchhoff governing equations, and closed form solutions can be found, as shown in [4].

In Kirchhoff theory the interlaminar shear strain is assumed to be zero. From the last two equations in (3.6) we get

$$\phi_x = \frac{\partial w_0}{\partial x}$$
$$\phi_y = \frac{\partial w_0}{\partial y} \qquad (3.10)$$

and introducing them into the first three equations in (3.6) we get

$$\epsilon_x(x, y, z) = \frac{\partial u_0}{\partial x} - z\frac{\partial^2 w_0}{\partial x^2} = \epsilon_x^0 + z\kappa_x$$

$$\epsilon_y(x, y, z) = \frac{\partial v_0}{\partial y} - z\frac{\partial^2 w_0}{\partial y^2} = \epsilon_y^0 + z\kappa_y$$

$$\gamma_{xy}(x, y, z) = \frac{\partial u_0}{\partial y} + \frac{\partial v_0}{\partial x} - 2z\frac{\partial^2 w_0}{\partial x \partial y} = \gamma_{xy}^0 + z\kappa_{xy} \tag{3.11}$$

Notice that the variables ϕ_x, ϕ_y have been eliminated and Kirchhoff theory only uses three variables $u_0(x, y)$, $v_0(x, y)$ and $w_0(x, y)$. This makes analytical solutions easier to find, but numerically Kirchhoff theory is more complex to implement. Since second derivatives of w_0 are needed to write the strains, the weak form (2.31) will have second derivatives of w_0. This will require that the interpolation functions (see Section 2.1.4) have C^1 continuity. That is, the interpolation functions must be such that not only the displacements but also the slopes be continuous across element boundaries. In other words, both the displacement w_0 and the slopes $\partial w_0/\partial x$, $\partial w_0/\partial y$ will have to be identical at the boundary between elements when calculated from either element sharing the boundary. This is difficult to implement.

Consider the case of beam bending. The ordinary differential equation (ODE) with an applied distributed load $\widehat{q}(x)$ is

$$EI\frac{d^4 w_0}{dx^4} = \widehat{q}(x) \tag{3.12}$$

The weak form is obtained as in (2.3)

$$0 = \int_{x_A}^{x_B} v\left[-EI\frac{d^4 w_0}{dx^4} + \widehat{q}(x)\right] dx \tag{3.13}$$

Integrating by parts twice

$$0 = \int_{x_A}^{x_B} \frac{d^2 v}{dx^2} EI\frac{d^2 w_0}{dx^2} dx + vEI\frac{d^3 w_0}{dx^3} - \frac{dv}{dx}EI\frac{d^2 w_0}{dx^2} - \int_{x_A}^{x_B} v\widehat{q}(x)dx$$

$$0 = B(v, w_0) + [vQ_x]_{x_A}^{x_B} - \left[\frac{dv}{dx}M_x\right]_{x_A}^{x_B} - \int_{x_A}^{x_B} v\widehat{M}(x)dx$$

$$0 = B(v, w_0) + L(v) \tag{3.14}$$

When the elements are assembled as in Section 2.1.6, it turns out that adjacent elements i and $i + 1$ that share a node have identical deflection but opposite shear force Q_x and bending moment M_x at their common node, as follows

$$w^i = w^{i+1}$$

$$Q^i = -Q^{i+1}$$

$$M^i = -M^{i+1} \tag{3.15}$$

For the shear forces to cancel as in (2.25), it is only required to have $v^i = v^{i+1}$, which is satisfied by C^0 continuity elements having $w^i = w^{i+1}$ at the common node. For the bending moments to cancel as in (2.25), it is required that $dv^i/dx = dv^{i+1}/dx$. This can only be done if the elements have C^1 continuity. That is, the slopes $dw^i/dx = dw^{i+1}/dx$ must be identical at the common node. Such elements are difficult to work with ([5, page 276]).

In FSDT theory, only first derivatives are used in the strains (3.6). So, the weak form (2.31) has only first derivatives and, like (2.25), all the internal generalized forces cancel at common nodes with only C^0 element continuity.

3.2 FE Analysis of Laminates

Stress and deformation analysis of composites can be done at different levels (Figure 3.4). The level of detail necessary for description of the material depends on the level of post-processing desired.

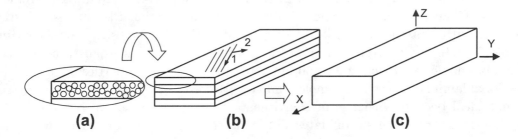

Fig. 3.4: (a) Micromechanics, (b) lamina level and (c) laminate level approach.

When a great level of detail is necessary, the strain and stress are computed at the constituent level, i.e. fiber and matrix. In this case, it is necessary to describe the micro-structure, including the fiber shape and geometrical distribution, and the material properties of the constituents. More details are given in Chapter 6 where micro-mechanical modeling is used to generate properties for any combination of fibers and matrix.

With a less detailed approach, the composite material can be considered as a homogeneous equivalent material. In this case, its structural behavior can be analyzed by using orthotropic properties shown in Chapter 1.

Sometimes it is not possible to use the shell simplification explained in Section 3.1. For example, when the composite material is a woven fabric, or the laminate is very thick, or when studying localized phenomena such as free edges effects (Chapter 5). In these cases, the composite should be analyzed as solid, as shown in Chapter 5. In any case, most of the laminated structures can be analyzed using the plates and shell simplifications explained in Section 3.1.

If the whole laminate is analyzed as a homogeneous equivalent shell, using the macro-scale level approach (see Figure 3.4.c), the stress distribution in the laminate cannot be obtained. However, this very simple description of the laminate is sufficient when only displacements, buckling loads and modes, or vibration frequencies and modes are required. In these cases, only the laminate stiffness (3.8) is needed (see Section 3.2.2). In certain cases, even a simpler material description will suffice. For example, when the laminate is only unidirectional, or if the laminate is balanced and symmetric (see [1] pp. 153-154), the laminate can be modeled as a single lamina of orthotropic material (see Section 3.2.3).

On the other hand, if stress and strains are to be calculated, then the actual laminate stacking sequence (LSS) must be input to the program (see Section 3.2.4). In this case, the elastic properties of each lamina, as well as thickness and fiber orientation of every lamina must be given. This method is usually called the meso-scale level approach (see Figure 3.4.b).

For the sake of generality, laminae are always modeled as orthotropic materials, while in fact a unidirectional lamina is transversely isotropic. Then, it suffices to enter $E_3 = E_2$, and $G_{23} = E_3/2(1 + \nu_{23})$ into the equations of the orthotropic material.

The elastic properties of a unidirectional lamina can be computed using microme-chanics (Chapter 6) or with experimental data of unidirectional laminates. Elastic and strength material properties of some unidirectional composites are shown in Table 3.1. In the analysis of most composite structures, it is usual to avoid the micromechanics approach and to obtain experimentally the properties of the unidi-rectional laminae, or even the whole laminate. However, the experimental approach is not ideal because a change of constituents or volume fraction of reinforcement during the design process invalidates all the material data and requires a new exper-imental program for the new material. It is better to calculate the elastic properties of the lamina using micromechanics formulas, such as those presented in Section 6.1. Unfortunately, micromechanics formulas are not accurate to predict strength, so experimental work cannot be ruled out completely.

Table 3.1: Material properties of unidirectional carbon/epoxy composites

Property	Unit	AS4D/9310	T300/5208
E_1	[GPa]	133.86	136.00
$E_2 = E_3$	[GPa]	7.706	9.80
$G_{12} = G_{13}$	[GPa]	4.306	4.70
G_{23}	[GPa]	2.76	5.20
$\nu_{12} = \nu_{13}$		0.301	0.280
ν_{23}		0.396	0.150
ρ	[g/cm^3]	1.52	1.54
F_{1t}	[MPa]	1830	1550
F_{1c}	[MPa]	1096	1090
$F_{2t} = F_{3t}$	[MPa]	57	59
$F_{2c} = F_{3c}$	[MPa]	228	59
F_6	[MPa]	71	75

3.2.1 Shell Element Types in FE codes

In commercial FE codes, various plates and shells theories are implemented and differentiated by element types, called shell elements [6, 7]. Laminate properties can be specified in two ways: defining the constitutive matrices or specifying individual layer properties.

When the average properties of the laminate are used, the shell element cannot distinguish between different laminae. It can only relate generalized forces and moments to generalized strains and curvatures. Some shell elements, called layered shell elements, have the capability to compute the laminate properties using the LSS and the laminae properties. Then, it is possible to analyze the model at the meso-scale level.

Choosing the proper element type is very important when using the finite element method to analyze composites. The element type should be chosen based on the problem and the desired results. The main non-layered and layered shell elements types in ANSYS are shown in Tables 3.2 and 3.3. These elements allow one to model thin to moderately thick shells, up to a side-to-thickness ratio of 10. Some of them have 4 nodes and others have 8 nodes using interpolation functions of higher degree. These shell elements are defined in 3D space and have six degrees of freedom (DOF) at each node (translations in the nodal x, y, and z directions and rotations about the nodal x, y, and z axis). The 6th DOF (rotation about the z axis) is included in the shell formulation to allow modeling of folded plates, but it would not be necessary if the shell surface is smooth.

Modeling of different types of laminates with various levels of detail is explained in the next few sections.

3.2.2 A-B-D-H Input Data for Laminate FEA

As previously mentioned, macro-scale level (laminate level) is enough if only deflections, modal analysis or buckling analysis is to be performed, with no requirement for detailed stress analysis. Then it is not necessary to specify the laminate stacking sequence (LSS), the thickness, and the elastic properties of each lamina of the laminate. Only the elastic laminate properties (A, B, D and H matrices) defined in (3.9) are required. This is convenient because it allows one to input the aggregate composite material behavior with few parameters. The reduction of the complexity of the input data allows modeling of laminates with an unlimited number of laminae, using only four matrices.

When the A, B, D and H matrices are used to define the FE analysis, the computer model knows the correct stiffness but it does not know the LSS. Therefore, the software can compute the deformation response (including buckling and vibrations) and even the strain distribution through the thickness of the shell, but it cannot compute the stress components because it does not know where the lamina material properties change.

The A-B-D-H input data can be found by using (3.9). Then, these are input in the FE software, as illustrated in Example 3.1.

Table 3.2: Non-layered shell element types in ANSYS

SHELL43: 4-Node Plastic Large Strain Shell
Element type well suited to model moderately thick shell structures (FSDT). It permits nonlinear material formulations (e.g. plasticity or damage), but it does not have large strain capability. • ANSYS recommends using the SHELL181 for nonlinear analysis (with better formulation and large strain capability).
SHELL63: 4-Node Structural Shell
This element type has thin shell capability. Shear deflection is not included in this thin-shell element, so it should not be used with moderately thick structures. The element permits nonlinear material capability, but does not support large strain formulation. • We recommend using the SHELL93 instead of SHELL63 whenever possible.
SHELL93: 8-Node Structural Shell
Element type particularly well suited to model curved shells. For its midside node capability, the deformation shapes are quadratic in both in-plane directions. Shear deflection is included in this shell element (FSDT). The element has nonlinear and large strain capabilities.

Table 3.3: Shell element types with layered option in ANSYS

SHELL91: 8-Node Nonlinear Layered Structural Shell
This element type may be used for layered applications of a structural shell model or for modeling thick laminates and sandwich structures (uses FSDT). Up to 100 different layers are permitted. The element has nonlinear and large strain capabilities.   • If applicable, SHELL99 is usually more efficient (smaller formulation time) than SHELL91. However, SHELL91 is more robust for nonlinear analysis.
SHELL99: 8-Node Linear Layered Structural Shell
The element type SHELL99 may be used for layered applications of a structural shell model. It allows a total of 250 layers, and if more than 250 layers are required, a user-input constitutive matrix is available. The element does not have nonlinear capabilities. It also has the option to offset the nodes to the top, middle or bottom surface.   • SHELL99 allows more layers than SHELL91 but not all types of nonlinear behavior. It usually has a smaller element formulation time.   • With KEYOPT(2) = 2, a user-input constitutive matrix is used, then it does not need the definition of a LSS.
SHELL181: 4-Node Finite Strain Shell
This element type is suitable for analyzing thin to moderately thick shell structures. It may be used for layered applications for modeling laminated composite shells (use FSDT) or sandwich construction. A maximum of 250 layers is supported. The element is well suited for linear and nonlinear applications. It has large strain capabilities. This element is able to input the LLS using section definitions rather than real constants sets.   • We recommend the use of SHELL181 instead of SHELL43.   • ANSYS recommends the use of KEYOPT(3) = 2 for most composite analysis (necessary to capture the stress gradients).

Example 3.1 *Consider a simply supported square plate $a_x = a_y = 2000$ mm, thickness $t = 10$ mm laminated with AS4D/9310 (Table 3.1) in a $[0/90]_n$ configuration. Tabulate the center deflection perpendicular to the plate surface when the number of layers is $n = 1, 5, 10, 15$ and 20. The plate is loaded in compression with and edge load $N_x = -1$ N/mm and ($N_y = N_{xy} = M_x = M_y = M_{xy} = 0$). Use symmetry to model 1/4 of the plate. Generate the A, B, D and H matrices and enter them into ANSYS.*

Solution to Example 3.1 *The matrices A, B, D and H are calculated using (3.9), which are implemented in CADEC [2].*

In ANSYS the only shell element that allows data input as A, B, D and H matrices is SHELL99 with $KEYOPT(2) = 2$. Then, the laminate stiffness matrices must be placed in the input file using three matrices: E_0, E_1 and E_2. Each of these 6×6 symmetric matrices is defined by 21 values.

The upper triangular part of the extensional stiffness $[E_0]$ matrix is

$$[E_0] = \begin{bmatrix} A_{11} & A_{12} & 0 & A_{16} & 0 & 0 \\ \cdot & A_{22} & 0 & A_{26} & 0 & 0 \\ \cdot & \cdot & 0 & 0 & 0 & 0 \\ \cdot & \cdot & \cdot & A_{66} & 0 & 0 \\ \cdot & \cdot & \cdot & \cdot & H_{44} & H_{45} \\ \cdot & \cdot & \cdot & \cdot & \cdot & H_{55} \end{bmatrix} \tag{3.16}$$

The upper triangular part of the bending-extension coupling stiffness $[E_1]$ matrix is

$$[E_1] = \begin{bmatrix} B_{11} & B_{12} & 0 & B_{16} & 0 & 0 \\ \cdot & B_{22} & 0 & B_{26} & 0 & 0 \\ \cdot & \cdot & 0 & 0 & 0 & 0 \\ \cdot & \cdot & \cdot & B_{66} & 0 & 0 \\ \cdot & \cdot & \cdot & \cdot & 0 & 0 \\ \cdot & \cdot & \cdot & \cdot & \cdot & 0 \end{bmatrix} \tag{3.17}$$

and the upper triangular part of the bending stiffness $[E_2]$ matrix is

$$[E_2] = \begin{bmatrix} D_{11} & D_{12} & 0 & D_{16} & 0 & 0 \\ \cdot & D_{22} & 0 & D_{26} & 0 & 0 \\ \cdot & \cdot & 0 & 0 & 0 & 0 \\ \cdot & \cdot & \cdot & D_{66} & 0 & 0 \\ \cdot & \cdot & \cdot & \cdot & 0 & 0 \\ \cdot & \cdot & \cdot & \cdot & \cdot & 0 \end{bmatrix} \tag{3.18}$$

Using SHELL99 element with $KEYOPT(2) = 2$, enter the values of the non-zero matrix coefficients using R and RMODIF commands in ANSYS, as shown in the following command lines

```
R,1,A11,A12,0,A16,0,0          ! real set #1, 1st row of E0 matrix
RMODIF,1,7,A22,0,A26,0,0       ! real set #1, 2nd row
RMODIF,1,16,A66,0,0            ! real set #1, 4th row
RMODIF,1,19,H44,H45            ! real set #1, 5th row
RMODIF,1,21,H55                ! real set #1, 6th row
RMODIF,1,22,B11,B12,0,B16,0,0  ! real set #1, 1st row of E1 matrix
RMODIF,1,28,B22,0,B26,0,0      ! real set #1, 2nd row
RMODIF,1,37,B66,0,0            ! real set #1, 4th row
```

```
RMODIF,1,43,D11,D12,0,D16,0,0    ! real set #1, 1st row of E2 matrix
RMODIF,1,49,D22,0,D26,0,0        ! real set #1, 2nd row
RMODIF,1,58,D66,0,0,             ! real set #1, 4th row
RMODIF,1,76,dens,thick  ! real set #1, Average density & thickness
```

In this case the input data do not need material properties (these are included in the constitutive A, B, D and H matrices). The complete input file, which is available on the Web site [12], is listed below for $n = 1$. See Appendix E for an introduction to the software interface.

Note that in ANSYS a **pressure** applied on the boundary acts **on** the element; that is, opposite to the outside normal to the boundary and thus a compressive edge load in this case is applied as SFL,2,PRES,1.0.

```
/TITLE,Simply Supported [0/90] Plate Nx=1 N/mm -  - SHELL99
! Material is AS4D/9310 - laminate [0/90]n, n=1
/UNITS,MPA               ! Units are in mm, MPa, and Newtons
/PREP7                   ! Pre-processor module
! This input data does not need material properties

ET,1,SHELL99,,2          ! Chooses SHELL91 element for analysis
                         ! Set KEYOPT(2)=2, then supply ABDH matrix
KEYOPT,1,10,2            ! Set KEYOPT(10)=2, print ABDH matrix file.abd

! Real constant set #1, ABDH matrix definition
R,1,711563,23328.8,0,0,0,0              ! set #1, A11,A12,0,A16,0,0
RMODIF,1,7,711563,0,0,0,0               ! set #1, A22,0,A26,0,0
RMODIF,1,16,43600,0,0                   ! set #1, A66,0,0
RMODIF,1,19,29666.7,0                   ! set #1, H44,H45
RMODIF,1,21,29666.7                     ! set #1, H55
RMODIF,1,22,-1.58515e+006,0,0,0,0,0     ! set #1, B11,B12,0,B16,0,0
RMODIF,1,28,1.58515e+006,0,0,0,0        ! set #1, B22,0,B26,0,0
RMODIF,1,37,0,0,0                       ! set #1, B66,0,0
RMODIF,1,43,5.92969e+006,194407,0,0,0,0 ! set #1, D11,D12,0,D16,0,0
RMODIF,1,49,5.92969e+006,0,0,0,0        ! set #1, D22,0,D26,0,0
RMODIF,1,58,363333,0,0                  ! set #1, D66,0,0
RMODIF,1,76,,10          ! set #1, Average density and thick

! Geometry and mesh
RECTNG,0,1000,0,1000     ! Creates a rectangle with x=1 m and y=1 m
ESIZE,100                ! Element size 100 mm
AMESH,all                ! Mesh the area
FINISH                   ! Exit pre-processor module

/SOLU                    ! Solution module
ANTYPE,STATIC            ! Set static analysis
DL,2,1,uz,0              ! Impose Simple Supported BC
DL,3,1,uz,0
DL,1,1,symm              ! Impose Symmetry BC
DL,4,1,symm
!d,all,rotz              ! Constrain rotations about z axes (optional)
```

```
SFL,2,PRES,1.0              ! Apply uniform edge load in N/mm

SOLVE                       ! Solve current load state FINISH
! Exit solution module

/POST1                      ! Post-processor module
PLDISP                      ! Display deformed shape
PLNSOL,u,z                  ! Display contour of displacements z
!plesol,s,x                 ! notice this cannot be done, try it
FINISH                      ! Exit post-processor module
```

The solution is tabulated in Table 3.4. Bending extension coupling produces a lateral deflection, which diminishes as the number of layers grow.

Table 3.4: Lateral deflection vs. number of layers in Example 3.1

n	δ [mm]
1	0.2191
5	0.0211
10	0.0104
15	0.0069
20	0.0052

3.2.3   Equivalent Orthotropic Input for Laminate FEA

Some FEA codes do not have laminated elements and do not accept the A, B, D and H matrices as explained in Section 3.2.2. However, if they have orthotropic elements, it is still possible to perform deformation, vibration, and buckling analysis for laminated composites.

Unidirectional Laminate FEA

To model a unidirectional laminate, standard shell elements can be used, even if they are not layered shell, and still obtain correct results of displacements, strains, and stress. The geometry of shells is a surface that represents the mid-surface of the real shell, located halfway through the thickness. The positive thickness coordinate points along a normal to the shell mid-surface (local z-direction which coincides with the 3-direction). This is the normal definition of shells and it is used in shell elements, as shown in Example 3.2.

Example 3.2 *Use ANSYS to model a simply supported rectangular plate with dimensions $a_x = 4000$ mm, $a_y = 2000$ mm, thickness $t = 10$ mm. Apply a uniform transverse load $q_0 = 0.12 \times 10^{-3}$ MPa. The material is a unidirectional laminate AS4D/9310 (Table 3.1), with the fibers oriented in the x-direction.*

Solution to Example 3.2 *The thickness coordinate is eliminated from the governing equations so that the 3D problem simplifies to 2D. In the process, the thickness becomes a parameter, which is known and supplied to the modeling software. Most software packages differentiate between material properties and parameters but both are supplied as known input data. For example, the shell thickness is supplied to ANSYS as a real constant set (R command), while material properties are entered separately (MP command).*

SHELL93, an 8-node shell element, is used in this example. Symmetry with respect to the x-z and y-z planes is used to model 1/4 of the plate. See input file next [12].

```
/TITLE,Simply Supported Plate under q=0.12e-3 MPa - SHELL93
! Material is UD AS4D/9310 Carbon/Epoxy - 8 layers 1.25 mm thick
/UNITS,MPA               ! Units are in mm, MPa, and Newtons

/PREP7                   ! Pre-processor module
! Material properties FOR AS4D/9110 orthotropic laminate
uimp,1,ex,ey,ez,133.86E3,7.706E3,7.706E3
uimp,1,gxy,gyz,gxz,4.306E3,2.76E3,4.306E3
uimp,1,prxy,pryz,prxz,0.301,0.396,0.301

ET,1,SHELL93             ! Chooses SHELL93 element for analysis
R,1,10 ! Real constant set #1, thickness of 10 mm

! Geometry and mesh
RECTNG,0,2000,0,1000     ! Creates a rectangle with x=2 m and y=1 m
ESIZE,250                ! Element size 250 mm
AMESH,all                ! Mesh the area

FINISH                   ! Exit pre-processor module

/SOLU                    ! Solution module
ANTYPE,STATIC            ! Set static analysis
DL,2,1,uz,0              ! Impose Simple Supported BC
DL,3,1,uz,0
DL,1,1,symm              ! Impose Symmetry BC
DL,4,1,symm
!d,all,rotz              ! Constraint rotations about z axes (optional)

SFA,all,2,PRES,1.2e-4,   ! Apply uniform pressure in MPa

SOLVE                    ! Solve current load state
FINISH                   ! Exit solution module

/POST1                   ! Post-processor module
PLDISP,1                 ! plots displaced plate
PLNSOL,u,z               ! contour plot of z direction displacements
PLESOL,s,x               ! contour plot of x direction stress
FINISH                   ! Exit post-processor module
```

Symmetric Laminate FEA

If a multidirectional laminate is balanced and symmetric, the apparent laminate orthotropic properties can be found as explained in Section 1.15. The apparent laminate properties represent the stiffness of an equivalent (fictitious) orthotropic plate that behaves like the actual laminate under in-plane loads. These apparent properties should not be used to predict bending response. When the only important response is bending, e.g. a thick cantilever plate under bending, the formulation shown in [1, (6.33)] should be used to obtain the apparent laminate properties. However, in most of the structural design using composite shell, the laminate works under in-plane loads and the formulation in Section 1.15 should be used.

If the laminate is symmetric but not balanced, the axes of orthotropy are rotated with respect to the global coordinate system, but still the laminate is equivalent to an orthotropic material as per Section 1.15. For example, a unidirectional laminate oriented at an angle θ with respect to global axes, should be modeled on a coordinate system oriented along the fiber direction (see Section 3.2.7).

Example 3.3 *Use ANSYS to model a simply supported rectangular plate with dimensions $a_x = 2000 \ mm$, $a_y = 2000 \ mm$, for a laminate $[\pm 45/0]_S$. Apply a tensile edge load $N_x = 200 \ N/mm$. Each layer is 1 mm thick with the following properties*

$$E_1 = 37.88 \ GPa \quad G_{12} = 3.405 \ GPa \quad \nu_{12} = 0.299$$
$$E_2 = 9.407 \ GPa \quad G_{23} = 3.308 \ GPa \quad \nu_{23} = 0.422$$

Solution to Example 3.3 *Since the laminate is balanced symmetric, compute the averaged laminate properties E_x, E_y and so on using Section 1.15 to yield*

$$E_x = 20.104 \ GPa \quad G_{xy} = 8.237 \ GPa \quad \nu_{xy} = 0.532$$
$$E_y = 12.042 \ GPa \quad G_{yz} = 3.373 \ GPa \quad \nu_{yz} = 0.203$$
$$E_z = 10.165 \ GPa \quad G_{xz} = 3.340 \ GPa \quad \nu_{xz} = 0.307$$

The input file is shown below [12]. Note that in ANSYS a pressure applied on the boundary acts on the element; that is, opposite to the outside normal to the boundary and thus a tensile load in this case is applied as SFL,2,PRES,-200.0.

```
/TITLE,Simply Supported Plate Nx=200 N/mm - equivalent [45/-45/0]s
! [45/-45/0]s laminate of E-Glass/Vinyl, vf=0.5 and th=1*6 mm
/UNITS,MPA                ! Units are in mm, MPa, and Newtons

/PREP7                    ! Pre-processor module
! Equivalent orthotropic material properties for the LAMINATE
uimp,1,ex,ey,ez,20.104E3,12.042E3,10.165E3
uimp,1,gxy,gyz,gxz,8.237E3,3.373E3,3.34E3
uimp,1,prxy,pryz,prxz,0.532,0.203,0.307

ET,1,SHELL93              ! Chooses Shell 93 element for analysis
R,1,10                    ! Real constant set #1, thickness of 6 mm

! Geometry and mesh
RECTNG,0,1000,0,1000      ! Creates a rectangle with x=1 m and y=1 m
```

```
ESIZE,100              ! Element size 100 mm
AMESH,all              ! Mesh the area
/PNUM,LINE,1
LPLOT
FINISH                 ! Exit pre-processor module

/SOLU                  ! Solution module
ANTYPE,STATIC          ! Set static analysis
DL,2,1,uz,0            ! Impose Simple Supported BC
DL,3,1,uz,0
DL,1,1,symm            ! Impose Symmetry BC
DL,4,1,symm
!d,all,rotz            ! Constraint rotations about z axes (optional)

SFL,2,PRES,-200        ! Apply uniform linear load in N/mm

SOLVE                  ! Solve current load state
FINISH                 ! Exit solution module

/POST1                 ! Post-processor module
/VIEW,,1,1,1
!PLDISP,2                ! Plots displaced plate
PLNSOL,u,x             ! Contour plot of x direction displacements
!plesol,s,x             ! Notice the stress results are incorrect
FINISH                 ! Exit post-processor module
```

The resulting maximum horizontal displacement on a quarter-plate model is 0.995 mm. The planes $x = 0$ and $y = 0$ are not symmetry planes for a $[\pm 45/0]_S$ but once the laminate is represented by equivalent orthotropic properties, as is done in this example, the lack of symmetry at the lamina level is lost and it does not have any effect on the mid-surface displacements. Therefore, one-quarter of the plate represents well the entire plate as long as no stress analysis is performed. Furthermore, displacement and mid-surface strain analysis can be done with the laminate replaced by an equivalent orthotropic material. However, even if the full plate were to be modeled, the stress values in the equivalent orthotropic material are not the actual stress values of the laminate. While the material analyzed in this example is not homogeneous, but laminated, the material in Example 3.2 is a homogeneous unidirectional material. Therefore, the stress values are not correct in this example but they are correct in Example 3.2.

Asymmetric Laminate FEA

If the laminate is not symmetric, bending-extension coupling must be considered. Strictly speaking, such material is not orthotropic and should not be modeled with an equivalent laminate material. Even then, if only orthotropic shell elements are available and the bending-extension coupling effects are not severe, the material could be approximated by an orthotropic material by neglecting the matrices B and D. The ratios defined in [1, (6.34)-(6.36)] can be used to assess the quality of the approximation obtained using apparent elastic properties. Care must be taken for unbalanced laminates that the A and H matrix are formulated in a coordinate system coinciding with the axes of orthotropy of the laminate.

3.2.4   LSS for Multidirectional Laminate FEA

For computation of strain and stress at the meso-scale (lamina level), it is necessary to know the description of the laminate and the properties of each lamina. The description of the multidirectional laminate includes the LSS, which specifies the angle of each lamina with respect to the x-axis of the laminate, the thickness, and the elastic material properties of each lamina. Then, the software computes the matrices A, B, D, and H internally. In this way, the software can compute the stress components in each layer. This approach is illustrated in the following example.

Example 3.4 *Consider a simply supported square plate $a_x = a_y = 2000$ mm, $t = 10$ mm thick, laminated with AS4D/9310 (Table 3.1) in a $[0/90/\pm45]_S$ symmetric laminate configuration. Tabulate the center deflection perpendicular to the plate surface. The plate is loaded with a tensile load $N_x = 100$ N/mm and $(N_y = N_{xy} = M_x = M_y = M_{xy} = 0)$.*

Solution to Example 3.4 *Two solutions using different shell elements, SHELL91 and SHELL181, are presented here. Each element uses different instructions to introduce the LSS and lamina properties. The specifications of LSS for SHELL91 and SHELL99 are identical.*

> ### Solution Using SHELL91
> *Half of the LSS is entered by specifying that the laminate is symmetric through the real constant (R and RMORE commands, see input file listed below). Note the LSS is given starting at layer #1 at the bottom.*

```
/TITLE,Simply Supported [0/90/45/-45]s - uniform load - SHELL91
! Material is AS4D/9310 - [0/90/45/-45]s, Th=1.25 mm per lamina
/UNITS,MPA                 ! Units are in mm, MPa, and Newtons

/PREP7                     ! Pre-processor module
! Material properties FOR AS4D/9310 orthotropic laminate
uimp,1,ex,ey,ez,133.86E3,7.706E3,7.706E3
uimp,1,gxy,gyz,gxz,4.306E3,2.76E3,4.306E3
uimp,1,prxy,pryz,prxz,0.301,0.396,0.301

ET,1,shell91,,1            ! Chooses SHELL91 element for analysis
                           ! Set KEYOPT(2)=1, then supply 12+(6*NL) const.
KEYOPT,1,1,8               ! Set KEYOPT(1)=8, Max number of layers = 8
KEYOPT,1,5,1               ! Set KEYOPT(5)=1, Element output: Middle layer
KEYOPT,1,8,1               ! Set KEYOPT(8)=1, Storage data: All layers
KEYOPT,1,11,0              ! Set KEYOPT(11)=0, Nodes on midshell surface

! Real constant set #1, [0/90/45/-45]s, NL=8, lamina thick 1.25 mm
R,1,8,1,                   ! 8 layers symmetrical
RMORE,,,,,,
RMORE,1,0,1.25             ! 1st layer: mat. #1,   0 deg, Th=1.25 mm
RMORE,1,90,1.25            ! 2nd layer: mat. #1,  90 deg, Th=1.25 mm
RMORE,1,45,1.25            ! 3nd layer: mat. #1, +45 deg, Th=1.25 mm
```

```
RMORE,1,-45,1.25          ! 4rt layer: mat. #1, -45 deg, Th=1.25 mm

! Geometry and mesh
RECTNG,-1000,1000,-1000,1000    ! Creates a rectangle
ESIZE,250                 ! Element size 250 mm
AMESH,all                 ! Mesh the area

FINISH                    ! Exit pre-processor module

/SOLU                     ! Solution module
ANTYPE,STATIC             ! Set static analysis
DL,2,1,uz,0               ! Impose Simple Supported BC
DL,3,1,uz,0
DL,1,1,uz,0
DL,4,1,uz,0
!d,all,rotz               ! Constraint rotations about z (optional)
CEN_NODE=NODE(0,0,0)      ! Center node
D,CEN_NODE,UX       ! Constraint nodes to avoid rigid body motion
LEFT_NODE=NODE(-1000,0,0)    ! Middle node in left edge
D,LEFT_NODE,UY
RIGH_NODE=NODE(1000,0,0)     ! Middle node in right edge
D,RIGH_NODE,UY

SFL,2,PRES,-100           ! Apply uniform pressure in N/mm
SFL,4,PRES,-100

SOLVE                     ! Solve current load state
FINISH                    ! Exit solution module

/post1                    ! Post-processor module
PLDISP,1                  ! Display the displaced plate
FINISH                    ! Exit post-processor module
```

Note that KEYOPTION(11)=0 is used to place the nodes on the mid surface. The deformation of the layer is in-plane only because the laminate is symmetric, balanced, and in-plane loaded. The distributed edge load is applied at the mid-surface. However, the same model is re-executed with KEYOPTION(11)=1 (nodes on bottom face) or with KEYOPTION(11)=2 (nodes on top face) the edge load will be applied on the bottom edge or on the top edge of the laminate. In these cases, out-plane deformations will appear.

Solution Using SHELL181
A different family of commands is used next to define a section (SECTYPE, SECDATA and SECOFFSET, see input file listed below [12]). Note the LSS is also given starting at layer #1 at the bottom surface.

```
/TITLE,Simply Supported [0/90/45/-45]s - uniform load - SHELL181
! Material is AS4D/9310 - [0/90/45/-45]s, Th=1.25 mm per lamina
/UNITS,MPA                ! Units are in mm, MPa, and Newtons

/PREP7                    ! Pre-processor module
```

```
! Material properties FOR AS4D/9310 orthotropic laminate
uimp,1,ex,ey,ez,133.86E3,7.706E3,7.706E3
uimp,1,gxy,gyz,gxz,4.306E3,2.76E3,4.306E3
uimp,1,prxy,pryz,prxz,0.301,0.396,0.301

ET,1,SHELL181,,,2        ! Chooses SHELL181 element for analysis
! Set KEYOPT(3)=2, full integration (recommended in composites)

SECTYPE,1,SHELL,,La1     ! set #1, [0/90/45/-45]s, label=La1
SECDATA, 1.25,1,0.0,3    ! 1st layer: mat. #1,   0 deg, Th=1.25 mm
SECDATA, 1.25,1,90,3     ! 2nd layer: mat. #1,  90 deg, Th=1.25 mm
SECDATA, 1.25,1,45,3     ! 3nd layer: mat. #1, +45 deg, Th=1.25 mm
SECDATA, 1.25,1,-45,3    ! 4rt layer: mat. #1, -45 deg, Th=1.25 mm
SECDATA, 1.25,1,-45,3    ! Same layers in symmetrical order
SECDATA, 1.25,1,45,3
SECDATA, 1.25,1,90,3
SECDATA, 1.25,1,0.0,3
SECOFFSET,MID            ! Nodes on the laminate middle thickness

! Geometry and mesh
RECTNG,-1000,1000,-1000,1000    ! Creates a rectangle
ESIZE,250               ! Element size 250 mm
AMESH,all               ! Mesh the area

FINISH                  ! Exit pre-processor module

/SOLU                   ! Solution module
ANTYPE,STATIC           ! Set static analysis
DL,2,1,uz,0             ! Impose Simple Supported BC
DL,3,1,uz,0
DL,1,1,uz,0
DL,4,1,uz,0
!d,all,rotz             ! Constraint rotations about z (optional)
CEN_NODE=NODE(0,0,0)    ! Center node
D,CEN_NODE,UX           ! Constraint nodes to avoid rigid body motion
LEFT_NODE=NODE(-1000,0,0)    ! Middle node in left edge
D,LEFT_NODE,UY
RIGH_NODE=NODE(1000,0,0)     ! Middle node in right edge
D,RIGH_NODE,UY

SFL,2,PRES,-100         ! Apply uniform pressure in N/mm
SFL,4,PRES,-100

SOLVE                   ! Solve current load state
FINISH                  ! Exit solution module

/post1                  ! Post-processor module
PLDISP,1                ! Display the displaced plate
FINISH                  ! Exit post-processor module
```

Other model definition aspects that can be controlled include: the position of the bottom and top surfaces of the laminate (i.e., the direction of the vector normal to the surface of the shell), the relative position of the shell surface through the real laminate thickness (at the bottom, at the middle, or at the top), the orientation of laminate reference axis (as is shown in Section 3.2.7), and so on.

3.2.5 FEA of Ply Drop-Off Laminates

Sometimes it is convenient to set the reference surface at the bottom (or top) of the shell. One such case is when the laminate has ply drop-offs, as shown in Figure 3.5. When the design calls for a reduction of laminate thickness, plies can be gradually terminated from the thick to the thin part of the shell. As a rule of thumb, ply drop-off should be limited to a 1:16 to 1:20 ratio ($Th : L$ ratio in Figure 3.5) unless detailed analysis and/or testing supports a steeper drop-off ratio. For this case, it is convenient to specify the geometry of the smooth surface, or tool surface.

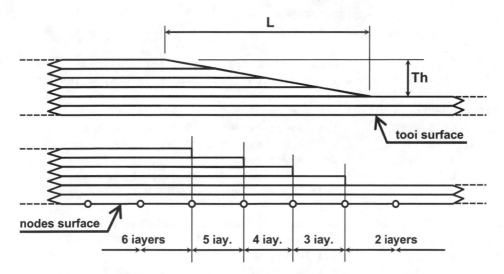

Fig. 3.5: Ply drop-off of length (L) and thickness (Th) and FE model simplifications.

Then, every time a ply or set of plies is dropped, the material and thickness for those elements is changed. This is illustrated in the next examples. Not all software has this capability and it may be necessary to assume that the mid-surface is smooth while in reality only the tool surface is smooth. As long as the thickness is small compared to the other two dimensions of the structure, such assumption is unlikely to have a dramatic effect in the results of a global analysis, such as deformation, buckling, and even membrane stress analysis. The exact description of the thickness geometry begins to pay a role when detailed 3D stress analysis of the ply drop-off region is required, but at that point, a 3D local model is more adequate.

Example 3.5 *A ply drop-off is defined between the laminate A, $[90/0]_S$, and the laminate B, $[90/0]$. The ply drop-off ratio is 1:20 The lamina thickness is 0.75 mm. Consider a* *composite strip* 120 *mm long and* 100 *mm wide under tension* $N = 10$ *N/mm applied to the bottom edges on the strip. Use symmetry to model 1/2 of the tape.*

Solution to Example 3.5 *Using shell elements SHELL181, three different sections are defined, one for A, one for B, and one section to model the ply drop-off between them.*

The thickness difference between both laminates is $0.75 \cdot 2 = 1.5$ *mm. Therefore, the total length of ply drop-off is* $1.5 \cdot 20 = 30$ *mm. Every 15 mm there is a section change.*

The bottom layer is designated as layer #1, and additional layers are stacked from bottom to top in the positive normal direction of the element coordinate system.

```
/TITLE,Tape with Ply Drop-off between [90/0]s and [90/0]
! Material is AS4D/9310 - Th=0.75 mm per lamina - SHELL181
/UNITS,MPA               ! Units are in mm, MPa, and Newtons

/PREP7                   ! Pre-processor module
! Material properties FOR AS4D/9310 orthotropic laminate
uimp,1,ex,ey,ez,133.86E3,7.706E3,7.706E3
uimp,1,gxy,gyz,gxz,4.306E3,2.76E3,4.306E3
uimp,1,prxy,pryz,prxz,0.301,0.396,0.301

ET,1,SHELL181,,,2        ! Chooses SHELL181 element for analysis
    ! Set KEYOPT(3)=2, full integration (recommended in composites)

SECTYPE,1,SHELL,,A       ! Section shell set #1, [90/0]s, A
SECDATA, 0.75,1,90,3     ! 1st layer: mat. #1, 90 deg, Th=0.75 mm
SECDATA, 0.75,1,0,3      ! 2nd layer: mat. #1,  0 deg, Th=0.75 mm
SECDATA, 0.75,1,0,3      ! 3rd layer: mat. #1,  0 deg, Th=0.75 mm
SECDATA, 0.75,1,90,3     ! 4th layer: mat. #1, 90 deg, Th=0.75 mm
SECOFFSET,BOT            ! Nodes on the laminate BOTTOM thickness

SECTYPE,2,SHELL,,DROP    ! Section shell set #2, [90/0/0], DROP
SECDATA, 0.75,1,90,3     ! 1st layer: mat. #1, 90 deg, Th=0.75 mm
SECDATA, 0.75,1,0,3      ! 2nd layer: mat. #1,  0 deg, Th=0.75 mm
SECDATA, 0.75,1,0,3      ! 3rd layer: mat. #1,  0 deg, Th=0.75 mm
SECOFFSET,BOT            ! Nodes on the laminate BOTTOM thickness

SECTYPE,3,SHELL,,B       ! Section shell set #2, [90/0], B
SECDATA, 0.75,1,90,3     ! 1st layer: mat. #1, 90 deg, Th=0.75 mm
SECDATA, 0.75,1,0,3      ! 2nd layer: mat. #1,  0 deg, Th=0.75 mm
SECOFFSET,BOT            ! Nodes on the laminate BOTTOM thickness

! Geometry and mesh
RECTNG,0,60,0,50              ! Laminate A x=60 mm and y=50 mm
RECTNG,60,(60+15),0,50       ! Laminate Drop x=15 mm and y=50 mm
RECTNG,(60+15),(120+15),0,50 ! Laminate B x=60 mm and y=50 mm
AGLUE,all               ! Glue all areas
ESIZE,5                 ! Element size 5 mm
SECNUM,1
AMESH,1                 ! Mesh the area number 1
SECNUM,2
```

```
AMESH,4                  ! Mesh the area number 2
SECNUM,3
AMESH,5                  ! Mesh the area number 3
FINISH                   ! Exit pre-processor module

/SOLU                    ! Solution module
ANTYPE,STATIC            ! Set static analysis
DL,4,1,all,0             ! Impose clamped BC
DL,1,1,symm              ! Impose Symmetry BC
DL,13,4,symm
DL,15,5,symm

SFL,10,PRES,-10          ! Apply uniform tensile edge load in N/mm

SOLVE                    ! Solve current load state
FINISH                   ! Exit solution module

/post1                   ! Post-processor module
PLDISP,1                 ! Display the displaced plate
FINISH                   ! Exit post-processor module
```

When an inner lamina extends over only part of the geometry, it is convenient that the remaining laminae maintain their numbering through the entire model. Otherwise, if the continuity in numbering of layers is lost, the post-processing and result visualization will be extremely difficult. A model with a few layers, some of which are dropped over part of the laminate, is shown in Figure 3.6.

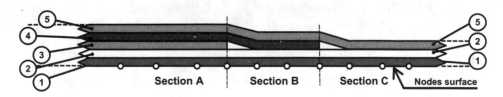

Fig. 3.6: Laminate with dropped laminae.

Example 3.6 *Define in ANSYS the different sections of the laminate shown in Figure 3.6. The laminate in section A is a $[+45/-45/0/90/0]$. The thickness of each lamina is 1.2 mm.*

Solution to Example 3.6 *Using shell elements SHELL181, the different sections are defined as shown in the command list below. The bottom layer is designated as layer #1, and additional layers are stacked from bottom to top in the positive normal direction of the element coordinate system. The dropped laminae are modeled using zero thickness in order to maintain continuous numbering of the remaining laminae.*

```
ET,1,SHELL181,,,2        ! Chooses SHELL181 element for analysis
   ! Set KEYOPT(3)=2, full integration (recommended in composites)
```

```
SECTYPE,1,SHELL,,A          ! Section shell set #1, section A
SECDATA,1.2,1,45,3          ! 1st layer: mat. #1, +45 deg, Th=1.2 mm
SECDATA,1.2,1,-45,3         ! 2nd layer: mat. #1, -45 deg, Th=1.2 mm
SECDATA,1.2,1,0,3           ! 3rd layer: mat. #1,   0 deg, Th=1.2 mm
SECDATA,1.2,1,90,3          ! 4th layer: mat. #1,  90 deg, Th=1.2 mm
SECDATA,1.2,1,0,3           ! 5th layer: mat. #1,   0 deg, Th=1.2 mm
SECOFFSET,BOT               ! Nodes on the laminate BOTTOM thickness

SECTYPE,1,SHELL,,B          ! Section shell set #1, section B
SECDATA,1.2,1,45,3          ! 1st layer: mat. #1, +45 deg, Th=1.2 mm
SECDATA,1.2,1,-45,3         ! 2nd layer: mat. #1, -45 deg, Th=1.2 mm
SECDATA,0,1,0,3             ! 3rd layer: Th=0 mm, do not compute
SECDATA,1.2,1,90,3          ! 4th layer: mat. #1,  90 deg, Th=1.2 mm
SECDATA,1.2,1,0,3           ! 5th layer: mat. #1,   0 deg, Th=1.2 mm
SECOFFSET,BOT               ! Nodes on the laminate BOTTOM thickness

SECTYPE,1,SHELL,,C          ! Section shell set #1, section C
SECDATA,1.2,1,45,3          ! 1st layer: mat. #1, +45 deg, Th=1.2 mm
SECDATA,1.2,1,-45,3         ! 2nd layer: mat. #1, -45 deg, Th=1.2 mm
SECDATA,0,1,0,3             ! 3rd layer: Th=0 mm, do not compute
SECDATA,0,1,90,3           ! 4th layer: Th=0 mm, do not compute
SECDATA,1.2,1,0,3           ! 5th layer: mat. #1,   0 deg, Th=1.2 mm
SECOFFSET,BOT               ! Nodes on the laminate BOTTOM thickness
```

3.2.6   FEA of Sandwich Shells

Some laminates can be considered *sandwich* when specifically designed for sandwich construction with thin faceplates and a thick, relatively weak, core. The faceplates are intended to carry all, or almost all, of the bending and in-plane normal load. Conversely, the core is assumed to carry all of the transverse shear. Example 3.7 shows how to define and calculate a sandwich cantilever beam.

The following assumptions are customarily made for a sandwich shell:

- The terms H_{ij} in (3.9) depend only on the middle layer (core) and they can be calculated as

$$H_{ij} = \left(\overline{Q}_{ij}^*\right)_{core} t_{core}; i,j = 4,5 \tag{3.19}$$

- The transverse shear moduli (G_{23} and G_{13}) are set to zero for the top and bottom layers (non-core layers).

- The transverse shear strains and stresses in the face plate (non-core) layers are neglected or assumed to be zero.

- The transverse shear strains and shear stresses in the core are assumed constant through the thickness.

Example 3.7 *Calculate the maximum deflection of a sandwich cantilever beam subject to an end load $F_z = -100$. The beam is made of a sandwich of two outer aluminum plates (with thickness 1 mm each, $E = 69\,GPa, \nu = 0.3$) and an inner core of foam (with thickness 50 mm, $E = 300\,MPa, \nu = 0.1$).*

Solution to Example 3.7 *ANSYS element SHELL91 is used in this example. This element allows the sandwich option. It is activated by KEYOPTION(9) = 1. Both faceplates are assumed to have the same number of layers, up to seven layers each. With the sandwich option, use of KEYOPT(5) = 1 is recommended since the best results are obtained at the midplane.*

Another element with capabilities to analyze sandwich structures is SHELL181, which models the transverse shear deflection using an energy equivalence method that renders the sandwich option unnecessary.

```
/title,Cantilever Beam with Sandwich
[Al/foam/Al] material ! Material is Aluminum (th=1mm) and FOAM
(Th=50 mm) - SHELL91 /UNITS,MPA              ! Units are in mm, MPa,
and Newtons

/PREP7                    ! Pre-processor module
! Material definition
MP,EX,1,69e3              ! Aluminum Young's modulus
MP,NUXY,1,0.3             ! Aluminum Poisson ratio
MP,EX,2,300              ! Foam Young's modulus
MP,NUXY,2,0.1            ! Foam Poisson ratio

ET,1,SHELL91,,1           ! Chooses SHELL91 element for analysis
                          ! Set KEYOPT(2)=1, then supply 12+(6*NL) const.
KEYOPT,1,1,3             ! Set KEYOPT(1)=3, Max number of layers = 3
KEYOPT,1,5,1             ! Set KEYOPT(5)=1, Element output: Middle layer
KEYOPT,1,8,1             ! Set KEYOPT(8)=1, Storage data: All layers
KEYOPT,1,9,1             ! Set KEYOPT(9)=1, Sandwich option activated
KEYOPT,1,11,0            ! Set KEYOPT(11)=0, Nodes on midshell surface

! Real constant set #1, [Al/Foam/Al]T, NL=3, laminate thick=52 mm
R,1,3                    ! 3 layers, no symmetry RMORE,,,,,,,
RMORE,1,0,1              ! 1st layer: mat. #1,    0 deg, Th=1 mm
RMORE,2,90,50            ! 2nd layer: mat. #2,    0 deg, Th=50 mm
RMORE,1,0,1              ! 3nd layer: mat. #1,    0 deg, Th=1 mm

! Geometry and mesh
RECTNG,0,3e3,0,600       ! Creates a rectangle with x=3 m and y=600 mm
ESIZE,200                ! Element size 200 mm
AMESH,all                ! Mesh the area

FINISH                   ! Exit pre-processor module

/SOLU                    ! Solution module
ANTYPE,STATIC            ! Set static analysis
DL,4,1,all,0             ! Impose Clamped BC

NSEL,S,LOC,x,3e3
CP,1,UZ,ALL              ! Coupling DOF set, vertical displacement
NSEL,R,LOC,y,300
F,all,FZ,-100            ! Apply force in a end line node
```

```
NSEL,all

SOLVE                   ! Solve current load state
FINISH                  ! Exit solution module

/POST1                  ! Post-processor module
PLDISP,1
FINISH                  ! Exit post-processor module
```

3.2.7   Element Coordinate System

In the pre-processor, during the definition of the laminate, it is very important to know the orientation of the laminate coordinate system. Material properties, the relative lamina orientation with respect to the laminate axis, and other parameters and properties are defined in the laminate coordinate system, unless specified otherwise. Also, it can be used to obtain the derived results (strains and stress) in these directions. In FEA, the laminate coordinate system is associated to the element coordinate system, with a unique right-handed orthogonal system associated to each element.

The element coordinate system orientation is associated with the element type. For bar or beam elements the orientation of the x-axis is generally along the line defined by the end nodes of the element. For solid elements in two and three dimensions, the orientation is typically defined parallel to the global coordinate system. For shell elements this is not useful. Axes x and y need to be defined on the element surface, with the z-axis always normal to the surface. The default orientation of x and y axes depends on the commercial code and the element type.

There are various ways to define the default orientation of x and y in shell elements. Two of them are shown in Figure 3.7. In Figure 3.7.(a) the x-axis is aligned with the edge defined by the first and second nodes of each element, the z-axis normal to the shell surface (with the outward direction determined by the right-hand rule), and the y-axis perpendicular to the x- and z-axis (ANSYS uses this rule as default). Other packages, such as MSC-MARC^{TM} calculate the orientation of the x-axis from the lines defined by the middle points of the edges as shown in Figure 3.7(b).

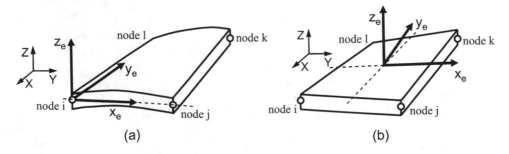

Fig. 3.7: Default orientations of element (or material) coordinate systems in shell elements: (a) ANSYS, (b) MSC-MARC.

In Examples 3.2-3.7, only rectangular plates with rectangular elements are analyzed. All of them have the first and the second node aligned with the global X-axis. Therefore, the material axes have been chosen parallel to the global axis. But this doesn't need to be the case. Most commercial codes have utilities to change the element coordinate system. Example 3.8 illustrates how to change the element coordinate system orientation in a plate. Example 3.9 illustrates how it can be done in a shell with curvature. Example 3.10 illustrates how different orientations can be used in different locations of the structure.

Example 3.8 *Use a local coordinate system and SHELL99 to model the plate of Example 3.2 if the orthotropic material is rotated +30 degrees with respect the x-direction.*

Solution to Example 3.8 *In ANSYS, a local coordinate system is defined using* LOCAL *commands, which can be cartesian, cylindrical, or spherical. Then, each element is linked to a previously defined local coordinate system using the element property* ESYS. *The objective is to orient the x-axis element coordinate system parallel to the x-axis material coordinate system. Also, it is possible to define element coordinate system orientations by user written subroutines [3].*

Element coordinate systems may be displayed as a triad with the /PSYMB *command or as an* ESYS *number (if specified) with the* /PNUM *command.*

This example illustrates the use of local coordinate system and the use of SHELL99 in a rectangular plate. See input file listed below [12].

```
/TITLE,Simply supported plate under uniform load q=1.2E-4 MPa - SHELL99
! Material is AS4D/9310 Carbon/Epoxy - 8 layers of 1.25 mm thickness
/UNITS,MPA              ! Units are in mm, MPa, and Newtons

/PREP7                  ! Pre-processor module
! Material properties FOR AS4D/9310 orthotropic laminate
uimp,1,ex,ey,ez,133.86E3,7.706E3,7.706E3
uimp,1,gxy,gyz,gxz,4.306E3,2.76E3,4.306E3
uimp,1,prxy,pryz,prxz,0.301,0.396,0.301

ET,1,SHELL99,,1,      ! KEYOPT(2)=1, supply 12+(6*NL)const.
KEYOPT,1,5,1          ! Set KEYOPT(5)=1, Stress results
KEYOPT,1,8,1          ! Set KEYOPT(8)=1, Storage data: All layers
R,1,8,1,,,,           ! Real constant set #1, layers properties
RMORE,,,,,,,
RMORE,1,0,1.25        ! 1st layer: mat. #1, 0 deg, Th=1.25 mm
RMORE,1,0,1.25        ! 2nd layer: mat. #1, 0 deg, Th=1.25 mm
RMORE,1,0,1.25        ! 3rd layer: mat. #1, 0 deg, Th=1.25 mm
RMORE,1,0,1.25        ! 4th layer: mat. #1, 0 deg, Th=1.25 mm

LOCAL,11,0,,,,30,0,0  ! Define local coord. system, XYrot=30 deg
ESYS,11               ! Set coord. system for elements meshed

! Geometry and mesh
RECTNG,0,2000,0,1000  ! Creates a rectangle with x=2 m and y=1 m
ESIZE,250             ! Element size 250 mm
AMESH,all             ! Mesh the area
```

```
CSYS,0                    ! Go back to default coord. system
/psymb,esys,1             ! Set on display laminate orientation
eplot                     ! Display elements
FINISH                    ! Exit pre-processor module

/SOLU                     ! Solution module
DL,2,1,uz,0               ! Impose Simple Supported BC
DL,3,1,uz,0
DL,1,1,symm               ! Impose Symmetry BC
DL,4,1,symm
d,all,rotz                ! Constraint rotations about z axes (optional)
SFA,all,2,PRES,1.2E-4     ! Apply uniform pressure in MPa
SOLVE                     ! Solve current load state
FINISH                    ! Exit solution module

/POST1                    ! Post-processor module
PLDISP,1                  ! Plots displaced plate
RSYS,SOLU                 ! Activate results in solution coord. system
LAYER,2                   ! Set layer #2, top face (by default)
PLESOL,s,x                ! Contour plot stress 1(fiber), layer2, top face
FINISH                    ! Exit post-processor module
```

Example 3.9 *Align the laminate coordinate system with the global Y-axis of a 3D curved shell.*

Solution to Example 3.9 *For shells defined in 3D, the ESYS orientation uses the projection of the local system on the shell surface. The element x-axis is determined from the projection of the local x-axis on the shell surface. The z-axis is determined normal to the shell surface (with the outward direction determined by the right-hand rule), and the y-axis perpendicular to the x- and z-axis. For elements without midside nodes (linear interpolation functions), the projection is evaluated at the element centroid and it is assumed constant in direction throughout the element. For elements with midside nodes (quadratic interpolation functions), the projection is evaluated at each integration point and may vary in direction throughout the element.*

   See input file listed below (available in [12]), to align the element x-axis with the global Y-axis.

```
/TITLE,Curved surface - SHELL99
! Material is AS4D/9310 Carbon/Epoxy - 6 layers of 1.05 mm thickness
! Units are in mm, MPa, and Newtons

/PREP7                      ! Pre-processor module
! Material properties FOR AS4D/9310 orthotropic laminate
uimp,1,ex,ey,ez,133.86E3,7.706E3,7.706E3
uimp,1,gxy,gyz,gxz,4.306E3,2.76E3,4.306E3
uimp,1,prxy,pryz,prxz,0.301,0.396,0.301

ET,1,SHELL99,,1,,,1       ! Chooses SHELL99 element for analysis
R,1,6,1,,,,               ! Real constant set #1, layers properties
RMORE,,,,,,,
RMORE,1,90,1.05           ! 1st layer: mat. #1,90 deg, Th=1.05 mm
```

```
RMORE,1,45,1.05          ! 2nd layer: mat. #1, 45 deg, Th=1.05 mm
RMORE,1,-45,1.05         ! 3rd layer: mat. #1,-45 deg, Th=1.05 mm

! Create geometry by solid modeling
K,1,300,0,135
K,2,0,0,235
K,3,-300,0,135
K,4,200,200,0
K,5,0,200,135
K,6,-200,200,0
L,1,4
L,3,6
BSPLIN,1,2,3
BSPLIN,4,5,6
AL,ALL
! Mesh geometry
LOCAL,11,0,,,,90,0,0     ! Define local coord. system, XYrot=90 deg
ESYS,11                  ! Set coord. system for elements meshed
ESIZE,50                 ! Define element size
AMESH,1                  ! Mesh the area
CSYS,0                   ! Go back to default coord. system
/PSYMB,esys,1            ! Set on display laminate orientation
/TYPE,1,0                ! Not hidden surfaces
EPLOT                    ! Display elements
FINISH                   ! Exit pre-processor module
```

Example 3.10 *Model in ANSYS a flanged tube with axial and radial laminate orientation. In the cylindrical part, the reference axis will be in the longitudinal direction. In the flange, the reference axis will be radial (see Figure 3.8).*

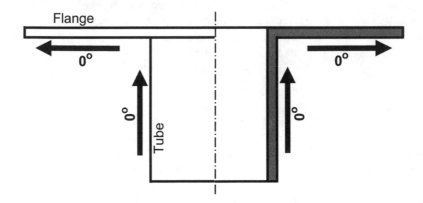

Fig. 3.8: Reference axis in a flange tube.

Solution to Example 3.10 *Different orientation systems are needed for different model locations. Therefore, two local reference axes are defined and activated using the **ESYS** command. The input file listed below aligns the elements on the cylinder in the axial direction and aligns the elements on the flange in the radial direction [12].*

```
/TITLE,Flange tube with axial and radial laminate orientation - SHELL99
! Material is AS4D/9310 Carbon/Epoxy - 6 layers of 1.05 mm thickness
! Units are in mm, MPa, and Newtons

/PREP7                      ! Pre-processor module
! Material properties FOR AS4D/9310 orthotropic laminate
uimp,1,ex,ey,ez,133.86E3,7.706E3,7.706E3
uimp,1,gxy,gyz,gxz,4.306E3,2.76E3,4.306E3
uimp,1,prxy,pryz,prxz,0.301,0.396,0.301

ET,1,SHELL99,,1,,,1         ! Chooses SHELL99 element for analysis
R,1,6,1,,,,                 ! Real constant set #1, layers properties
RMORE,,,,,,,
RMORE,1,0,1.05              ! 1st layer: mat. #1, 0 deg, Th=1.05 mm
RMORE,1,45,1.05            ! 2nd layer: mat. #1, 45 deg, Th=1.05 mm
RMORE,1,-45,1.05          ! 3rd layer: mat. #1,-45 deg, Th=1.05 mm

! Create geometry by solid modeling
CYL4,0,0,350,,,,300
CYL4,0,0,350,,550
AGLUE,3,4,5                 ! Glue areas, area 5 become area 6

! Mesh geometry
LOCAL,11,0,,,,0,0,90       ! Define rotation=90 deg around Y (cylinder)
LOCAL,12,1,,,,0,0,0        ! Define polar coordinate system (flange)
ESIZE,50                    ! Define element size
ESYS,11                     ! Set coord. system for elements meshed
AMESH,3,4                   ! Mesh the cylindrical areas (areas 3 and 4)
ESYS,12                     ! Set coord. system for elements meshed
AMESH,6                     ! Mesh the flange area (area 6)
CSYS,0                      ! Go back to default coord. system
/PSYMB,esys,1               ! Set on display laminate orientation
/TYPE,1,0                   ! Not hidden surfaces
EPLOT                       ! Display elements
FINISH                      ! Exit pre-processor module
```

Example 3.11 *Create a FE model for a pultruded composite column under axial compression load $P = 11,452$ N[13] and calculate the end axial displacement $u(L/2)$, where $x = 0$ is located at the half-length of the column. The column is simply supported (pinned) at both ends $x = (-L/2, L2)$. Its length is $L = 1.816$ m. The cross section of the column is that of a wide-flange I-beam (also called H-beam) with equal outside height and width, $H = W = 304.8$ mm. The thickness of both the flange and the web is $t_f = t_w = 12.7$ mm. The material properties are given by the A, B, D, and H matrices, with units [mm MPa], [mm^2 MPa], [mm^3 MPa], and [mm MPa], respectively. For the flange:*

$$[A] = \begin{bmatrix} 335,053 & 47,658 & 0 \\ 47,658 & 146,155 & 0 \\ 0 & 0 & 49,984 \end{bmatrix} \quad ; \quad [B] = \begin{bmatrix} -29,251 & -1,154 & 0 \\ -1,154 & -5,262 & 0 \\ 0 & 0 & -2,274 \end{bmatrix}$$

$$[D] = \begin{bmatrix} 4{,}261{,}183 & 686{,}071 & 0 \\ 686{,}071 & 2{,}023{,}742 & 0 \\ 0 & 0 & 677{,}544 \end{bmatrix} \quad ; \quad [H] = \begin{bmatrix} 34{,}216 & 0 \\ 0 & 31{,}190 \end{bmatrix}$$

For the web:

$$[A] = \begin{bmatrix} 338{,}016 & 44{,}127 & 0 \\ 44{,}127 & 143{,}646 & 0 \\ 0 & 0 & 49{,}997 \end{bmatrix} \quad ; \quad [B] = \begin{bmatrix} -6{,}088 & -14{,}698 & 0 \\ -14{,}698 & -6{,}088 & 0 \\ 0 & 0 & 0 \end{bmatrix}$$

$$[D] = \begin{bmatrix} 4{,}769{,}538 & 650{,}127 & 0 \\ 650{,}127 & 2{,}155{,}470 & 0 \\ 0 & 0 & 739{,}467 \end{bmatrix} \quad ; \quad [H] = \begin{bmatrix} 34{,}654 & 0 \\ 0 & 31{,}623 \end{bmatrix}$$

Solution to Example 3.11 *The* `.log` *file is shown next. A number of modeling techniques are illustrated, which are very useful for FEA of composite structures. Solid modeling is based on areas, and the use of lines to effectively impose boundary conditions and to control the mesh refinement is illustrated. Only one-half of the length of the column is modeled using symmetry boundary conditions. The loaded end is constrained to move as a rigid body using* **CERIG** *so that the pinned boundary condition is properly simulated. The* **PSTRESS,ON** *is included to keep the stress solution for buckling analysis in Example 4.4. The model is set up parametrically so that all the geometric parameters of the column, such as the length, as well as mesh refinement can be easily changed. Only displacements can be displayed because the model is set up with A-B-D-H matrices, but the* `.log` *file can be easily modified to enter the LSS along with lamina material properties as in Example 3.4.* **LOCAL** *and* **ESYS** *coordinate systems are used so that all the local and the element coordinate systems are oriented similarly; this is necessary to facilitate the specification of directionally dependent materials properties and also interpretation of stress and strain results. The axial displacement at the loaded end of the column can be read from the GUI as* **DMX=.036518.**

```
/TITLE, I-COLUMN, Ref: CST 58 (1998) 1335-1341
!Ref: COMPOSITE SCIENCE AND TECHNOLOGY 58 (1998) 1335-1341
/UNITS,MPA              !UNITS ARE mm, MPa, Newtons
WPSTYLE,,,,,,,,1        !WORKPLANE VISIBLE
/VSCALE,1,2.5,0 !2.5 LARGER ARROWS
!                 ARROW & TRIAD COLORS: WHITE=X, GREEN=Y, BLUE=Z
/TRIAD,LBOT       !MOVE COORDINATE LABELS TO LEFT-BOTTOM
/VIEW,,1,2,3      !OBLIQUE VIEW
/PNUM,KP,0        !THESE ARE ALL THE NUMBERING OPTIONS
/PNUM,LINE,0      !ENTITY NUMBERING OFF=0, ON=1
/PNUM,AREA,1
/PNUM,VOLU,0
/PNUM,NODE,0
/PNUM,TABN,0
/PNUM,SVAL,0
/NUMBER,0

/PBC,ALL,,1       !THESE ARE ALL THE BC DISPLAY OPTIONS
/PBC,NFOR,,0      !DISPLAY ALL APPLIED BC, OFF=0, ON=1
```

```
/PBC,NMOM,,0
/PBC,RFOR,,0
/PBC,RMOM,,0
/PBC,PATH,,0

/PSYMB,CS,1        !THESE ARE ALL THE SYMBOL DISPLAY OPTIONS
/PSYMB,NDIR,0
/PSYMB,ESYS,0
/PSYMB,LDIV,1      !SHOW LESIZE ON LINES
/PSYMB,LDIR,1      !SHOW LDIR TO DECIDE ON LESIZE BIAS
/PSYMB,ADIR,0
/PSYMB,ECON,0
/PSYMB,XNODE,0
/PSYMB,DOT,1
/PSYMB,PCONV,
/PSYMB,LAYR,0
/PSYMB,FBCS,0


/PREP7             !ENTER THE PREPROCESSOR
!DEFINE PARAMETRIC VALUES===================
LOAD=11452         !APPLIED LOAD              [N]
L2=1816/2          !COLUMN HALF LENGTH        [mm]
WW=304.8           !OUTER WEB WIDHT           [mm]
FW=WW              !OUTER FLANGE WIDTH
WT=12.7            !WEB THICKNESS             [mm]
FT=WT              !FLANGE THICKNESS
WW1=WW-FT          !MID-PLANE WEB WIDTH
FW1=FW-WT          !MID-PLANE FLANGE WIDTH
WW2=WW1/2
FW2=FW1/2
CSAREA=2*FW*FT+(WW-FT)*WT        !CROSS SECTION AREA
NELEN=10           !NUMBER OF ELEMENTS ALONG THE LENGTH
NEWEB=4            !NUMBER OF ELEMENTS ON THE WEB
NEFLA=2            !NUMBER OF ELEMENTS ON 1/2 FLANGE
!
ET,1,SHELL99,,2    ! KEYOPT(2)=2: USES ABDH MATRICES
!KEYOPT,1,10,2     ! Set KEYOPT(10)=2, print ABDH matrix in file.abd
! NOTE HOW THE A-B-D-H MATRICES ARE GIVEN LIKE THIS
! R,1,A11,A12,0,A16
! RMODIF,1,7,A22,0,A26
! RMODIF,1,16,A66
! RMODIF,1,19,H44,H45
! RMODIF,1,21,H55
! RMODIF,1,22,B11,B12,0,B16
! RMODIF,1,28,B22,0,B26
! RMODIF,1,37,B66
! RMODIF,1,43,D11,D12,0,D16
! RMODIF,1,49,D22,0,D26
! RMODIF,1,58,D66
! RMODIF,1,76,AVERAGE DENSITY, LAMINATE THICKNESS
```

```
!
!FLANGE [mm,Mpa,Newton]
R,1,    3.35E+05,   4.77E+04,   0,  0.00E+00
RMODIF,1,7, 1.46E+05,   0,  0.00E+00
RMODIF,1,16,   5.00E+04,
RMODIF,1,19,   3.42E+04,   0.00E+00
RMODIF,1,21,   3.12E+04
RMODIF,1,22,   -2.93E+04,  -1.15E+03,  0,  0.00E+00
RMODIF,1,28,   -5.26E+03,  0,  0.00E+00
RMODIF,1,37,   -2.27E+03,
RMODIF,1,43,   4.26E+06,   6.86E+05,   0,  0.00E+00
RMODIF,1,49,   2.02E+06,   0,  0.00E+00
RMODIF,1,58,   6.78E+05
RMODIF,1,76,     ,   12.7
!
! WEB [mm, Mpa, Newton]
R,2,    3.38E+05,   0.00E+00,   0,  0.00E+00
RMODIF,2,7, 1.44E+05,   0,  0.00E+00
RMODIF,2,16,   5.00E+04,
RMODIF,2,19,   3.47E+04,   0.00E+00
RMODIF,2,21,   3.16E+04
RMODIF,2,22,   -6.09E+03,  -1.47E+04,  0,  0.00E+00
RMODIF,2,28,   -6.09E+03,  0,  0.00E+00
RMODIF,2,37,   -6.47E-11,
RMODIF,2,43,   4.77E+06,   6.50E+05,   0,  0.00E+00
RMODIF,2,49,   2.16E+06,   0,  0.00E+00
RMODIF,2,58,   7.39E+05,
RMODIF,2,76,     ,   12.7
!
!DEFINE SOLID MODEL USING AREAS======================
RECTNG,0,L2,-WW2,WW2,   !WEB
WPAVE,0,WW2,0             !MOVE WORKPLANE TO TOP FLANGE
WPRO,,-90.000000,        !ROTATE WORKPLANE AS FLANGE
RECTNG,0,L2,-FW2,FW2,   !TOP FLANGE
WPAVE,0,-WW2,0
RECTNG,0,L2,-FW2,FW2,   !BOTTOM FLANGE
AOVLAP,all       !JOINS AREAS CREATING INTERSECTIONS IF NEEDED
NUMCMP,AREA      !COMPRESS AREA NUMBERS
/REPLOT
WPSTYLE,,,,,,,,0         !HIDE WORKPLANE
!
LPLOT              !PLOT LINES
LSEL,S,LOC,X,0  !SELECT SYMMETRY END
/REPLOT
/PBC,ALL,,1       !DISPLAY ALL APPLIED BC, OFF=0, ON=1
DL,ALL, ,SYMM   !APPLY SYMMETRY BC
!NOTE SYMM DISPLAYED AS S ON LINES, BUT WILL NOT SHOW ON NODES LATER
ALLSEL,ALL       !RESELECT ALL ENTITIES
/REPLOT
!
```

```
!MESHING=========================================================
LOCAL,11,0,,,, , ,0      !DEFINE LOCAL COORD SYS TO ALIGN W/MAT PROPS
ESYS,11                  !USE IT FOR ALL ELEMS
/VSCALE,1,2.5,0          !2.5 LARGER ARROWS
/PSYMB,ESYS,1            !DISPLAY IT
/PSYMB,ESYS,0            !DO NOT DISPLAY IT
/VSCALE,1,1.0,0          !RESET TO DEFAULT ARROW LENGTH
MSHAPE,0,2D              !QUADRILATERAL 0, MESHING 2D
MSHKEY,1                 !MAPPED MESHING 1 (FREE WOULD BE 0)
!
LESIZE,ALL,,,NEFLA       !ALL LINES DIVIDED IN NEFLA ELEMENTS
LESIZE, 2,,,NEWEB,0      !LINE 2, DIV NEWEB, NO BIAS, OVERRIDES PREVIOUS
LESIZE, 4,,,NEWEB,0
LESIZE, 1,,,NELEN,1/2    !LINE 1, DIV 10, BIAS 1/2 TOWARDS LINE END
LESIZE, 3,,,NELEN,2      !LINE 3, DIV 10, BIAS 2 TOWARDS LINE ORIGIN
LESIZE, 5,,,NELEN,1/2
LESIZE, 7,,,NELEN,2
LESIZE, 9,,,NELEN,1/2
LESIZE,11,,,NELEN,2
/PNUM,REAL,1     !COLOR AND NUMBER ELEMENTS BY REAL SET
ASEL,ALL         !SELECT ALL AREAS
ASEL,S,,,1       !SELECT AREA 1 (WEB)
AATT,,2,         !USE REAL CONST 2 FOR THE WEB
AMESH,ALL        !MESH ALL AREAS CURRENTLY SELECTED (I.E., WEB)
ASEL,S,,,ALL     !SELECT ALL
ASEL,U,,,1       !UNSELECT AREA 1 TO KEEP THE FLANGES
AATT,,1          !USE REAL CONST 1 FOR THE FLANGE
AMESH,ALL        !MESH ALL AREAS CURRENTLY SELECTED (I.E., FLANGE)
ASEL,ALL
/PNUM,REAL,0     !SUPRESS NUMBERING
CSYS,0           !RETURN TO GLOBAL COORD SYSTEM TO DISPLAY STRESSES
!
NSEL,S,LOC,X,0   !PREVENT RIGID BODY TRANSLATION
NSEL,R,LOC,Y,0
D,ALL,UY,0
D,ALL,UZ,0
D,ALL,ROTX,0     !PREVENT RIGID BODY TWIST
NSEL,ALL
!
NSEL,S,LOC,X,L2 !SELECT LOAD END
NSEL,R,LOC,Y,0  !SELECT CENTER NODE ONLY
/PNUM,NODE,1
NPLOT
*GET,MYNODE,NODE,,NUM,MIN         !GET LABEL OF CENTER NODE
NSEL,S,LOC,X,L2                   !SELECT LOAD END AGAIN
/PNUM,NODE,0                      !TURN OFF NODE NUMBER DISPLAY
/REPLOT
! APPLY RIGID BC AT LOADED END
CERIG,MYNODE,ALL,UXYZ, , , ,      !MYNODE MASTER, ALL OTHER SLAVES
F,MYNODE,FX,-1*LOAD               !APPLY COMPRESSION LOAD
```

```
!
ALLSEL,ALL                  !RESELECT EVERYTHING
FINISH                      !EXIT PREPROCESSOR
!
/SOLU                       !ENTER SOLUTION MODULE
ANTYPE,STATIC               !STATIC ANALYSIS
PSTRESS, ON                 !KEEP STRESS FOR BUCK ANAL LATER
SOLVE
FINI                        !EXIT SOLUTION MODULE
!
/POST1                      !ENTER POSTPROCESSOR MODULE
/VIEW,,1,1,1                !ISOMETRIC VIEW
PLDISP,2                    !PLOT DEFORMED SHAPE AND UNDEF OUTLINE
!FINI
```

3.3   Failure Criteria

Failure criteria are curve fits of experimental data that attempt to predict failure
under multiaxial stress based on experimental data obtained under uniaxial stress.
All failure criteria described in this section predict the first occurrence of failure in
one of the laminae but are unable to track failure propagation until complete laminate failure. Continuum damage mechanics is used in Chapter 8 to track damage
evolution up to laminate failure. The truncated-maximum-strain criterion estimates
laminate failure without tracking damage evolution by making certain approximations and assumptions about the behavior of the laminate [1].

In this section, failure criteria are presented using the notion of failure index,
which is used for several FEA packages, and it is defined as

$$I_F = \frac{stress}{strength} \tag{3.20}$$

Failure is predicted when $I_F \geq 1$. The strength ratio ([1] Section 7.1.1) is the
inverse of the failure index

$$R = \frac{1}{I_F} = \frac{strength}{stress} \tag{3.21}$$

Failure is predicted when $R \leq 1$.

3.3.1   2D Failure Criteria for Unidirectional Laminae

Strength-based failure criteria are commonly used in FEA to predict failure events
in composite structures. Numerous criteria exist for unidirectional (UD) laminae
subjected to a state of plane stress ($\sigma_3 = 0$). The most commonly used are described
in [1]. Another failure criterion worth mentioning is the Puck failure criterion [10],
because it distinguishes between fiber failure (FF) and inter fiber failure (IFF). In
the case of plane stress, the IFF criteria discriminates three different modes. The
IFF mode A is when perpendicular transversal cracks appear in the lamina under
transverse tensile stress with or without in-plane shear stress. The IFF mode B

also denotes perpendicular transversal cracks, but in this case they appear under in-plane shear stress with small transverse compression stress. The IFF mode C indicates the onset of oblique cracks (typically with an angle of 53° in carbon epoxy laminates) when the material is under significant transversal compression.

The FF and the three IFF modes yield separate failure indexes. The Puck criterion assumes that FF only depends on longitudinal tension. Therefore, the failure index for FF is defined as

$$I_{FF} = \begin{cases} \sigma_1/F_{1t} & \text{if } \sigma_1 > 0 \\ -\sigma_1/F_{1c} & \text{if } \sigma_1 < 0 \end{cases} \tag{3.22}$$

The IFF failure indexes have different expressions depending on the stress state and the IFF mode that becomes active. For IFF with positive transverse stress, mode A is active. The failure index in this case is defined as

$$I_{IFF,A} = \sqrt{\left(\frac{\sigma_6}{F_6}\right)^2 + \left(1 - p_{6t}\frac{F_{2t}}{F_6}\right)^2 \left(\frac{\sigma_2}{F_{2t}}\right)^2} + p_{6t}\frac{\sigma_2}{F_6} \quad \text{if } \sigma_2 \geq 0 \tag{3.23}$$

where p_{6t} is a fitting parameter. Lacking experimental values, it is assumed that $p_{6t} = 0.3$ [10].

Under negative transverse stress, either mode B or mode C is active, depending on the relationship between in-plane shear stress and transversal shear stress. The limit between mode B and C is defined by the relation F_{2A}/F_{6A}, where

$$F_{2A} = \frac{F_6}{2p_{6c}}\left[\sqrt{1 + 2p_{6c}\frac{F_{2c}}{F_6}} - 1\right] \tag{3.24}$$

$$F_{6A} = F_6\sqrt{1 + 2p_{2c}} \tag{3.25}$$

and p_{2c} is defined as

$$p_{2c} = p_{6c}\frac{F_{2A}}{F_6} \tag{3.26}$$

and p_{6c} is another fitting parameter. Lacking experimental values, it is assumed that $p_{6c} = 0.2$ [10].

Finally, the failure indexes in mode B and C are defined as

$$I_{IFF,B} = \frac{1}{F_6}\left[\sqrt{\sigma_6^2 + (p_{6c}\sigma_2)^2} + p_{6c}\sigma_2\right] \quad \text{if } \begin{cases} \sigma_2 < 0 \\ \left|\frac{\sigma_2}{\sigma_6}\right| \leq \frac{F_{2A}}{F_{6A}} \end{cases} \tag{3.27}$$

and

$$I_{IFF,C} = -\frac{F_{2c}}{\sigma_2}\left[\left(\frac{\sigma_6}{2(1+p_{2c})F_6}\right)^2 + \left(\frac{\sigma_2}{F_{2c}}\right)^2\right] \quad \text{if } \begin{cases} \sigma_2 < 0 \\ \left|\frac{\sigma_2}{\sigma_6}\right| \geq \frac{F_{2A}}{F_{6A}} \end{cases} \tag{3.28}$$

3.3.2   3D Failure Criteria

Failure criteria presented here are 3D generalizations of the ones presented in ([1] Section 7.1). The user of FEA packages should be careful because some packages use only the in-plane stress components for the computation of the failure index (e.g., ABAQUS), even though all six stress components may be available from the analysis. In those cases the interlaminar and thickness components of stress should be evaluated separately to see if they lead to failure.

In this section, the numerical subscript denotes the directions of 1) fiber, 2) in-plane transverse to the fibers, and 3) through the thickness of the lamina. The letter subscript denotes t) tensile and c) compressive. Contracted notation is used for the shear components as described in Section 1.5.

Maximum Strain Criterion

The failure index is defined as

$$
I_F = \max \begin{cases}
\epsilon_1/\epsilon_{1t} & \text{if } \epsilon_1 > 0 \text{ or } -\epsilon_1/\epsilon_{1c} \text{ if } \epsilon_1 < 0 \\
\epsilon_2/\epsilon_{2t} & \text{if } \epsilon_2 > 0 \text{ or } -\epsilon_2/\epsilon_{2c} \text{ if } \epsilon_2 < 0 \\
\epsilon_3/\epsilon_{3t} & \text{if } \epsilon_3 > 0 \text{ or } -\epsilon_3/\epsilon_{3c} \text{ if } \epsilon_3 < 0 \\
abs(\gamma_4)/\gamma_{4u} \\
abs(\gamma_5)/\gamma_{5u} \\
abs(\gamma_6)/\gamma_{6u}
\end{cases}
\tag{3.29}
$$

The quantities in the denominator are the ultimate strains of the unidirectional lamina. Note that compression ultimate strains in (3.29) are positive numbers.

Maximum Stress Criterion

The failure index is defined as

$$
I_F = \max \begin{cases}
\sigma_1/F_{1t} & \text{if } \sigma_1 > 0 \text{ or } -\sigma_1/F_{1c} \text{ if } \sigma_1 < 0 \\
\sigma_2/F_{2t} & \text{if } \sigma_2 > 0 \text{ or } -\sigma_2/F_{2c} \text{ if } \sigma_2 < 0 \\
\sigma_3/F_{3t} & \text{if } \sigma_3 > 0 \text{ or } -\sigma_3/F_{3c} \text{ if } \sigma_3 < 0 \\
abs(\sigma_4)/F_4 \\
abs(\sigma_5)/F_5 \\
abs(\sigma_6)/F_6
\end{cases}
\tag{3.30}
$$

The letter F is used here to denote a strength value for a unidirectional lamina as in [9]. Note that compression strength in (3.30) are positive numbers.

Tsai-Wu Criterion

Using the Tsai-Wu criterion the failure index is defined as

$$
I_F = \frac{1}{R} = \left[-\frac{B}{2A} + \sqrt{\left(\frac{B}{2A}\right)^2 + \frac{1}{A}} \right]^{-1}
\tag{3.31}
$$

98

Finite Element Analysis of Composite Materials

with

$$A = \frac{\sigma_1^2}{F_{1t}F_{1c}} + \frac{\sigma_2^2}{F_{2t}F_{2c}} + \frac{\sigma_3^2}{F_{3t}F_{3c}} + \frac{\sigma_4^2}{F_4^2} + \frac{\sigma_5^2}{F_5^2} + \frac{\sigma_6^2}{F_6^2}$$
$$+ c_4 \frac{\sigma_2\sigma_3}{\sqrt{F_{2t}F_{2c}F_{3t}F_{3c}}} + c_5 \frac{\sigma_1\sigma_3}{\sqrt{F_{1t}F_{1c}F_{3t}F_{3c}}} + c_6 \frac{\sigma_1\sigma_2}{\sqrt{F_{1t}F_{1c}F_{2t}F_{2c}}} \quad (3.32)$$

and

$$B = \left(F_{1t}^{-1} - F_{1c}^{-1}\right)\sigma_1 + \left(F_{2t}^{-1} - F_{2c}^{-1}\right)\sigma_2 + \left(F_{3t}^{-1} - F_{3c}^{-1}\right)\sigma_3 \quad (3.33)$$

where c_i, $i = 4..6$, are the Tsai-Wu coupling coefficients, that by default are taken to be -1. Note that compression strength in (3.32) and (3.33) are here positive numbers.

The through-the-thickness strength values F_{3t} and F_{3c} are seldom available in the open literature, so it is common practice to use the corresponding in-plane transverse values of strength. Also, the interlaminar strength F_5 is commonly assumed to be equal to the in-plane shear strength. Lacking experimental data for the remaining interlaminar strength F_4, it can be estimated as the shear strength of the matrix.

Example 3.12 *Compute the failure index I_F in each layer of Example 3.4 using the maximum stress failure criterion and the Tsai-Wu criterion. The lamina strength values are given in Table 3.1. Determine the strength ratio of the laminate using both criteria.*

Solution to Example 3.12 *ANSYS has two standard ways to introduce and use failure criteria. The first option uses FC commands, and it can be applied on any shell element type. The second way uses FB commands with the option FAIL, but it can be used only with shell element types SHELL91 and SHELL99.*

Using FC Commands

Once the model is solved, inside /POST1 module, the FC commands (FC, FCDELE, FCLIST, etc.) can be used to define the failure criteria parameters. Before defining the failure criteria parameters, ANSYS must show the solution in "results" coordinate system using RSYS, SOLU command. The LAYER command is used to select the layer where the failure criterion is to be calculated.

After Example 3.4, include the commands below to compute the I_F of each lamina. Note that in ANSYS compression strength must be introduced using negative numbers. If the compression strength value is not given, ANSYS takes the compression strength equal to the negative value of the tensile strength. Also note that ANSYS uses x, y, z to denote the material coordinates that are denoted as $1, 2, 3$ in this textbook.

```
/POST1               ! Post-processor module
RSYS,SOLU            ! Show the results in 'solution' reference system

!Failure criteria definition
FC,1,s,xten, 1830    ! F1t strength
FC,1,s,xcmp,-1096    ! F1c strength
FC,1,s,yten, 57      ! F2t strength
FC,1,s,ycmp,-228     ! F2c strength
FC,1,s,zten, 1e6     ! F3t=F3c strength (large value, do not compute)
FC,1,s,xy,71         ! F6 strength
```

```
FC,1,s,yz,1e4          ! F4 strength (large value, do not compute)
FC,1,s,xz,1e4          ! F5 strength (large value, do not compute)
FC,1,s,XYCP,-1         ! c6 coefficient. Defaults to -1.0
FC,1,s,YZCP,-1         ! c4 coefficient. Defaults to -1.0
FC,1,s,XZCP,-1         ! c5 coefficient. Defaults to -1.0

LAYER, 1               ! Select layer #1
PRNSOL,S,FAIL          ! Print table with FAIL index, where:
     ! MAXF is Maximum Stress Criterion failure index
     ! TWWI is Tsai-Wu 'strength index' (we recommend not to use it)
     ! TWSR is Tsai-Wu failure index (inverse of strength ratio R)
PLNSOL,S,MAXF          ! Maximum Stress Criterion failure index
PLNSOL,S,TWSR          ! Tsai-Wu failure index

! Repeat this with the remaining layers
!LAYER, 2 ....

FINISH                 ! End Post-process module
```

The MAXF *and the* TWSR *are the failure index defined in Eqs. (3.30) and (3.31) respectively. The* TWSI, *called Tsai-Wu "strength index," is the addition of the value A in Eq. (3.32) and value B in Eq. (3.33), i.e.* $TWSI = A + B$. *This "index" does not have engineering interpretation and we recommend not to use it.*

Using TB Commands

The TB *data definition allows, using the option* FAIL, *to define up to six failure criteria. Using this option predefined failure criteria can be used (maximum strain failure criterion, maximum stress failure criterion, and Tsai-Wu criterion), or other failure criteria can be defined by user written subroutines (using USRFC1 to USRFC6 [3]).*

The parameters for the failure criteria must be specified in the /PREP7 module, using a set of commands. The definition starts with TB,FAIL *and* TBTEMP,,CRIT *commands. Next, one of the six failure criteria is activated using the* TBDATA *command with one of the failure criteria "keys." A typical sequence of commands to specify failure criteria "keys" is shown below.*

```
TB,FAIL,#mat,1  ! Data table for failure criterion, material #mat,
TBTEMP,,CRIT    ! Failure criterion key
TBDATA,1,#key1,#key2,#key3,#key4,#key5,#key6
        ! #key1 Maximum Strain Failure Criterion
        ! #key2 Maximum Stress Failure Criterion
        ! #key3 Tsai-Wu Failure Criterion
        ! #key4...#key6 User defined Failure Criterion
```

where the #mat *is the material reference number, and the constants from* \#key1 *to* \#key6 *entered on the* TBDATA *command are described in Table 3.5.*

Then, the different failure criteria constants are defined using TBDATA *as:*

```
TBTEMP   ! Temperature for subsequent properties (use default value)
TBDATA,1,#e1t,#e1t,#e1t,#e1t,#e1t,#e1t  ! ultimate normal strains
TBDATA,7,#e6u,#e4u,#e5u                 ! ultimate shear strains
TBDATA,10,#F1t,#F1c,#F2t,#F2c,#F3t,#F3c ! normal strengths
```

Table 3.5: Key values to activate failure criteria using TB commands

Key	Value	Included or not Included
#key1	0	Do not include this predefined criterion
	1	Predefined maximum strain failure criterion (uses constants 1-9). Output as FC1
	-1	Include user-defined criterion with subroutine USRFC1
#key2	0	Do not include this predefined criterion
	1	Predefined maximum stress failure criterion (uses constants 10-18). Output as FC2
	-1	Include user-defined criterion with subroutine USRFC2
#key3	0	Do not include this predefined criterion
	2	Predefined Tsai-Wu failure criterion, using failure index (uses constants 10-21). Output as FC3
	-1	Include user-defined criterion with subroutine USRFC3
#key4		User-defined failure criteria - Output as FC4 TO FC6
to	0	Do not include this criterion
#key6	-1	User-defined criteria with subroutines USRFC4, USRFC5, USRFC6, respectively

```
TBDATA,16,#F6,#F4,#F5                  ! shear strengths
TBDATA,16,#c6,#c4,#c5                  ! Tsai-Wu coupling coeff.
```

The tension ultimate strains must be positive. The compression ultimate strains must be negative. If a compression value is not given, ANSYS takes the compression ultimate strain to be equal to the negative value of the tensile ultimate strain. The compression strengths must be introduced using negative numbers. If a compression value is not specified, ANSYS takes the compression strength to be equal to the negative value of the tensile strength. The coupling coefficients for Tsai-Wu (c_4, c_5 and c_6) default to -1 (values between -1 and 0 are recommended). For 2D analysis, set the values with indexes 3, 4 and 5 to a value several orders of magnitude larger than the values with 1, 2 and 6 indexes, and set c_4 and c_5 to zero.

For this example, the strength properties are shown below. These commands can be introduced in the /PREP7 module (see comment line marked with (*), after material definition, in the solution to Example 3.4).

```
TB,FAIL,1,1                   ! Define Failure Criteria, mat #1
TBTEMP,,CRIT
TBDATA,1,,1,2                 ! To include Maximum Stress and Tsai-Wu
TBTEMP
TBDATA,10,1830,-1096,57,-228,1e6     ! F1t,F1c,F2t,F2c,F3t,F3c
TBDATA,16,71,1E4,1E4          ! F6, F4, and F5 shear strengths
TBDATA,19,-1,-1,-1            ! c6, c4, c5, default are -1 in Tsai-Wu
TBLIST                        ! list the Failure Criteria constants
```

The failure indexes are given by ANSYS in the results file. To read them, the ETABLE command can be used in the /POST1 module, taking the reference values shown in Table 3.6.

Table 3.6: Sequence numbers to a failure index with ETABLE command

Output Quantity	Item	Sequence number
FCMAX (over all layers), Failure criterion number where maximum occurs	NMISC	1
VALUE, Value for this criterion	NMISC	2
LN, Layer # where maximum occurs	NMISC	3
FC, Failure criterion number	NMISC	$(4NL) + 8 + 15(N-1) + 1$
VALUE, Maximum value for this criterion	NMISC	$(4NL) + 8 + 15(N-1) + 2$

In Table 3.6, row 4, the output quantity FC is the failure criterion identifier (1 = maximum strain, 2 = maximum stress and 3 = Tsai-Wu criterion). The VALUE will be the failure index obtained. NL is the maximum layer number. N is the failure number as stored on the results file in compressed form. When only the maximum stress and the Tsai-Wu failure criteria are included, the maximum stress criteria will be stored first (N = 1) and the Tsai-Wu failure criteria will be stored second (N = 2). In addition, if more than one criterion is requested, the maximum value over all criteria is stored last (N = 3 for this example).

After that, it is possible to print or display non-averaged contour plots using **PRETABLE** *or* **PLETABLE** *commands respectively. For example, to read the failure index values in LAYER #2 in the present example, use the following commands in the post-processor module.*

```
/POST1                           ! Post-processor module
RSYS,SOLU
ETABLE,IFMAX,NMISC,2             ! Maximum failure index (over all layers)
ETABLE,LNMAX,NMISC,3             ! Layer where maximum occurs
ETABLE,FCMAX,NMISC,1             ! Failure criteria number of the maximum v.
LAYER,2                          ! Read 2nd LAYER Failure Index values
ETABLE,STRX2,SX                  ! Read stress in material direction 1
ETABLE,STRY2,SY                  ! Read stress in material direction 2
ETABLE,STRXY2,SXY                ! Read shear stress in material directions
ETABLE,IF12,NMISC,(4*8)+8+15*(1-1)+2   ! Max. stress failure index
SEXP,RS12,IF12,,-1                      ! Max. stress strength ratio
ETABLE,FC12,NMISC,(4*8)+8+15*(1-1)+1   ! Max. stress criterion FC2
ETABLE,IF22,NMISC,(4*8)+8+15*(2-1)+2   ! Tsai-Wu failure index
SEXP,RS22,IF22,,-1                      ! Tsai-Wu strength ratio
ETABLE,FC22,NMISC,(4*8)+8+15*(2-1)+1   ! Tsai-Wu criterion FC3
! Print
PRETAB,IFMAX,LNMAX,FCMAX                ! maximum (over all layers)
PRETAB,STRX2,STRY2,STRXY2,IF12,IF22     ! stress and I_F layer #2
PRETAB,IF12,RS12,FC12,IF22,RS22,FC22    ! I_F, R, and Failure crit.
FINISH                                  ! Exit post-processor module
```

The solution is tabulated in Table 3.7, showing the failure indexes and the strength ratios obtained for maximum stress criterion and Tsai-Wu criterion in each layer.

Example 3.13 *Compute the 2D Tsai-Wu failure index I_F in each layer of Example 3.4 using the APDL language of ANSYS. The lamina strength values are given in Table 3.1.*

Table 3.7: Failure indexes and strength ratios for each layer in Example 3.12

Layer		Maximum Stress		Tsai-Wu	
		I_F	R	I_F	R
#1,	0°	0.0144	69.34	0.0144	69.38
#2,	90°	0.0243	41.16	0.0294	34.04
#3,	$+45^o$	0.0157	63.84	0.0199	50.18
#4,	-45°	0.0157	63.84	0.0199	50.18

Solution to Example 3.13 *Using the APDL scripting language of ANSYS [11], it is possible to compute any user defined criterion. With APDL, it is possible to automate common tasks (macros) or even build parametric models (in terms of parameters or variables). To create macros, a set of commands can be saved in a text file using extension* ∗.mac *in the ANSYS working directory. Then, these commands can be executed by using the name of the file.*

In this example, APDL is used in the \POST1 *processor to compute a failure criterion, as follows: (i) activate the solution reference axes with* RSYS,SOLU, *(ii) select the layer to compute the failure index (e.g.* LAYER,1, *for layer number 1), and (iii) execute the macro file using its name. In this example, the commands to compute, print, and plot the 2D Tsai-Wu failure index are saved in a file named* TSAIWU2D.mac. *Then, the set of commands can be recalled invoking* TSAIWU2D *within ANSYS.*

```
! Tsai-Wu failure criterion, Using APDL macro language

! Failure criteria definition parameters
F1t= 1830    ! F1t strength
F1c= 1096    ! F1c strength
F2t= 57      ! F2t strength
F2c= 228     ! F2c strength
F6 = 71      ! F6 strength
c6 = -1      ! Tsai-Wu coefficient

! Initialize arrays
*get,nelem,elem,,num,max     ! get number of elements
*get,nnode,node,,num,max     ! get number of nodes

*dim,I_F,,nnode              ! set up arrays for element nodes
*dim,sel,,nnode              ! set up array for select vector

NSLE,S,CORNER                ! select only nodes in element corners
*vget,sel(1),node,1,nsel     ! mask for compute only corners

! Compute Tsai-Wu failure criterion
*do,in,1,nnode
    *if,sel(in),gt,0,then    ! Read only selected nodes
        *get,s_1,node,in,s,x ! Get stress each node
        *get,s_2,node,in,s,y
        *get,s_6,node,in,s,xy
        A1= s_1**2/(F1t*F1c) ! Compute Failure Index
```

```
        A2= s_2**2/(F2t*F2c)
        A6= s_6**2/(F6)**2
        A12= c6*s_1*s_2/(F1t*F1c*F2t*F2c)**0.5
        A = A1+A2+A6+A12
        B= (1/F1t-1/F1c)*s_1+(1/F2t-1/F2c)*s_2
        R_tw=-B/(2*A)+((B/2/A)**2+1/A)**0.5
        I_F(in)=1/R_tw
    *endif
*enddo

*VPUT,I_F,NODE,1,EPSW,  ! Write Failure Index in results database
PRNSOL,EPSW             ! Print Failure Index in a list PLNSOL,EPSW
! Contour plot Failure Index
```

The solution is tabulated in Table 3.8, showing the failure indexes obtained by using the Tsai-Wu criterion in each layer.

Table 3.8: Failure indexes and strength ratios for each layer in Example 3.13

Layer		I_F	R
#1,	0°	0.0144	69.34
#2,	90°	0.0243	41.16
#3,	$+45^o$	0.0157	63.84
#4,	-45°	0.0157	63.84

Example 3.14 *Compute the Tsai-Wu failure index I_F on each layer of Example 3.8 using a USERMAT subroutine (*`usermatps.f`* for shell elements). The lamina strength values are given in Table 3.1.*

Solution to Example 3.14 *See user material subroutine* `usermatps.f` *and model file* `FEAcomp_Ex314.log` *on the Web site [12]. Refer to Appendix E for program compilation and execution details.*

Suggested Problems

Problem 3.1 *Compute the maximum bending moment per unit cross-sectional area m_u that can be applied to a beam of circular hollow cross section of outside radius r_o and inner radius r_i. The loading is pure bending, no shear. The material is homogeneous and failure occurs when the maximum stress reaches the strength σ_u of the material. The hollow section is filled with foam to prevent buckling. Derive an expression for the efficiency of the cross section as the ratio of m_u of the hollow beam by m_u of a solid rod of same outside radius. Faced with the problem of using a strong and relatively expensive material, would you recommend a small or large radius?*

Problem 3.2 *Compute the maximum outside radius for a cantilever beam of length L, loaded by a tip load P, otherwise similar to the beam in Problem 3.1 but subjected to pure shear loading. The shear strength is $\tau_u = \sigma_u/2$. Consider only shear. Buckling of the thin wall is likely to limit further the practical thickness of the wall.*

Problem 3.3 *Compute the maximum deflection per unit volume δ_V that can be applied to a beam of circular hollow cross section of outside radius r_o and inner radius r_i. This is a cantilever beam of length L, loaded by a tip load P. The hollow section is made of an homogeneous material with moduli E and $G = E/2.5$, filled with foam to prevent buckling. Derive an expression for the efficiency of the cross section as the ratio of δ_V between the hollow cross section and a solid rod of the same outside radius. Faced with the problem of using a relatively expensive and not quite stiff material, would you recommend a small or large radius?*

Problem 3.4 *Write a computer program to evaluate (3.9). The program data input is the LSS, the thickness of the laminae, and the material elastic properties. The output should be written in a file. Show all work in a report.*

Problem 3.5 *Using the program of Problem 3.4 compute the A,B, D and H matrices for the following laminates. The material is AS4D/9310 and all layers are 0.85 mm thick. Comment on the coupling of the constitutive equations for each case: (a) one layer $[0]$, (b) one layer $[30]$, (c) $[0/90]_2$, (d) $[0/90]_s$, (e) $[0/90]_8$, (f) $[\pm 45]_2 = [+45/-45/+45/-45]$, (g) $[\pm 45]_s = [+45/-45/-45/+45]$, (h) $[\pm 45/0/90/\pm 30]$. Show all work in a report.*

Problem 3.6 *Compute the value and location of the absolute maximum transverse shear strain γ_{23} in Example 3.2. At that location plot the distribution of γ_{23} through the thickness of the plate. Is that distribution a reasonable answer?*

Problem 3.7 *Recompute Example 3.2 with a doubly sinusoidal load $q(x,y) = q_0 \sin(\pi x/2a) \sin(\pi x/2b)$, where $2a, 2b$ are the plate dimensions in x and y, respectively. Compare the result with the exact solution at the center of the plate, that is $w_0 = 16q_0 b^4/[\pi^4(D_{11}s^4 + 2(D_{12} + 2D_{66})s^2 + D_{22})]$, where $s = b/a$ ([4, (5.2.8–5.2.10)]).*

Problem 3.8 *Calculate the first vibration frequency ϖ_{11} of the plate with the analytical solution $\varpi_{mn}^2 = \pi^4[D_{11}m^4s^4 + 2(D_{12} + 2D_{66})m^2n^2s^2 + D_{22}n^4]/(16\rho h b^4)$, where ρ, h are the density and thickness of the plate, respectively ([4, (5.7.8)]).*

Problem 3.9 *Using ANSYS FE code, generate a rectangular plate with $a_x = 1000$ mm and $b_y = 100$ mm. The laminae are made of AS4D/9310 (Table 3.1) 1.2 mm thick. Look up four different LSS laminates where appear: (a) bending extension coupling effect, (b) thermal expansion coupling effect, (c) torsion extension coupling effect, and (d) shear extension (these coupling effects are shown in [1, Figure 6.7]). Model (i) one half of the plate, 500 × 100 mm, and (ii) one quarter of the plate, 500 × 50 mm, applying symmetry conditions and report when it is correct or not to use each of theses reduced models. Show all work in a report.*

Problem 3.10 *Using a program (e.g. MATLAB) to plot the failure limits (with $I_f = 1$) of maximum stress, Tsai-Wu, and Puck failure criteria in the plane $\sigma_1 - \sigma_2$, and in the plane $\sigma_2 - \sigma_6$.*

Problem 3.11 *Compute the failure index I_F on each layer of Example 3.8 using the maximum stress failure criterion and the Tsai-Wu failure criterion. The lamina strength values are given in Table 3.1. (a) Calculate the failure indexes using the FC commands in ANSYS. (b) Write the nodal stress results at the top and bottom of each lamina in a file. Then, using an external program (e.g. MATLAB) compute the same failure indexes as in part (a) and compare them. Show all work in a report.*

Problem 3.12 *Compute the failure index I_F on each layer of Example 3.8 using the Puck failure criterion. The lamina strength values are given in Table 3.1. Calculate the failure indexes using: (a) APDL script in ANSYS and (b) a USERMAT subroutine (for shell elements* `usermatps.f`*). Show all work in a report.*

References

[1] E. J. Barbero, Introduction to Composite Materials Design, Taylor & Francis, Philadelphia, PA, 1998. http://www.mae.wvu.edu/barbero/icmd.html

[2] E. J. Barbero, Computer Aided Design Environment for Composites. http://www.mae.wvu.edu/barbero/cadec.html

[3] Guide to ANSYS User Programmable Features, ANSYS, Inc., Southpointe, PA, 2006.

[4] J. N. Reddy, Mechanics of Laminated Composite Plates and Shells, 2nd Ed., CRC Press, Boca Raton, FL, 2003.

[5] E. Hinton, and D. R. J. Owen, An Introduction to Finite Element Computations, Pineridge Press, Swansea, U.K., 1979.

[6] ANSYS v10.0 Theory Manual, ANSYS, Inc., Southpointe, PA, 2006.

[7] ABAQUS Theory Manual, Hibbitt, Karlsson & Sorensen, Inc., 2002.

[8] R. J. Roark, W. C. Young, Roark's Formulas for Stress and Strain, 6th Ed., McGraw-Hill, New York, NY, 1989.

[9] The Composite Materials Handbook, Web Resource, http://www.mil17.org

[10] A. Puck, and H. Schurmann, Failure Analysis of FRP Laminates by Means of Physically Based Phenomenological Models, Composites Science and Technology, 62 (2002) 1633-1662.

[11] ANSYS APDL Programmer's Guide, ANSYS, Inc., Southpointe, PA, 2006.

[12] E. J. Barbero, Web resource: http://www.mae.wvu.edu/barbero/feacm/

[13] Prediction and Measurement of Post-critical Behavior of Fiber-Reinforced Composite Columns, Composites Science and Technology, 58 (1998) 1335-1341.

Chapter 4

Buckling

Most composite structures are thin walled. This is a natural consequence of the following facts:

- Composites are stronger than conventional materials. Then, it is possible to carry very high loads with a small area, and thus small thickness in most components.

- Composites are expensive when compared to conventional materials. Therefore, there is strong motivation to reduce the volume, and thus the thickness as much as possible.

- The cost of polymer matrix composites increases with their stiffness. The stiffness in the fiber direction can be estimated by using the fiber dominated rule of mixtures, $E_1 = E_f V_f$. For example, when glass fibers are combined with a polymer matrix, the resulting composite stiffness is lower than that of aluminum. Using Aramid yields a stiffness comparable to aluminum. Carbon fibers yield composite stiffness lower than steel. Therefore, there is strong motivation to increase the moment of inertia of beams and stiffeners without increasing the cross-sectional area. The best option is to increase the moment of inertia by enlarging the cross section dimensions and reducing the thickness.

All the above factors often lead to design of composite structures with larger, thin walled cross sections, with modes of failure likely to be controlled by buckling.

4.1 Bifurcation Methods

Buckling is loss of stability due to geometric effects rather than material failure. But it can lead to material failure and collapse if the ensuing deformations are not restrained. Most structures can operate in a linear elastic range. That is, they return to the undeformed configuration upon removal of the load. Permanent deformations result if the elastic range is exceeded, as when matrix cracking occurs in a composite.

Consider a simply supported column of area A, length L, and moment of inertia I, made of homogeneous material with modulus E and strength F along the length of the column. The column is loaded by a compressive load P acting on the centroid of the cross section [1]. If the column geometry, loading, and material have no imperfections, the axial deformation is

$$u = PL/EA \qquad (4.1)$$

with no lateral deformation $w = 0$. The deformation of the structure (u, v, w) before buckling occurs is called the *primary path*. The slightest imperfection will make the column buckle when

$$P_{CR} = \pi^2(EI)/L^2 \qquad (4.2)$$

The load capacity for long slender columns will be controlled by buckling, as opposed to the crushing strength of the material. What happens after the column reaches its critical load depends largely on the support conditions. For the simply supported column, the lateral deflection

$$w = A\sin(\pi x/L) \qquad (4.3)$$

will grow indefinitely $(A \to \infty)$ when the load just barely exceeds P_{CR}. Such large lateral deflections will cause the material to fail and the column will collapse. The behavior of the structure after buckling has occurred is called *post-buckling*.

The column of the example above experiences no deformations in the shape of the buckling mode (4.3) before buckling actually happens. In this case, it is said that the structure has a *trivial* primary path. In the example above, this is a consequence of having a perfect structure with perfectly aligned loading. For these type of structures, buckling occurs at a bifurcation point. A bifurcation point is the intersection of the primary path with the secondary path, which in this example is the post-buckling path [2].

The bifurcation loads, one for every possible mode of buckling, are fairly easy to obtain using commercial software. The geometry of the structure is that of the perfect undeformed configuration, loaded with the nominal loads, and the material is elastic. Such analysis requires a minimum of effort on the part of the analyst. Commercial programs refer to this analysis as an eigenvalue buckling analysis because the critical loads are the eigenvalues λ_i of the discretized system of equations

$$([K] - \lambda[K_s])\{v\} = 0 \qquad (4.4)$$

where K and K_s are the stiffness and stress stiffness matrix, respectively, and v is the column of eigenvectors (buckling modes) [2].

Example 4.1 *Consider a simple supported plate, with side dimensions $a_x = 1000$ mm, $a_y = 500$ mm, edgewise loaded in compression with $N_x = N_y = 1$ N/mm. The plate is made of $[(0/90)_3]_S$, AS4/9310 composite with fiber volume fraction 0.6 and total thickness $t_T = 10.2$ mm. The lamina elastic properties are shown in Table 4.1. Compute the critical load of the lowest four modes using eigenvalue analysis. Visualize the four modes.*

Table 4.1: Lamina elastic properties

Young's Moduli	Shear Moduli	Poisson's Ratio
$E_1 = 145880$ MPa	$G_{12} = G_{13} = 4386$ MPa	$\nu_{12} = \nu_{13} = 0.263$
$E_2 = E_3 = 13312$ MPa	$G_{23} = 4529$ MPa	$\nu_{23} = 0.470$

Solution to Example 4.1 *Since the laminate is symmetric, and stress computation lamina by lamina is not required, the critical loads can be obtained using three different approaches.*

In the first approach, the equivalent laminate moduli are calculated and used along with an orthotropic shell element. In this case, laminate moduli represent the stiffness of an equivalent orthotropic plate that behaves likes the actual laminate under in-plane loads, neglecting the bending loads (see Section 3.2.3). Laminate moduli can be found as explained in Section 1.15. Introduce the lamina properties (Table 4.1) into (1.92), rotate each layer (1.52), add then according to (1.103) get the laminate moduli (1.106) listed in Table 4.2. The portion of the ANSYS input file used to enter the laminate moduli is listed below and available on the Web site [10]. Element type SHELL63 *is used.*

Table 4.2: Equivalent laminate moduli

Young's Moduli	Shear Moduli	Poisson's Ratio
$E_x = 79985$ MPa	$G_{xy} = 4386$ MPa	$\nu_{xy} = 0.044$
$E_y = 79985$ MPa	$G_{yz} = 4458$ MPa	$\nu_{yz} = 0.415$
$E_z = 16128$ MPa	$G_{xz} = 4458$ MPa	$\nu_{xz} = 0.415$

```
/TITLE,Orthotropic plate, edge load, Bifurcation Analysis, SHELL93
/UNITS,MPA               ! Units are in mm, MPa, and Newton

/PREP7                   ! Pre-processor module
! Equivalent Material properties
uimp,1,ex,ey,ez,79985,79985,16128
uimp,1,gxy,gyz,gxz,4386,4458,4458
uimp,1,prxy,pryz,prxz,0.044,0.415,0.415

ET,1,SHELL93             ! Chooses SHELL93 element for analysis
R,1,10.2                  ! Real constant set #1, thickness of 10.2 mm

! Geometry and mesh
RECTNG,0,500,0,250       ! Creates a rectangle with x=500 mm and y=250 mm
ESIZE,,25                ! 25 divisions for edge
AMESH,all                ! Mesh the area

FINISH                   ! Exit pre-processor module
```

In the second approach, $A - B - D - H$ *matrices are used. To get the laminate properties (A, B, D and H matrices), introduce the lamina properties (Table 4.1) into (3.9). The resulting laminate matrices are*

$$
\begin{bmatrix} A & B \\ B & D \end{bmatrix} = \begin{bmatrix}
817036 & 35937.6 & 0 & 0 & 0 & 0 \\
35937.6 & 817036 & 0 & 0 & 0 & 0 \\
0 & 0 & 44737.2 & 0 & 0 & 0 \\
0 & 0 & 0 & 8.55845\ 10^6 & 311579 & 0 \\
0 & 0 & 0 & 311579 & 5.60896\ 10^6 & 0 \\
0 & 0 & 0 & 0 & 0 & 387872
\end{bmatrix}
$$

$$
[H] = \begin{bmatrix} 37812.8 & 0 \\ 0 & 37964.7 \end{bmatrix}
$$

The ANSYS input file used to define the laminate using SHELL99 *elements and A, B, C and H matrices is listed below.*

```
/TITLE,Orthotropic plate with edge load, Bifurcation Analysis, SHELL99
/UNITS,MPA                  ! Units are in mm, MPa, and Newtons

/PREP7                      ! Pre-processor module
! This input data does not need material properties

ET,1,SHELL99,,2             ! Chooses SHELL99 element for analysis
                            ! Set KEYOPT(2)=2, then supply ABDH matrix
KEYOPT,1,10,2               ! Set KEYOPT(10)=2, print ABDH matrix file.abd

R,1,817036,35937.6,0,0,0,0              ! real set #1, A11,A12,0,A16,0,0
RMODIF,1,7,817036,0,0,0,0               ! real set #1, A22,0,A26,0,0
RMODIF,1,16,44737.2,0,0                 ! real set #1, A66,0,0
RMODIF,1,19,37812.8,0                   ! real set #1, H44,H45
RMODIF,1,21,37964.7                     ! real set #1, H55
RMODIF,1,22,3.49246e-010,1.09139e-011,0,0,0,0   ! B11,B12,0,B16,0,0
RMODIF,1,28,2.6921e-010,0,0,0,0          ! real set #1, B22,0,B26,0,0
RMODIF,1,37,1.45519e-011,0,0            ! real set #1, B66,0,0
RMODIF,1,43,8.55845e+006,311579,0,0,0,0 !    D11,D12,0,D16,0,0
RMODIF,1,49,5.60896e+006,0,0,0,0        !         D22,0,D26,0,0
RMODIF,1,58,387872,0,0                  ! real set #1, D66,0,0
RMODIF,1,76,,10.2           ! real set #1, Average density and thick


! Geometry and mesh
RECTNG,0,500,0,250          ! Creates a rectangle with x=500 mm and y=250 mm
ESIZE,,25                   ! 25 divisions for edge
AMESH,all                   ! Mesh the area

FINISH                      ! Exit pre-processor module
```

In the third approach, *the LSS and the lamina properties (Table 4.1) are entered. The ANSYS input file commands to define the laminate are listed below. Element type* SHELL91 *is used.*

```
/TITLE,Orthotropic plate, edge load, Bifurcation Analysis, SHELL91
/UNITS,MPA                  ! Units are in mm, MPa, and Newton
```

```
/PREP7                    ! Pre-processor module
! Material properties for lamina
uimp,1,ex,ey,ez,145880,13312,13312
uimp,1,gxy,gyz,gxz,4386,4529,4386
uimp,1,prxy,pryz,prxz,0.263,0.470,0.263

ET,1,SHELL91,,1           ! Chooses Shell91 element for analysis
                          ! Set KEYOPT(2)=1, then supply 12+(6*NL) const.
KEYOPT,1,1,12             ! Set KEYOPT(1)=12, Max number of layers = 12
KEYOPT,1,5,1              ! Set KEYOPT(5)=1, Element output: Middle layer

! Real constant set #1, [(0/90)3]s, NL=12, lamina thick=0.85 mm
R,1,12,1,                 ! 12 layers symmetrical
RMORE,,,,,,,
RMORE,1,0,0.85            ! 1st layer: mat. #1,   0 deg, Th=0.85 mm
RMORE,1,90,0.85           ! 2nd layer: mat. #1,  90 deg, Th=0.85 mm
RMORE,1,0,0.85            ! 3nd layer: mat. #1,   0 deg, Th=0.85 mm
RMORE,1,90,0.85           ! 4th layer: mat. #1,  90 deg, Th=0.85 mm
RMORE,1,0,0.85            ! 5th layer: mat. #1,   0 deg, Th=0.85 mm
RMORE,1,90,0.85           ! 6th layer: mat. #1,  90 deg, Th=0.85 mm

! Geometry and mesh
RECTNG,0,500,0,250        ! Creates a rectangle with x=500 mm and y=250 mm
ESIZE,,25                 ! 25 divisions for edge
AMESH,all                 ! Mesh the area

FINISH                    ! Exit pre-processor module
```

The procedure for obtaining the solution of "Eigenvalue Buckling Analysis" in ANSYS has three steps: (i) solve the static solution using the PSTRESS,ON command to obtain the stress stiffness matrix, (ii) obtain the bifurcation loads using the eigenvalue buckling solution and (iii) expand the solution if the buckled mode shapes are needed. By running the code listed below, the critical load and buckling mode shape for every mode are obtained.

```
/SOLU                     ! Solution module, (i) STATIC ANALYSIS
ANTYPE,STATIC             ! Set static analysis
PSTRESS,ON                ! Calculate the stress stiffness matrix
DL,2,1,uz,0               ! Impose Simple Supported BC
DL,3,1,uz,0
DL,1,1,symm               ! Impose Symmetry BC
DL,4,1,symm
!d,all,rotz               ! Constraint rotations about z (optional)
!Load application
SFL,2,PRES,1              ! Apply uniform load in x=500 mm
SFL,3,PRES,1              ! Apply uniform load in y=250 mm
SOLVE                     ! Solve current load state
FINISH                    ! Exit solution module

/SOLU                     ! Solution module, (ii) BIFURCATION LOADS
ANTYPE,BUCK
BUCOPT,SUBSP,10           ! Find the first 10 bifurcations loads
```

```
SOLVE                   ! Solve
FINISH                  ! Exit solution module

/SOLU                   ! Solution module, (ii) BUCKLING MODES
EXPASS
MXPAND
SOLVE                   ! Solve
FINISH                  ! Exit solution module
```

Using the command SET,LIST *in the post-processor, a list with the critical buckling loads is obtained. With* SET,1,n, *where n is the mode number, it is possible to select different solutions corresponding to different mode shapes, which can be plotted using* PLDISP,1 *command, as indicated in the listing below.*

```
/POST1                  ! Post-processor module
SET,LIST                ! List the critical loads
SET,1,2                 ! Set mode number 2 shape
PLDISP,1                ! Display the mode 2 shape displacements
FINISH                  ! Exit post-processor module
```

The results of the first five modes are summarized in Table 4.3 for the equivalent lamina (SHELL93), ABDH matrix input (SHELL99), as well as for the elements using LSS (SHELL91). Values are shown for only five modes because lack of accuracy of results for modes above 1/2 the number of iteration vectors used in the subspace method.

Table 4.3: Bifurcation loads [N/mm]

Mode	1	2	3	4	5
SHELL93	252.70	570.55	1547.4	2150.8	2318.2
SHELL99 ABDH	209.53	639.98	1802.8	1822.5	1863.4
SHELL91 LSS	209.53	639.96	1803.0	1822.3	1863.6

4.1.1 Imperfection Sensitivity

To illustrate the influence of imperfections in buckling, let us consider the solid lines in Figure 4.1. The lateral deflection is zero for any load below the bifurcation load P_{CR}, that is on the primary path of the perfect structure. The primary path intersects the secondary path at the bifurcation point, for which the load is P_{CR}. The post-critical behavior of the column is indifferent and slightly stable. Indifferent means that the column can deflect right or left. Stable post-critical path means that the column can take a slightly higher load once it has buckled. For a column, this stiffening behavior is so small that one cannot rely upon it to carry any load beyond P_{CR}. In fact, the column will deform laterally so much that the material will fail and the system will collapse. Unlike columns, simply supported plates experience significant stiffening on the secondary path.

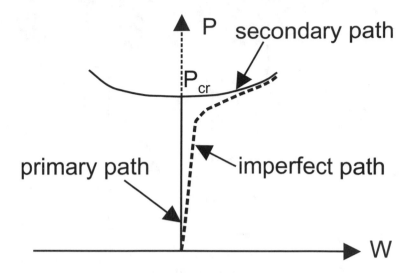

Fig. 4.1: Equilibrium paths for the perfect column.

4.1.2 Asymmetric Bifurcation

Consider the frame illustrated in Figure 4.2. An eigenvalue analysis using one finite element per bar ([2] Section 5.9 and 7.8) reveals the bifurcation load

$$P_{CR} = 8.932(10^{-6})AE \tag{4.5}$$

but gives no indication about the nature of the critical state: whether it is stable or not, whether the post-critical path is symmetric or not, and so on. We shall see later on that the frame has an asymmetric, and thus unstable post-critical path, as represented in Figure 4.2. That is, the post-critical path has a slope

$$P^{(1)} = 18.73(10^{-9})AE \ (1/\text{rad}) \tag{4.6}$$

in the force-rotation diagram in Figure 4.2, where θ is the rotation of the joint at the load point.

In general, the problem with eigenvalue analysis is that it provides no indication as to the nature of the post-critical path. If the post-critical path is stiffening and symmetric as in Figure 4.1, the real structure may have a load capacity close to the bifurcation load. But if the post-critical path is unstable and/or asymmetric, as in Figure 4.2 or if there is mode interaction [3, 4, 5, 6, 7, 8], the real structure may have a load capacity much smaller then the bifurcation load. In order to use the information provided by eigenvalue analysis, it is necessary to understand and quantify the post-buckling behavior.

4.1.3 Post-Critical Path

One way to investigate the post-buckling behavior is to perform a continuation analysis of the imperfect structure [9]. This is perfectly possible, but complicated

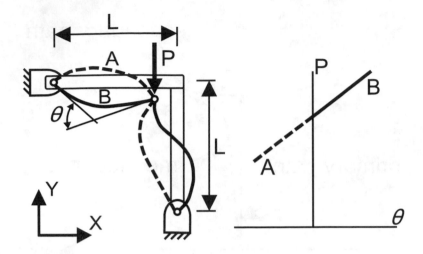

Fig. 4.2: Two bar frame.

and time consuming, as will be seen later in this chapter. A more expedient solution can be obtained using software capable of predicting the nature of the post-critical path, including symmetry, curvature, and mode interaction. If the secondary path is stable and symmetric, the bifurcation load can be used as a good estimate of the load capacity of the structure. The curvature of the post-critical path gives a good indication of the post-buckling stiffening and it can be used to a certain extent to predict post-buckling deformations.

The bifurcation load, slope, and curvature of the post-critical path emerging from the bifurcation (4.1) can be computed with BMI3 [4, 5, 6] available in [10]. The post-buckling behavior is represented by the following formula

$$\Lambda = \Lambda^{(cr)} + \Lambda^{(1)}s + \frac{1}{2}\Lambda^{(2)}s^2 + \dots \qquad (4.7)$$

where s is the perturbation parameter, which is chosen as one component of the displacement of one node, Λ_{CR} is the bifurcation multiplier, $\Lambda^{(1)}$ is the slope, and $\Lambda^{(2)}$ is the curvature of the post-critical path ([3, (43)], see also [4, 5, 6, 7, 8]). When the slope is zero, the post-critical path is symmetric. Therefore, buckling is indifferent, and the real structure will buckle to either side. There is no way to predict which way it is going to buckle, unless of course one knows the shape of the imperfections on the real plate, which is seldom the case. A positive curvature denotes stiffening during post-buckling, and a negative one indicates that the stiffness decreases.

Example 4.2 *Consider the simple supported plate of Example 4.1. Compute the bifurcation multiplier Λ_{CR}, the critical load N_{CR}, the slope $\Lambda^{(1)}$, and the curvature $\Lambda^{(2)}$ of the post-critical path. Estimate the load when the maximum lateral deflection is equal to the thickness of the plate. As perturbation parameter, use the largest displacement component of the buckling mode with lowest buckling load.*

Solution to Example 4.2 *The program BMI3 [10] is used in this case to compute the bifurcation multiplier Λ_{CR}, the slope $\Lambda^{(1)}$, and the curvature $\Lambda^{(2)}$ of the post-critical path.*

Refer to Appendix E for a description of the software interface and operation procedure. BMI3 is used from within the ANSYS GUI in this example. Since BMI3 requires the A-B-D-H matrices, the ANSYS input file is a slightly modified version of that used for the **second approach** *in Example 4.1. The model in ANSYS and solved by using the "Eigen-value buckling analysis" procedure for obtaining the bifurcation loads* $\Lambda^{(cr)}$. *Using the file* FEAcomp_Ex402_ABDH.log *(available in [10]), the critical lowest load is displayed on the ANSYS GUI as* FREQ=210.2 *for* STEP=1, SUB=1. *Since the buckling mode is scaled to a maximum amplitude of 1.0, we get* DMX=1. *A list of buckling loads can be recalled by the command* SET,LIST.

Within the ANSYS GUI, Run the APDL macro ans2i *simply by entering* ans2i *in the ANSYS command line [10] to calculate parameters of the post-critical path. BMI3 will be executed. In the "ANSYS Output Window", manually introduce the following responses to the prompts: (i) "..sort (0=none, 1=x, 2=y, 3=z)" 1, and (ii) "..for perturbation analysis (y/n)?"* n. *By responding* n, *BMI3 chooses as perturbation parameter the largest displacement component of the buckling mode with lowest buckling load. In this case, that corresponds to the first buckling mode, the node in the middle of the plate, and the deflection direction δ. Therefore, the following results are obtained:*

$\Lambda^{(cr)} = -209.0418$, $\Lambda^{(1)} \approx 0$, $\Lambda^{(2)} = -0.2308$. *Since BMI3 solves the problem using reversed loads (see Appendix E), then (4.7) becomes*

$$-N = \Lambda^{(cr)} + \Lambda^{(1)}s + \tfrac{1}{2}\,\Lambda^{(2)}s^2$$
$$N = -\Lambda^{(cr)} - \Lambda^{(1)}s - \tfrac{1}{2}\,\Lambda^{(2)}s^2$$

and, in this case the perturbation direction is $s = -\delta$, *so*

$$N = -\Lambda^{(cr)} - \Lambda^{(1)}(-\delta) - \tfrac{1}{2}\,\Lambda^{(2)}(-\delta)^2$$
$$N = -\Lambda^{(cr)} + \Lambda^{(1)}\delta - \tfrac{1}{2}\,\Lambda^{(2)}\delta^2$$

Therefore, using the results from BMI3, the secondary path of bifurcation analysis is

$$N = -(-209.0418) + (0)\,\delta - (-0.1154)\,\delta^2 = 209.0418 + 0.1154\,\delta^2$$

Since the slope $\Lambda(1)$ *is zero, the post-critical path is symmetric. The post-buckling load when the lateral deflection (w) is equal to the thickness (s = Th = 10.2 mm) is equal to 221 N/mm, as shown in Figure 4.3.*

4.2   Continuation Methods

The strain to failure of polymer matrix composites (PMC) is high. Compare 1.29% for AS4/3501 and 2.9% for S-glass/epoxy with only 0.2% for steel and 0.4% for aluminum. That means that buckling deformations can go into post-buckling regime while the material remains elastic. However, great care must be taken that no matrix dominated degradation mode takes place, in which case the material will not remain elastic (see Chapter 8). Eigenvalue buckling analysis is relatively simple as long as the material remains elastic because classical theory of elastic stability can be used, as was done in Section 4.1. Material nonlinearity is one reason that motivates an incremental analysis. Another reason to use an incremental analysis is to evaluate the magnitude of the buckling load for an *imperfection sensitive* structure.

In an incremental analysis, also called continuation analysis, the load is increased gradually step by step. At each step, the deformation, and possibly the changing

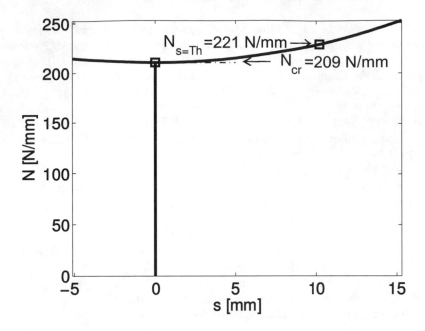

Fig. 4.3: Equilibrium paths for a perfect plate.

material properties, are evaluated. Incremental analysis must include some type of imperfection, in the geometry, material, or alignment of loads. Lacking any imperfection, incremental analysis will track the linear solution, revealing no bifurcations or limit points.

Continuation methods are a form of geometrically nonlinear analysis. The system must have a nontrivial fundamental path, such as a flat plate with asymmetric laminate stacking sequence (LSS) under edge loads.

If the system has a trivial fundamental path, such a flat plate with symmetric LSS under edge load, the nontrivial fundamental path can be forced by introducing an imperfection. Several types of imperfections are possible, including material imperfections (e.g., unsymmetrical LSS), geometric imperfections, or load eccentricity are used.

Since the real geometric imperfections are seldom known, the preferred artificial geometric imperfection is in the form of the bifurcation mode having the lowest bifurcation load. This is true in most cases; however, in some cases, a second mode that is associated to imperfections that are more damaging to the structure should be used [11]. Also, if the structure has an asymmetric post-buckling path, as the two-bar example in Figure 4.2, care must be taken not to force the structure along the stiffening path.

FEA codes allow the user to modify a mesh by superposing an imperfection in the shape of any mode from a previous bifurcation analysis onto the perfect geometry (see Example 4.3).

Example 4.3 *Using ANSYS, apply a geometric imperfection $w_p(x, y) = \delta_0 \ \phi(x, y)$ to Example 4.1 and plot the load-multiplier vs. maximum lateral deflection for an imperfection* *magnitude* $\delta_0 = Th/10$ *and* $\delta_0 = Th/100$, *where* Th *is the total laminate thickness, and* $\phi(x,y)$ *is the buckling mode corresponding to the lowest bifurcation load found in Example 4.1.*

Solution to Example 4.3 *First the buckling modes are found using the bifurcation method (execute the commands shown in Example 4.1).*

Then the nodal positions are updated using the UPGEOM *command. Using this command the displacements from a previous analysis can be added in order to update the geometry of the finite element model to that of the deformed configuration.*

Since the displacements have been obtained from a mode shape, the maximum displacement in the results file is 1.0. The UPGEOM *allows the user to define a multiplier for displacements being added to the nodal coordinates. In this case, the multiplier factors chosen are* $\delta_0 = Th/10$ *and* $\delta_0 = Th/100$. *Therefore, an initial deflection equal to the first mode of buckling with a central deflection* δ_0 *is forced on the structure.*

Using a continuation method with this non-perfect geometry, the continuation equilibrium paths shown in Figure 4.4 are obtained. It can be seen that eventually the continuation solution approaches the secondary path of the perfect structure, shown by dashed lines in Figure 4.4. For smaller imperfections, the continuation solution follows more closely the primary path, then the secondary path. A structure with large imperfections deviates more from the behavior of the perfect structure, as show by the solution corresponding to an imperfection $\delta_0 = Th/10$.

```
/PREP7                    ! Pre-processor module
ftr=(10.8/10)             ! Multiplicator shape factor (Th/10)
UPGEOM,ftr,1,1,file,rst
   ! ftr: Multiplier for displacements added to coordinates
   ! 1,1 : Load step 1, substep=1, equivalent to mode =1
   ! file,rst: results file to obtain displacements
FINISH                    ! Exit pre-processor module

/SOLU                     ! Solution module, Continuation loads
ANTYPE,STATIC             ! Set static analysis
NLGEOM,1                  ! Use large displacements analysis
OUTRES,ALL,ALL            ! Keep results of each substep

mult=225                  ! Apply loads until N = 225 N/mm
SFL,2,PRES,1*mult         ! Apply uniform pressure in x=500 mm
SFL,3,PRES,1*mult         ! Apply uniform pressure in y=250 mm

ARCLEN,1,10,0.1           ! Use ARCLENG method to obtain solution
NSUBST,50,0,0             ! #Substeps
SOLVE                     ! Solve current load state
FINISH                    ! Exit solution module

/POST26                   ! Post-processor module
LINES,1000                ! List without breaks between pages
NSOL,2,1,U,Z,UZ_node1     ! Load deflexion in central plate node
PLVAR,2                   ! DISPLAY VARIABLES evolution
PRVAR,2                   ! PRINT VARIABLES evolution
FINISH
```

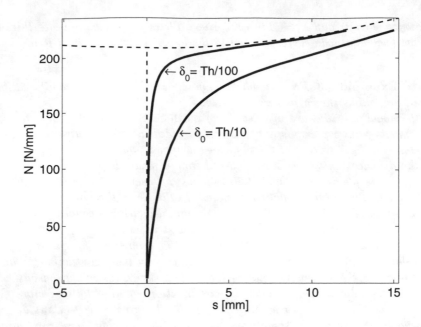

Fig. 4.4: Equilibrium paths for a $[(0/90)_6]_S$ plate, with $\delta_0 = Th/10$ and $\delta_0 = Th/100$.

Example 4.4 *Find the bucling load multiplier and the first mode shape for Example 3.11.*

Solution to Example 4.4 *The solution is found simply by executing the* .log *file shown below after executing the* .log *file in Example 3.11. The load multiplier Λ^{cr} can be read from the GUI as* FREQ=54.288 *on the same screen that shows the mode shape for mode one (*STEP=1, SUB=1*). The buckling load is simply the product of the load multiplier by the applied load $P^{cr} = 54.288 \times 11,452 = 621,706\ N$.*

```
!Buckling analysis for I-COLUMN CST 58 (1998) 1335-1341
/SOLU
ANTYPE,BUCK
BUCOPT,SUBSP,10
SOLVE
FINI
/POST1          !POSTPROCESSOR MODULE
SET,1,1         !SELECT 1ST LOAD CASE, 1ST EIGENVALUE
*GET,LCR,TIME   !GET THE EIGENVALUE IN USER DEFINED VARIABLE LCR
PLDISP,2        !PLOT MODE SHAPE AND OUTLINE OF UNDEFORMED SHAPE
```

Suggested Problems

Problem 4.1 *Compute the bifurcation load P^c of the two-bar frame in Figure 4.2 using one beam element per bar. Each bar has length $L = 580\ mm$, area $A = 41\ mm^2$, inertia $I = 8.5\ mm^4$, height $H = 10\ mm$, and modulus $E = 200\ GPa$.*

Problem 4.2 *Perform a convergence study on the bifurcation load P^c of the two-bar frame in Problem 4.1 by increasing the number of elements per bar N until the bifurcation load converges within 2%. Plot P^c vs. N.*

Problem 4.3 *Recalculate Example 4.2 when the LSS changes to $[(0/90)_6]_T$, thus becoming asymmetric. Do not introduce any imperfection but rather analyze the perfect system, which in this case is asymmetric.*

Problem 4.4 *Recalculate Example 4.2 with $[(0/90)_6]_T$, and $N_x = 1, N_y = N_{xy} = 0$. Do not introduce any imperfection but rather analyze the perfect system, which in this case is asymmetric.*

Problem 4.5 *Using a FEA code, plot the continuation solution for $\delta_0 = Th/100$ as in Figure 4.4, for a cylindrical shell with distributed axial compression on the edges. The cylinder has a length of $L = 965$ mm and a mid-surface radius of $a = 242$ mm. The LSS is $[(0/90)_6]_S$, with layer thickness $t = 0.127$ mm. The laminae are of E-glass/epoxy with $E_1 = 54$ GPa, $E_2 = 18$ GPa, $G_{12} = 9$ GPa, $\nu_{12} = 0.25$, and $\nu_{23} = 0.38$.*

Problem 4.6 *Compute the Tsai-Wu criterion failure index I_f of Problem 4.5 at $P = \Lambda^{(cr)}$. The strength values are $F_1 t = 1034$ MPa, $F_1 c = 1034$ MPa, $F_2 t = 31$ MPa, $F_2 c = 138$ MPa, and $F_6 t = 41$ MPa.*

Problem 4.7 *Plot the imperfection sensitivity of the cylindrical shell of Problem 4.5, for imperfections in the range $(Th/200) < s < Th$.*

References

[1] F. P. Beer, E. R. Johnston, and J. T. DeWolf, Mechanics of Materials, 3rd. Ed., McGraw Hill, NY, 2001.

[2] L. A. Godoy, Theory of Stability-Analysis and Sensitivity, Taylor & Francis, Philadelphia, PA, 2000.

[3] E. J. Barbero, L. A. Godoy, and I. Raftoyiannis, Finite Elements For Three-Mode Interaction in Buckling Analysis, International Journal for Numerical Methods in Engineering, 39 (1996) 469-488.

[4] L. A. Godoy, E. J. Barbero, and I. Raftoyiannis, Finite Elements for Post-Buckling Analysis–Part I: W-Formulation, Computer and Structures, 56(6) (1995), 1009-1017.

[5] E. J. Barbero, I. Raftoyiannis, and L. A. Godoy, Finite Elements for Post-Buckling Analysis–Part II: Applications to Composite Plate Assemblies, Computer and Structures, 56(6) (1995) 1019-1028.

[6] I. Raftoyiannis, L. A. Godoy, and E. J. Barbero, Buckling Mode Interaction in Composite Plate Assemblies, Applied Mechanics Reviews, 48(11, Part 2) (1995) 52-60.

[7] E. J. Barbero, and J. Trovillion, Prediction and Measurement of Post-Critical Behavior of Fiber Reinforced Composite Columns, Composite Science and Technology, special issue on Civil Infrastructural Applications of Composite Materials, 58(8) (1998) 1335-1341.

[8] E. J. Barbero, Prediction of Buckling Mode Interaction in Composite Columns, Mechanics of Composite Materials and Structures, 7 (2000) 269-284.

[9] D. Bushnell, Computerized Buckling Analysis of Shells, Martinus Nijhoff, Dordrecht, The Netherlands, 1985.

[10] E. J. Barbero, Web resource: http://www.mae.wvu.edu/barbero/feacm/

[11] S. Yamada, and J.G.A. Croll, Buckling Behavior Pressure Loaded Cylindrical Panels, ASCE Journal of Engineering Mechanics, 115(2) (1989) 327-344.

Chapter 5

Free Edge Stresses

In-plane loading N_x, N_y, N_{xy}, of symmetric laminates induces only in-plane stress σ_x, σ_y, σ_{xy}, in the interior of the laminate. Near the free edges, interlaminar stresses σ_z, σ_{yz}, σ_{xz}, are induced due to the imbalance of the in-plane stress components at the free edge.

For illustration, consider a long laminated strip of length $2L$, width $2b << 2L$, and thickness $2H < 2b$ (Figure 5.1). The laminate is symmetric. The strip is loaded by an axial force N_x only. Since the laminate is symmetric, the mid-plane strains and curvatures (see (3.6)) are uniform over the entire crosssection and given by

$$
\begin{aligned}
\epsilon_x^0 &= \alpha_{11} N_x \\
\epsilon_y^0 &= \alpha_{12} N_x \\
\gamma_{xy}^0 &= 0 \\
k_x = k_y &= k_{xy} = 0
\end{aligned}
\tag{5.1}
$$

where α_{11}, α_{22}, are in-plane laminate compliances, which are obtained by inverting (3.8); see also ([1, (6.18)]). From the constitutive equation ([1, (6.21)]) for layer k, we get

$$
\begin{aligned}
\sigma_x^k &= \left(\overline{Q}_{11}^k \alpha_{11} + \overline{Q}_{12}^k \alpha_{12} \right) N_x \\
\sigma_y^k &= \left(\overline{Q}_{12}^k \alpha_{11} + \overline{Q}_{22}^k \alpha_{12} \right) N_x \\
\sigma_{xy}^k &= \left(\overline{Q}_{16}^k \alpha_{11} + \overline{Q}_{26}^k \alpha_{12} \right) N_x \\
\sigma_z^k &= \sigma_{xz}^k = \sigma_{yz}^k = 0
\end{aligned}
\tag{5.2}
$$

A piece of laminate taken out of the interior of the laminate will have balanced σ_y and σ_{xy} on opposite faces; the free body diagram (FBD) is in equilibrium without the need for any additional forces. In this case we say the stress components are self-equilibrating. At the free edge in Figure 5.1, $\sigma_y = \sigma_{xy} = \sigma_{yz} = 0$. If σ_y and σ_{xy} are not zero in the interior of the laminate, but are zero at the free edge, then some other stresses must equilibrate them.

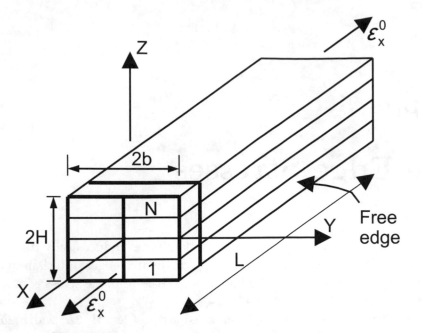

Fig. 5.1: Tensile coupon[1].

5.1   Poisson's Mismatch

A layer subjected to tensile loading in one direction will contract in the direction perpendicular to the load. If two or more layers with different Poisson's ratios are bonded together, interlaminar stress will be induced to force all layers to deform equally at the interfaces (Figure 5.2). Over the entire laminate thickness, these stresses add up to zero since there is no transverse loading N_y applied. In other words, they are self-equilibrating in such a way that

$$\int_{z_0}^{z_N} \sigma_y dz = 0 \qquad (5.3)$$

where z_0 and z_N are the coordinates of the bottom and top surfaces, respectively.

5.1.1   Interlaminar Force

As noted in (5.3), the in-plane stress σ_y calculated with classical lamination theory (CLT, [1, Chapter 6]) are self-equilibrating when added through the whole thickness of the laminate. But on a portion of the laminate (above z_k in Figure 5.3), the stresses σ_y may not be self-equilibrating. Therefore, the contraction or expansion of one or more layers must be equilibrated by interlaminar shear stress σ_{yz}. Since there is no shear loading on the laminate, the integral of σ_{yz} over the entire width of the sample must vanish. Over half the width of the laminate, however, an interlaminar

---

[1]Reprinted from Mechanics of Fibrous Composites, C. T. Herakovich, Fig. 8.1, copyright (1998), with permission from John Wiley & Sons, Inc.

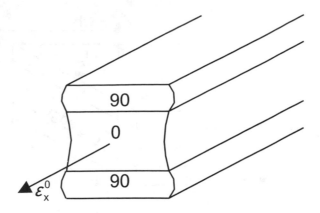

Fig. 5.2: Poisson's effect[2].

shear force exists if the stress σ_y above or below the surface is not self-equilibrating. The magnitude of these per unit length forces can be estimated by integrating the interlaminar shear stress σ_{yz} over half the width of the laminate ($0 < y < b$). By equilibrium

$$F_{yz}(z_k) = \int_0^b \sigma_{yz(z=z_k)} dy = -\int_{z_k}^{z_N} \sigma_y dz \qquad (5.4)$$

The interlaminar shear stress σ_{yz} is not available from classical lamination theory but the transverse stress σ_y is. Therefore, the magnitude of the interlaminar shear force can be computed anywhere through the thickness of a laminate in terms of the known transverse stress distribution σ_y.

The in-plane stress σ_y in a balanced laminate under tensile load is constant in each layer. Therefore, when the interlaminar force is evaluated at an interface (located at z=z_k), the integration above reduces to

$$F_{yz}(z_k) = -\sum_{i=k}^{N} \sigma_y^i t_i \qquad (5.5)$$

The magnitude of the interlaminar shear force F_{yz} can be used to compare different stacking sequences in an effort to minimize the free-edge interlaminar shear stress σ_{yz}. However, the force does not indicate how large the actual stress is. Therefore, it can be used to compare different LSS but not as a failure criterion.

5.1.2 Interlaminar Moment

The interlaminar shear stress σ_{yz} produces shear strain γ_{yz}, which must vanish at the center line of the sample because of symmetry. Therefore, $\sigma_{yz} = 0$ at the center line. Also, at the free edge, σ_{yz} must vanish because σ_{zy} vanishes there. But for any

---

[2]Reprinted from Mechanics of Fibrous Composites, C. T. Herakovich, Fig. 8.14, copyright (1998), with permission from John Wiley & Sons, Inc.

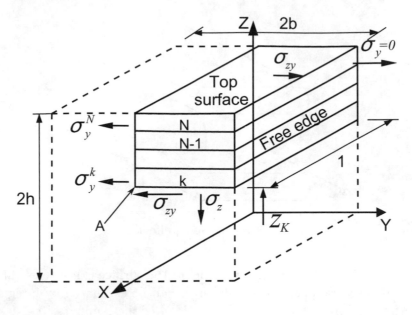

Fig. 5.3: Free body diagram of sublaminate for computation of Poison-induced forces and moments $\sigma_{zy} = \sigma_{yz}$.

position z_k above which σ_y is not self-equilibrating, σ_{yz} must be non-zero somewhere between the edge and the center line. A numerical solution of σ_{yz} is plotted in Figure 5.5 in terms of the distance y/b from the free edge. It reveals that σ_{yz} grows rapidly near the free edge and then tapers out at the interior of the laminate.

A not self-equilibrating distribution of stress yields both a force F_{yz} (5.5) and a moment. To compute the moment M_z, take moments of the stress σ_y with respect to point A in Figure 5.3. A non-vanishing moment produced by σ_y can only be equilibrated by a moment produced by transverse stress σ_z. Therefore, the moment M_z is defined as

$$M_z(z_k) = \int_0^b \sigma_{z(z=z_k)} y \, dy = \int_{z_k}^{z_N} (z - z_k)\sigma_y dz \qquad (5.6)$$

The existence of σ_z is corroborated by free-edge delamination during a tensile test, at a much lower load than the failure load of the laminate. The magnitude of the moment can be used to compare different stacking sequences in an effort to minimize the thickness stress σ_z. However, the moment does indicate how large the actual thickness stress is and where the maximum value occurs.

The in-plane stress σ_y in a balanced laminate under tensile load is constant in each layer. Therefore, when the interlaminar moment is evaluated at an interface (located at $z = z_k$), the integration above reduces to

$$M_z(z_k) = \sum_{i=k}^{N} \sigma_y^i \left(z_i t_i + \frac{t_i^2}{2} - z_k t_i \right) \qquad (5.7)$$

Since σ_z is a byproduct of σ_{yz} and σ_{xz} vanishes at $y = 0$ due to symmetry, then σ_z must vanish at the center line of the specimen ($y = 0$) but it is large near the edge. Since no vertical load is applied, the integral of σ_z must be zero. Therefore, it must be tensile (positive) on some regions and compressive (negative) at others. A numerical solution reveals that σ_z grows rapidly near the free edge, dips to negative values and then tapers out at the interior of the laminate. A numerical solution of σ_z is plotted in Figure 5.5 in terms of the distance y/b from the free edge. However, $\sigma_z \to \infty$ as $y \to b$. This is a singularity that is not handled well by FEA. Therefore the results, even for $y < b$, will be very dependent on the mesh refinement. Furthermore, since $\sigma_z \to \infty$, the results cannot be used in a failure criterion without further consideration.

Example 5.1 *Compute F_{yz} and M_z at all interfaces of a balanced $[0_2/90_2]_s$ symmetric laminate (Figure 5.1) loaded with $N_x = 175$ KN/m. Use unidirectional lamina carbon/epoxy properties $E_1 = 139$ GPa, $E_2 = 14.5$ GPa, $G_{12} = G_{13} = 5.86$ GPa, $G_{23} = 2.93$ GPa, $\nu_{12} = \nu_{13} = 0.21$, $\nu_{23} = 0.38$. The lamina thickness is $t_k = 0.127$ mm.*

Solution to Example 5.1 *The in-plane stress distribution σ_y through the thickness can be obtained by the procedure described in ([1, Section 6.2]), which is implemented in CADEC [3]. Stress values are displayed under "Macromechanics," "Global stress," "Total stresses." The data in the scrollable window can be retrieved from the file TOSTRSGI.OUT. The stress values are shown in Table 5.1.*

To calculate F_{yz}, compute the contribution of all layers above a given interface using (5.5). The in-plane stress σ_y in a balanced laminate under in-plane load is constant in each layer, so (5.5) applies. For other cases, (5.4) can be integrated exactly since σ_y is linear in z, or F_{yz} can be approximated by (5.5) using the average σ_y in each layer.

Since the laminate is balanced and loaded with in-plane loads only, M_z can be computed using (5.7). Otherwise, use (5.6) or approximate M_z by using the average σ_z in each layer into (5.7).

The results are shown Table 5.1 and Figure 5.4.

Example 5.2 *Plot σ_{yz} and σ_z vs y for $0 < y < b$ at the 90/0 interface above the middle surface of a $[0/90]_s$ laminate with properties $E_1 = 139$ GPa, $E_2 = 14.5$ GPa, $G_{12} = G_{13} = 5.86$ GPa, $G_{23} = 2.93$ GPa, $\nu_{12} = \nu_{13} = 0.21$, $\nu_{23} = 0.38$. Take $2b = 20$ mm, length of the sample $2L = 80$ mm, thickness of each layer $t_k = 1.25$ mm. Load the sample with a uniform strain $\epsilon_x = 0.01$ by applying a uniform displacement at $x = L$. Use orthotropic solid elements on each layer. Refine the mesh towards the free edge. Use at least two quadratic elements through the thickness of each layer and an element aspect ratio approximately one near the free edge.*

Solution to Example 5.2 *Note that it is not necessary to model the whole geometry. Symmetry can be used to model only the quadrant with $x > 0, y > 0, z > 0$; that is one-eighth of the plate, as shown in Figure 5.3. Since any cross section $y - z$ has the same behavior, only a short segment between $x = 0$ and $x = L\star$ needs to be modeled. Since free edge effects also occur at $x = 0$ and $x = L\star$, take $L\star = 8h$ and plot the results at $x = L \star /2$ to avoid free edge effects at the two loaded ends of the model. The solution is shown in Figure 5.5.*

*See the command input file below. The **PATH** commands define, plot, and print the stress values shown in Figure 5.5.*

Table 5.1: Poisson's interlaminar force F_{yz}

k	Pos	σ_y [MPa]	t_k [mm]	z [mm]	F_{yz} [kN/m]	M_z [N m/m]
8	TOP	$5.55\ 10^{-3}$		0.508	0.000	
8	BOT	$5.55\ 10^{-3}$	0.127	0.381	-0.705	0.045
7	TOP	$5.55\ 10^{-3}$		0.381	-0.705	
7	BOT	$5.55\ 10^{-3}$	0.127	0.254	-1.410	0.179
6	TOP	$-5.55\ 10^{-3}$		0.254	-1.410	
6	BOT	$-5.55\ 10^{-3}$	0.127	0.127	-0.705	0.313
5	TOP	$-5.55\ 10^{-3}$		0.127	-0.705	
5	BOT	$-5.55\ 10^{-3}$	0.127	0.000	0.000	0.358
4	TOP	$-5.55\ 10^{-3}$		0.000	0.000	
4	BOT	$-5.55\ 10^{-3}$	0.127	-0.127	0.705	0.313
3	TOP	$-5.55\ 10^{-3}$		-0.127	0.705	
3	BOT	$-5.55\ 10^{-3}$	0.127	-0.254	1.410	0.179
2	TOP	$5.55\ 10^{-3}$		-0.254	1.410	
2	BOT	$5.55\ 10^{-3}$	0.127	-0.381	0.705	0.045
1	TOP	$5.55\ 10^{-3}$		-0.381	0.705	
1	BOT	$5.55\ 10^{-3}$	0.127	-0.508	0.000	0.000

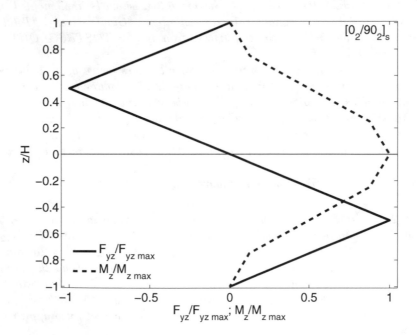

Fig. 5.4: Interlaminar force F_{yz} and moment M_z due to Poisson's effect.

```
/TITLE,Free Edge Analysis [0/90]s laminate
/PREP7                      ! Pre-processor module
*set,ThZ,1.25               ! Thickness of lamina in mm
*set,LX,8*ThZ               ! 1/2 Length of laminate in mm
*set,bY,10.0                ! 1/2 width of laminate in mm
*set,neX,8                  ! Number of elements in x/z direction
```

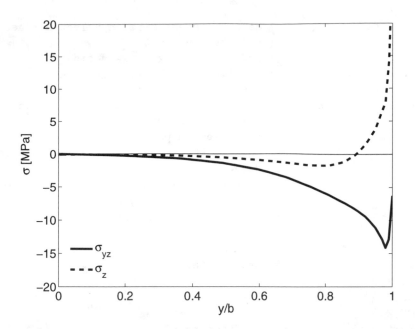

Fig. 5.5: Interlaminar stress σ_{yz} and σ_z at the 90/0 interface of a carbon/epoxy $[0/90]_S$ laminate (FEA).

```
*set,neY,14             ! Number of elements in y direction
*set,epsX,0.01          ! Uniform strain in x direction

! Equivalent Material properties
uimp,1,ex,ey,ez,139e3,14.5e3,14.5e3
uimp,1,gxy,gyz,gxz,5.86e3,2.93e3,5.86e3
uimp,1,prxy,pryz,prxz,0.21,0.38,0.21

ET,1,SOLID95            ! Chooses SOLID95 element for analysis

! Define material orientation by local Coordinate
local,11,,0,0,0,90      ! defines 90 degree local cs
local,12,,0,0,0,0       ! defines 0 degree local cs
CSYS,0                  ! set active cs to cart. system

! Generate Geometry
BLOCK,0,LX,0,bY,0,ThZ      ! 90 degrees layer
BLOCK,0,LX,0,bY,ThZ,2*ThZ  ! 0 degress layer
VGLUE,ALL                  ! Glue volumes

! Mesh Control and Mesh
lesize,all,,,neX        ! line number divisions = neX
lsel,s,loc,z,0          ! selects lines z=0
lsel,a,loc,z,ThZ        ! add lines z=Thz to selection
lsel,a,loc,z,2*ThZ      ! add lines z=2Thz to selection
lsel,r,loc,x,0          ! reselects lines x=0
```

```
LESIZE,ALL,,,neY,15,1,,,1     ! define element size in selected lines
lsel,s,loc,z,0                ! selects lines z=0
lsel,a,loc,z,ThZ              ! add lines z=Thz to selection
lsel,a,loc,z,2*ThZ            ! add lines z=2Thz to selection
lsel,r,loc,x,LX               ! reselects lines x=LX
LESIZE,ALL,,,NEy,(1/15),1,,,1 ! define ele. size in selected lines
lsel,all                      ! select all lines
MSHKEY,1                      ! Specifies mapped meshing
ESYS,11                       ! Selects 90 degrees material orientation
VMESH,1                       ! Meshes 90 degrees layer
ESYS,12                       ! Selects 0 degrees material orientation
VMESH,3                       ! Meshes 0 degree layer
FINISH              ! Exit pre-processor module

/SOLU               ! Solution module,
ANTYPE,STATIC       ! Set static analysis
ASEL,S,LOC,X,0
ASEL,A,LOC,Y,0
ASEL,A,LOC,Z,0
DA,ALL,SYMM         ! Impose Symmetry BC
ASEL,S,LOC,X,LX
DA,ALL,UX,(epsX*LX) ! Impose displacement on the end = epsX*LX
ALLSEL,ALL          ! Selects all areas
SOLVE               ! Solve current load state
FINISH              ! Exit solution module

/POST1              ! Post-processor module
RSYS,0              ! Set results in global coordinates system
PATH,INTERFACE,2,,100  ! Define a path between two points,
                       ! compute 100 values
PPATH,1,0,0,0,ThZ,0    ! 1st point of the path location
PPATH,2,0,0,bY,ThZ,0   ! 2nd point of the path location
PDEF,zero,EPSW,,AVG    ! Compute zero axis (optional)
PLPATH,Sz,Sxz,Syz,zero ! Plot Sz,Sxz,Syz
/page,1000,,1000       ! Define print list w/o skips between pages
PRPATH,Sz,Sxz,Syz      ! Print Sz,Sxz,Syz
FINISH                 ! Exit post-processor module


/POST1              ! Post-processor module
RSYS,0              ! Set results in global coordinates system
PATH,INTERFACE,2,,100  ! Define a path between two points,
                       ! compute 100 values
PPATH,1,0,LX/2,0,ThZ,0 ! 1st point of the path location
PPATH,2,0,LX/2,bY,ThZ,0 ! 2nd point of the path location
PDEF,Sz ,S,Z,AVG    ! Compute Sz
PDEF,Syz,S,YZ,AVG   ! Compute Syz
PDEF,zero,EPSW,,!AVG ! Compute zero axis (optional)
PLPATH,Sz,Syz,zero  ! Plot Sz,Syz
/page,1000,,1000    ! Define print list w/o skips between pages
```

```
PRPATH,Sz,Syz          ! Print Sz,Syz
FINISH                 ! Exit post-processor module
```

5.2   Coefficient of Mutual Influence

In classical lamination theory, it is assumed that the portion of the laminate being analyzed is far from the edges of the laminate. Stress resultants N and M are then applied to a portion of the laminate and these induce in-plane stress σ_x, σ_y, σ_{xy}, on each layer. In the interior of the laminate, interlaminar stress σ_{xz}, σ_{yz}, are induced only if shear forces are applied.

For uniaxial loading N_x, the transverse stresses generated in each layer as a result of Poisson's effect must cancel out to yield a null laminate force N_y. Also, the in-plane shear stress on off-axis layers must cancel out with those of other layers to yield zero shear force N_{xy} for the laminate. The situation is more complex near the edges as the various components of in-plane stress do not cancel each other across the lamina interfaces. For the time being, let us revisit the concept of laminate engineering properties. In material axes, the plane stress compliance equations are

$$\left\{ \begin{array}{c} \epsilon_1 \\ \epsilon_2 \\ \gamma_6 \end{array} \right\} = \left[\begin{array}{ccc} S_{11} & S_{12} & 0 \\ S_{12} & S_{22} & 0 \\ 0 & 0 & S_{66} \end{array} \right] \left\{ \begin{array}{c} \sigma_1 \\ \sigma_2 \\ \sigma_6 \end{array} \right\} \tag{5.8}$$

It is also known that the compliance coefficients can be written in terms of engineering properties as

$$[S] = \left[\begin{array}{ccc} 1/E_1 & -\nu_{12}/E_1 & 0 \\ -\nu_{12}/E_1 & 1/E_2 & 0 \\ 0 & 0 & 1/G_{12} \end{array} \right] \tag{5.9}$$

For an off-axis layer (oriented arbitrarily with respect to the global axes), we have

$$\left\{ \begin{array}{c} \epsilon_x \\ \epsilon_y \\ \gamma_{xy} \end{array} \right\} = \left[\begin{array}{ccc} \overline{S}_{11} & \overline{S}_{12} & \overline{S}_{16} \\ \overline{S}_{12} & \overline{S}_{22} & \overline{S}_{26} \\ \overline{S}_{16} & \overline{S}_{26} & \overline{S}_{66} \end{array} \right] \left\{ \begin{array}{c} \sigma_x \\ \sigma_y \\ \sigma_{xy} \end{array} \right\} \tag{5.10}$$

Here it can be seen that uniaxial load ($\sigma_y = \sigma_{xy} = 0$) yields shear strain as a result of the shear-extension coupling

$$\gamma_{xy} = \overline{S}_{16}\sigma_x \tag{5.11}$$

where

$$\overline{S}_{16} = (2S_{11} - 2S_{12} - S_{66})\sin\theta\cos^3\theta \tag{5.12}$$
$$- (2S_{22} - 2S_{12} - S_{66})\sin^3\theta\cos\theta$$

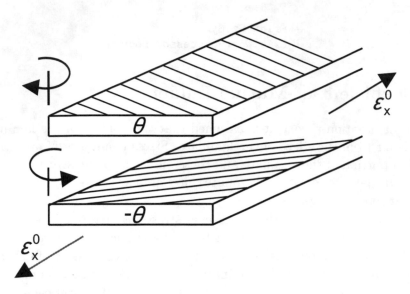

Fig. 5.6: Deformation caused by mutual influence[3].

Now, the coefficients of $\boxed{S}$ can be defined in term of the engineering properties for the off-axis layer as

$$\overline{S}_{11} = 1/E_x \; ; \; \overline{S}_{12} = -\nu_{xy}/E_x = -\nu_{yx}/E_y \tag{5.13}$$
$$\overline{S}_{22} = 1/E_y \; ; \; \overline{S}_{66} = 1/G_{xy}$$

To complete the definition of $\boxed{S}$ in (5.10), two new engineering properties describing shear-extension coupling, $\eta_{xy,x}$ and $\eta_{xy,y}$, are defined as

$$\overline{S}_{16} = \frac{\eta_{xy,x}}{E_x} \; ; \; S_{26} = \frac{\eta_{xy,y}}{E_x} \tag{5.14}$$

The engineering properties $\eta_{xy,x}$ and $\eta_{xy,y}$ are called coefficients of mutual influence and they represent the shear caused by stretching. Their formal definition is obtained by imposing an axial stress and measuring the resulting shear strain

$$\eta_{ij,i} = \frac{\gamma_{ij}}{\epsilon_i} \tag{5.15}$$

Alternatively, two other coefficients of mutual influence could be defined to represent the stretching caused by shear

$$\overline{S}_{16} = \frac{\eta_{x,xy}}{G_{xy}} \; ; \; \overline{S}_{26} = \frac{\eta_{y,xy}}{G_{xy}} \tag{5.16}$$

These are defined by imposing a shear stress and measuring the axial strain

$$\eta_{i,ij} = \frac{\epsilon_i}{\gamma_{ij}} \tag{5.17}$$

---

[3]Reprinted from Mechanics of Fibrous Composites, C. T. Herakovich, Fig. 8.14, copyright (1998), with permission from John Wiley & Sons, Inc.

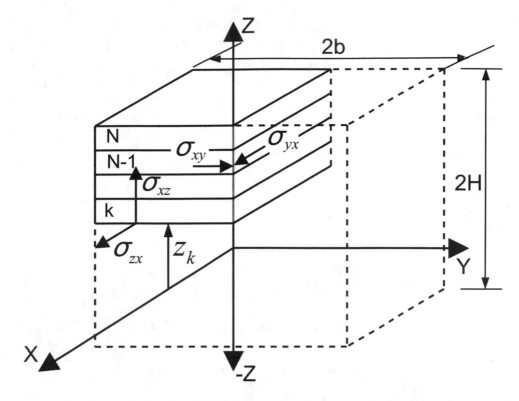

Fig. 5.7: Free body diagram of sublaminate to compute the interlaminar force due to mutual influence.

5.2.1 Interlaminar Stress due to Mutual Influence

Off-axis laminae induce in-plane shear stress when subject to axial loading because the natural shear deformations that would occur on an isolated lamina (Figure 5.6) are constrained by the other laminae. Through the whole thickness of the laminate, these stresses cancel out, but over unbalanced sublaminates (e.g., the top layer in Figure 5.6), they amount to a net shear.

That shear can only be balanced by interlaminar stress σ_{zx} at the bottom of the sublaminate (Figure 5.7). Then, summation of forces along x leads to a net force

$$F_{xz}(z_k) = \int_0^b \sigma_{zx_{(z=z_k)}} dy = -\int_{z_k}^{z_N} \sigma_{xy} dz \qquad (5.18)$$

Once again, the in-plane shear stress calculated with classical lamination theory (CLT) ([1, Chapter 6]) can be used to compute the interlaminar force per unit length F_{xz}. For in-plane loading, CLT yields constant shear stress in each layer. When the interlaminar force is evaluated at an interface (located at $z = z_k$), the integration above reduces to

$$F_{xz}(z_k) = -\sum_{i=k}^{N} \sigma_{xy}^i t_i \qquad (5.19)$$

The force F_{xz}, as well as the values of the coefficients of mutual influence can be used to qualitatively select the LSS with the least interlaminar stress. Actual values of interlaminar stresses can be found by numerical analysis. However, $\sigma_z \to \infty$ as $y \to b$. This is a singularity that is not handled well by FEA. Therefore the results, even for $y < b$, are very dependent on mesh refinement. Furthermore, since $\sigma_z \to \infty$, the results cannot be used in a failure criterion without further consideration. A numerical approximation of σ_{xz} for a $[\pm 45]_S$ laminate is plotted in Figure 5.9 in terms of the distance y' from the free edge.

Example 5.3 *Compute F_{xz} at all interfaces of a $[30_2/-30_2]_s$ balanced symmetric laminate (Figure 5.1) loaded with $N_x = 175$ KN/m. The material properties are given in Example 5.1. The lamina thickness is $t_k = 0.127$ mm.*

Solution to Example 5.3 *In plane shear stress σ_{xy} through the thickness of the laminate can be obtained following the same procedure used to obtain σ_y in Example 5.1.*

For a symmetric balanced laminate under in-plane loads, use (5.19). For a general laminate under general load, use (5.18) or approximate F_{xz} by (5.19) taking the average of σ_{xy} in each layer.

The results are obtained with a spreadsheet and shown in Table 5.2 and Figure 5.8.

Table 5.2: Interlaminar force F_{xz} due to mutual influence

k	Pos	σ_{xy} [MPa]	t_k [mm]	z [mm]	F_{xz} [kN/m]
8	TOP	$78.6\ 10^{-3}$		0.508	0.000
8	BOT	$78.6\ 10^{-3}$	0.127	0.381	-9.982
7	TOP	$78.6\ 10^{-3}$		0.381	-9.982
7	BOT	$78.6\ 10^{-3}$	0.127	0.254	-19.964
6	TOP	$-78.6\ 10^{-3}$		0.254	-19.964
6	BOT	$-78.6\ 10^{-3}$	0.127	0.127	-9.982
5	TOP	$-78.6\ 10^{-3}$		0.127	-9.982
5	BOT	$-78.6\ 10^{-3}$	0.127	0.000	0.000
4	TOP	$-78.6\ 10^{-3}$		0.000	0.000
4	BOT	$-78.6\ 10^{-3}$	0.127	-0.127	9.982
3	TOP	$-78.6\ 10^{-3}$		-0.127	9.982
3	BOT	$-78.6\ 10^{-3}$	0.127	-0.254	19.964
2	TOP	$78.6\ 10^{-3}$		-0.254	19.964
2	BOT	$78.6\ 10^{-3}$	0.127	-0.381	9.982
1	TOP	$78.6\ 10^{-3}$		-0.381	9.982
1	BOT	$78.6\ 10^{-3}$	0.127	-0.508	0.000

Example 5.4 *Plot σ_{xz} at the interface above the middle surface of a $[\pm 45]_S$ laminate using the material properties, geometry, and loading of Example 5.2.*

Solution to Example 5.4 *Note that in this case it is not possible to use the same symmetry conditions used in Example 5.2. Since the LSS is symmetric, it is possible to model half of the laminate $(z > 0)$. Since the LSS contains layers at angles other than 0 and 90, the plane $x = 0$ is not a symmetry plane, but rather a plane with $\epsilon_x = 0$. Also, the edge effects at the ends of the model in $x = 0$ and in $x = L\star$ are now important, so the results*

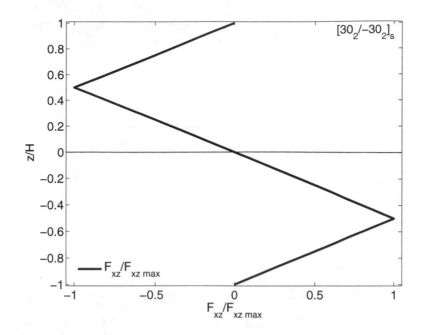

Fig. 5.8: Interlaminar shear force due to mutual influence F_{xz}.

must be plotted at $x = L \star /2$ *to avoid free edge effects at the loaded ends. The solution is shown in Figure 5.9.*

```
/TITLE,Free Edge Analysis [45/-45]s laminate
/PREP7                    ! Pre-processor module
*set,ThZ,1.25             ! Thickness of lamina in mm
*set,LX,8*ThZ             ! 1/2 Length of laminate in mm
*set,bY,10.0              ! 1/2 width of laminate in mm
*set,neX,8                ! Number of elements in x/z direction
*set,neY,14               ! Number of elements in y direction
*set,epsX,0.01            ! Uniform strain in x direction

! Equivalent Material properties
uimp,1,ex,ey,ez,139e3,14.5e3,14.5e3
uimp,1,gxy,gyz,gxz,5.86e3,2.93e3,5.86e3
uimp,1,prxy,pryz,prxz,0.21,0.38,0.21

ET,1,SOLID95             ! Chooses SOLID95 element for analysis

! Define material orientation by local Coordinate
local,11,,0,0,0,-45      ! defines -45 degree local cs
local,12,,0,0,0,45       ! defines +45 degree local cs
CSYS,0                   ! set active cs to cart. system

! Generate Geometry
BLOCK,0,LX,0,bY,0,ThZ    ! -45 degrees layer
BLOCK,0,LX,0,bY,ThZ,2*ThZ  ! +45 degress layer
```

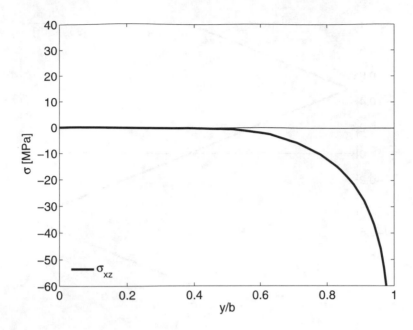

Fig. 5.9: Interlaminar shear stress σ_{xz} at the interface above the middle-surface of a carbon/epoxy $[\pm 45]_S$ laminate (FEA).

```
VSYMM,Y,ALL
VGLUE,ALL                      ! Glue volumes

! Mesh Control and Mesh
lesize,all,,,neX               ! line number divisions = neX
lsel,s,loc,z,0                 ! selects lines z=0
lsel,a,loc,z,ThZ               ! add lines z=Thz to selection
lsel,a,loc,z,2*ThZ             ! add lines z=2Thz to selection
lsel,r,loc,x,0                 ! reselects lines x=0
lsel,r,loc,y,0,2*By
LESIZE,ALL,,,neY,20,1,,,1      ! define element size in selected lines
lesize,all,,,neX               ! line number divisions = neX
lsel,s,loc,z,0                 ! selects lines z=0
lsel,a,loc,z,ThZ               ! add lines z=Thz to selection
lsel,a,loc,z,2*ThZ             ! add lines z=2Thz to selection
lsel,r,loc,x,0                 ! reselects lines x=0
lsel,r,loc,y,-2*By,0
LESIZE,ALL,,,neY,(1/20),1,,,1   ! element size in selected lines
lsel,s,loc,z,0                 ! selects lines z=0
lsel,a,loc,z,ThZ               ! add lines z=Thz to selection
lsel,a,loc,z,2*ThZ             ! add lines z=2Thz to selection
lsel,r,loc,x,LX                ! reselects lines x=LX
LESIZE,ALL,,,NEy,(1/20),1,,,1  ! define ele. size in selected lines
lsel,all                       ! select all lines
MSHKEY,1                       ! Specifies mapped meshing
ESYS,11                        ! Select -45 degrees material orientation
```

```
VMESH,1                          ! Meshes -45 degrees layer
VMESH,6                          ! Meshes -45 degrees layer
ESYS,12                          ! Select 45 degrees material orientation
VMESH,5                          ! Meshes 45 degree layer
VMESH,7                          ! Meshes 45 degree layer
FINISH                    ! Exit pre-processor module

/SOLU                     ! Solution module,
ANTYPE,STATIC             ! Set static analysis
ASEL,S,LOC,X,0
ASEL,A,LOC,Z,0
DA,ALL,SYMM               ! Impose Symmetry BC
ASEL,S,LOC,X,LX
DA,ALL,UX,(epsX*LX)       ! Impose displacement on the end = epsX*LX
ALLSEL,ALL                ! Selects all areas
DK,2,ALL                  ! Constraint node to avoid solid motion
SOLVE                     ! Solve current load state
FINISH                    ! Exit solution module

/POST1                    ! Post-processor module
RSYS,0                    ! Set results in global coordinates system
PATH,INTERFACE,2,,100     ! Define a path between two points,
                          ! compute 100 values
PPATH,1,0,LX/2,0,ThZ,0    ! 1st point of the path location
PPATH,2,0,LX/2,bY,ThZ,0   ! 2nd point of the path location
PDEF,Sxz,S,XZ,!AVG        ! Compute Sxz
PDEF,zero,EPSW,,!AVG      ! Compute zero axis (optional)
PLPATH,Sxz,zero           ! Plot Sxz
/page,1000,,1000          ! Define print list w/o skips between pages
PRPATH,Sxz                ! Print Sxz
FINISH                    ! Exit post-processor module
```

Suggested Problems

Problem 5.1 *Write a computer program to use tabulated data of σ_y (at the top and bottom of every layer) to compute F_{yz}, F_{xz}, and M_z, for all locations through the thickness of a laminate with any number of layers. Using the program, plot F_{yz}, F_{xz}, and M_z, through the thickness $-4t < z < 4t$ of a $[\pm45/0/90]_s$ laminate with layer thickness $t = 0.125$ mm, loaded with $N_x = 100$ kN/m. Use carbon/epoxy properties $E_1 = 139$ GPa, $E_2 = 14.5$ GPa, $G_{12} = G_{13} = 5.86$ GPa, $G_{23} = 2.93$ GPa, $\nu_{12} = \nu_{13} = 0.21$, $\nu_{23} = 0.38$. Submit a report including the source code of the program.*

Problem 5.2 *Repeat Problem 5.1 for $M_x = 1$ Nm/m. Submit a report including the source code of the program.*

Problem 5.3 *Plot σ_z/σ_{x0} and σ_{yz}/σ_{x0} vs. y/b ($0 < y/b < 1$) at $x = L/2$, at the first interface above the mid-surface for laminates, $[0/90]_s$ and with layer thickness $t = 0.512$ mm, loaded with $\epsilon_x = 0.01$. Compute the far-field uniform stress σ_{x0} in terms of the applied strain. Use quadratic solid elements and a mesh biased toward the free edge (bias 0.1) to model 1/8 of a tensile specimen (see Example 5.2), of width $2b = 25.4$ mm and length $2L = 20$ mm.*

Use carbon/epoxy properties $E_1 = 139$ GPa, $E_2 = 14.5$ GPa, $G_{12} = G_{13} = 5.86$ GPa, $G_{23} = 2.93$ GPa, $\nu_{12} = \nu_{13} = 0.21$, $\nu_{23} = 0.38$. Attempt to keep the aspect ratio of the elements near the free edge close to one. Submit the input command file to obtain the solution and the plot. In addition, submit the plot.

Problem 5.4 *For the laminate and loading described in Problem 5.3, plot σ_z/σ_{x0} and σ_{yz}/σ_{x0} versus z/t_k ($0 < z/t_k < 2$) above the mid-surface, at a distance $0.1t_k$ from the free edge and $x = L/2$. Study the effect of mesh refinement by providing four curves with different number of divisions along the $z-$direction. Attempt to keep the aspect ratio of the elements near the free edge close to one. Submit the input command file to obtain the solution and the plot. In addition, submit the plot.*

Problem 5.5 *Plot σ_{xz}/σ_{x0} as in Problem 5.3 for all the interfaces above the middle surface of a $[\pm 10_2]_S$ laminate.*

Problem 5.6 *Plot σ_{xz}/σ_{x0} as in Problem 5.4 for a $[\pm 10_2]_S$ laminate.*

Problem 5.7 *Use solid elements and a biased mesh to model 1/8 of a tensile specimen (see Example 5.2), of width $2b = 24$ mm and length $2L = 20$ mm. The laminate is $[\pm 45/0/90]_s$ with layer thickness $t = 0.125$ mm, loaded with $N_x = 175$ KN/m. Use carbon/epoxy properties $E_1 = 139$ GPa, $E_2 = 14.5$ GPa, $G_{12} = G_{13} = 5.86$ GPa, $G_{23} = 2.93$ GPa, $\nu_{12} = \nu_{13} = 0.21$, $\nu_{23} = 0.38$. Plot the three interlaminar stress components, from the edge to the center line of the specimen, at the mid-surface of each layer. Lump all four plots of the same stress into a single plot. Submit the input command file to obtain the solution and the three plots. In addition, submit the three plots.*

Problem 5.8 *Plot E_x/E_2, G_{xy}/G_{12}, $10\nu_{xy}$, $-\eta_{xy,x}$ and $-\eta_{x,xy}$ in the same plot vs θ in the range $-\pi/2 < \theta < \pi/2$ for a unidirectional single layer oriented at an angle θ. The material is S-glass/epoxy (Table 1.1 in [1]).*

Problem 5.9 *Using the plot from Problem 5.8 and considering a $[\theta_1/\theta_2]_S$ laminate, what is the worst combination of values θ_1, θ_2 for (a) Poisson's mismatch and (b) shear mismatch.*

Problem 5.10 *In a single plot, compare $-\eta_{xy,x}$ of E-glass/epoxy, Kevlar49/epoxy, and T800/3900-2 in the range $-\pi/2 < \theta < \pi/2$ (Table 1.1 in [1]).*

Problem 5.11 *Obtain contour plots of the three deformations u_x, u_y, u_z (independently) on the top surface of a $[\pm 45]_s$ laminate. Use dimensions, load, and material properties of Problem 5.7. Explain your findings.*

Problem 5.12 *Repeat Problem 5.11 for a $[0/90]_s$ laminate. Explain your findings.*

Problem 5.13 *Use solid elements and a biased mesh to model 1/4 of a tensile specimen (Figure 5.1) of a total width $2b = 12$ mm and length $2L = 24$ mm. Compare in the same plot σ_z vs z/H for $[\pm 15/\pm 45]_s$ and $[\pm(15/45)]$ of SCS-6/aluminum with 50% fiber volume. Use micromechanics (6.8) to predict the unidirectional composite properties. Layer thickness $t_k = 0.25$ mm. The laminate is loaded with $\epsilon_x = 0.01$.*

	Al-2014-T6	SCS-6
	([2] App. B)	([1] Table 2.1)
E [GPa]	75.0	427.0
G [GPa]	27.0	177.9

Problem 5.14 *Use a FEA model similar to Problem 5.13 to plot $\sigma_{xz}/\sigma_{zx_{\max}}$ vs θ ($0 < \theta < \pi/2$) for a $[\pm\theta]_s$ SCS-6/ Al laminate with $\epsilon_x = 0.01$.*

Problem 5.15 *Use the FEA model of Problem 5.13 to plot σ_z vs y/b ($0 < y < 0.95b$) at the mid-surface of the $[\pm 15/\pm 45]_s$ laminate. Note $\sigma_z \to \infty$ near $y = b$, so the actual value from FEA at $y = b$ is mesh dependent. Investigate mesh dependency at $y = 0.95b$ by tabulating the result using different mesh densities.*

Problem 5.16 *Use an FEA model similar to Problem 5.13 to plot σ_x, σ_{xy} and σ_{xz} vs y/b ($0 < y < b$) when a $[\pm\theta]_s$ SCS-6/Al laminate is subjected to 1% axial strain ($\epsilon_x = 0.01$).*

Problem 5.17 *A $[0/90]_s$ laminate with properties $E_1 = 139$ GPa, $E_2 = 14.5$ GPa, $G_{12} = G_{13} = 5.86$ GPa, $G_{23} = 2.93$ GPa, $\nu_{12} = \nu_{13} = 0.21$, $\nu_{23} = 0.38$ is shown in Figure 5.1. The strength properties of the lamina are $F_{1t} = 1550$ MPa, $F_{1c} = 1090$ MPa, $F_{2t} = F_{2c} = 59$ MPa, and $F_6 = 75$ MPa. Take $2b = 20$ mm, length of the sample $2L = 200$ mm, thickness of each layer $t_k = 1.25$ mm. Load the sample with a uniform strain $\epsilon_x = 0.01$ by applying a uniform displacement. Use symmetry to model only the quadrant with $x > 0$, $y > 0$, $z > 0$. Use orthotropic solid elements on each layer (element SOLID186 in ANSYS), with at least two quadratic elements through the thickness of each layer. Compute the 3D Tsai-Wu failure index I_F using a USERMAT subroutine for solid elements (`usermat3d.f`). Obtain the contour plot of I_F in each lamina (do not use results averaging). Show all work in a report.*

References

[1] E. J. Barbero, Introduction to Composite Materials Design, Taylor & Francis, Philadelphia, PA, 1998.

[2] C. T. Herakovich, Mechanics of Fibrous Composites, Wiley, New York, 1998.

[3] E. J. Barbero, Computer Aided Design for Composites
http://www.mae.wvu.edu/barbero/cadec.html

[4] E. J. Barbero, Web resource: http://www.mae.wvu.edu/barbero/feacm/

Chapter 6

Computational Micromechanics

In Chapter 1, the elastic properties of composite materials were assumed to be available in the form of elastic modulus E, shear modulus G, Poisson's ratio ν, and so on. For heterogeneous materials such as composites, a large number of material properties are needed, and experimental determination of these many properties is a tedious and expensive process. Furthermore, the values of these properties change as a function of the volume fraction of reinforcement and so on. An alternative, or at least a complement to experimentation, is to use homogenization techniques to predict the elastic properties of the composite in terms of the elastic properties of the constituents (matrix and reinforcements). Since homogenization models are based on more or less accurate modeling of the microstructure, these models are also called micromechanics models, and the techniques used to obtain approximate values of the composite's properties are called micromechanics methods or techniques [1]. Micromechanics models can be classified into empirical, semiempirical, analytical, and numerical. Accurate semiempirical models are described in [1].

This book deals only with strictly analytical or numerical models that do not require empirical adjusting factors, so that no experimentation is required. Since most of this book deals with 3D analysis, emphasis is placed on micromechanics models that can estimate the whole set of elastic properties using a single model, rather than using a disjoint collection of models based on different assumptions to assemble the set of properties needed, as is done for example in [2]. Many analytical techniques of homogenization are based on the equivalent eigenstrain method [3, 4], which considers the problem of a single ellipsoidal inclusion embedded in an infinite elastic medium. The Eshelby solution is used in [5] to develop a method that takes into account, approximately, the interactions among the inclusions. One of the more used homogenization techniques is the self-consistent method [6], which considers a random distribution of inclusions in an infinite medium. The infinite medium is assumed to have properties equal to the unknown properties sought. Therefore, an iterative procedure is used to obtain the overall moduli. Homogenization of composites with periodic microstructure has been accomplished by using various techniques including an extension of the Eshelby inclusion problem [3, 4], the Fourier series technique (see Section 6.1.3 and [7, 8]), and variational principles.

The periodic eigenstrain method was further developed to determine the overall relaxation moduli of linear viscoelastic composite materials (see Section 7.6 and [9, 10]). A particular case, the cell method for periodic media, considers a unit cell with a square inclusion [11].

The analytical procedures mentioned so far yield approximate estimates of the exact solution of the micromechanical problem. These estimates must lie between lower and upper bounds for the solution. Several variational principles were developed to evaluate bounds on the homogenized elastic properties of macroscopically isotropic heterogeneous materials [12]. Those bounds depend only on the volume fractions and the physical properties of the constituents.

In order to study the nonlinear material behavior of composites with periodic microstructure, numerical methods, mainly the finite element method, are employed. Nonlinear finite element analysis of metal matrix composite has been studied by looking at the behavior of the microstructure subjected to an assigned load history [13]. Bounds on overall instantaneous elastoplastic properties of composites have been derived by using the finite element method [14].

6.1   Analytical Homogenization

As discussed in the introduction, estimates of the average properties of heterogeneous media can be obtained by various analytical methods. Detailed derivations of the equations fall outside the scope of this book.

Available analytical models vary greatly in complexity and accuracy. Simple analytical models yield formulas for the stiffness $\mathbf{C}$ and compliance $\mathbf{S}$ tensors of the composite [11, (2.9) and (2.12)], such as

$$\mathbf{C} = \sum V_i \, \mathbf{C}^i \mathbf{A}^i \quad ; \quad \sum V_i \mathbf{A}^i = \mathbf{I}$$
$$\mathbf{S} = \sum V_i \, \mathbf{S}^i \mathbf{B}^i \quad ; \quad \sum V_i \mathbf{B}^i = \mathbf{I} \qquad (6.1)$$

where $V_i, \mathbf{C}^i, \mathbf{S}^i$, are the volume fraction, stiffness, and compliance tensors (in contracted notation)[1] of the i-th phase in the composite, respectively, and $\mathbf{I}$ is the 6×6 identity matrix. Furthermore, $\mathbf{A}^i, \mathbf{B}^i$, are the strain and stress concentration tensors (in contracted notation) of the i-th phase [11]. For fiber reinforced composites, $i = f, m$, represent the fiber and matrix phases, respectively.

6.1.1   Reuss Model

The *Reuss model* (also called rule of mixtures), assumes that the strain tensors[2] in the fiber, matrix, and composite are the same $\varepsilon = \varepsilon^f = \varepsilon^m$, so, the strain concentration tensors are all equal to the 6×6 identity matrix $\mathbf{A}^i = \mathbf{I}$. The rule of mixtures (ROM) formulas for E_1 and ν_{12} are derived and computed in this way.

---

[1]Fourth-order tensors with minor symmetry are represented by a 6×6 matrix taking advantage of contracted notation.

[2]Tensors are indicated by boldface type, or by their components using index notation.

6.1.2   Voigt Model

The *Voigt model* (also called inverse rule of mixtures), assumes that the stress tensors in the fiber, matrix, and composite are the same $\boldsymbol{\sigma} = \boldsymbol{\sigma}^f = \boldsymbol{\sigma}^m$, so, the stress concentration tensors are all equal to the 6×6 identity matrix $\mathbf{B}^i = \mathbf{I}$. The inverse rule of mixtures (IROM) formulas for E_2 and G_{12} are derived and computed in this way.

More realistic concentration tensors are given in Appendix B.

6.1.3   Periodic Microstructure Model

If the composite has periodic microstructure, or if it can be approximated as having such a microstructure (see Section 6.1.4), then Fourier series can be used to estimate all the components of the stiffness tensor of a composite. Explicit formulas for a composite reinforced by long circular cylindrical fibers, which are periodically arranged in a square array (Figure 6.1), are presented here [8]. The fibers are aligned with the x_1 axis, and they are equally spaced ($a_2 = a_3$). A generalization for transversely isotropic materials is presented in Section 6.1.4.

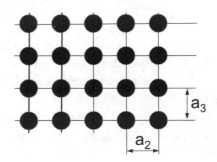

Fig. 6.1: Composite material with a periodic, square array of fibers.

Because the microstructure has square symmetry, the stiffness tensor has six unique coefficients given by

$$C_{11}^* = \lambda_m + 2\,\mu_m - \frac{V_f}{D}\left[\frac{S_3^2}{\mu_m^2} - \frac{2S_6 S_3}{\mu_m^2 g} - \frac{aS_3}{\mu_m\,c} + \frac{S_6^2 - S_7^2}{\mu_m^2 g^2} + \frac{aS_6 + bS_7}{\mu_m\,gc} + \frac{a^2 - b^2}{4\,c^2}\right]$$

$$C_{12}^* = \lambda_m + \frac{V_f}{D}b\left[\frac{S_3}{2c\mu_m} - \frac{S_6 - S_7}{2c\mu_m\,g} - \frac{a+b}{4\,c^2}\right]$$

$$C_{23}^* = \lambda_m + \frac{V_f}{D}\left[\frac{aS_7}{2\,\mu_m\,gc} - \frac{ba + b^2}{4\,c^2}\right]$$

$$C_{22}^* = \lambda_m + 2\,\mu_m - \frac{V_f}{D}\left[-\frac{aS_3}{2\,\mu_m\,c} + \frac{aS_6}{2\,\mu_m\,gc} + \frac{a^2 - b^2}{4\,c^2}\right]$$

$$C_{44}^* = \mu_m - V_f\left[-\frac{2\,S_3}{\mu_m} + (\mu_m - \mu_f)^{-1} + \frac{4\,S_7}{\mu_m(2 - 2\nu_m)}\right]^{-1}$$

$$C_{66}^* = \mu_m - V_f \left[-\frac{S_3}{\mu_m} + (\mu_m - \mu_f)^{-1} \right]^{-1} \tag{6.2}$$

where

$$D = \frac{aS_3^2}{2\,\mu_m^2 c} - \frac{aS_6 S_3}{\mu_m^2 gc} + \frac{a(S_6^2 - S_7^2)}{2\mu_m^2 g^2 c} + \\ + \frac{S_3(b^2 - a^2)}{2\,\mu_m\,c^2} + \frac{S_6(a^2 - b^2) + S_7(ab + b^2)}{2\,\mu_m gc^2} + \frac{(a^3 - 2b^3 - 3\,ab^2)}{8\,c^3} \tag{6.3}$$

and

$$\begin{aligned}
a &= \mu_f - \mu_m - 2\,\mu_f\,\nu_m + 2\,\mu_m\,\nu_f \\
b &= -\mu_m\,\nu_m + \mu_f\,\nu_f + 2\,\mu_m\,\nu_m\,\nu_f - 2\,\mu_1\,\nu_m\,\nu_f \\
c &= (\mu_m - \mu_f)(\mu_f - \mu_m + \mu_f\,\nu_f - \mu_m\,\nu_m + 2\,\mu_m\,\nu_f - 2\,\mu_f\,\nu_m + \\
&\quad + 2\,\mu_m\,\nu_m\,\nu_f - 2\,\mu_f\,\nu_m\,\nu_f) \\
g &= (2 - 2\nu_m)
\end{aligned} \tag{6.4}$$

The subscripts $()_m$, $()_f$ refer to matrix and fiber, respectively. Assuming the fiber and matrix are both isotropic (Section 1.12.5), Lame constants of both materials are obtained by using (1.79) in terms of the Young's modulus E, the Poisson's ratio ν, and the shear modulus $G = \mu$.

For a composite reinforced by long circular cylindrical fibers, periodically arranged in a square array (Figure 6.1), aligned with x_1 axis, with $a_2 = a_3$, the constants S_3, S_6, S_7 are given as follows [8]

$$\begin{aligned}
S_3 &= 0.49247 - 0.47603V_f - 0.02748V_f^2 \\
S_6 &= 0.36844 - 0.14944V_f - 0.27152V_f^2 \\
S_7 &= 0.12346 - 0.32035V_f + 0.23517V_f^2
\end{aligned} \tag{6.5}$$

The resulting tensor $\mathbf{C}^*$ has square symmetry due to the microstructural periodic arrangement in the form of a square array. The tensor $\mathbf{C}^*$ is therefore described by six constants. However, most composites have random arrangement of the fibers (see Figure 1.12), resulting in a transversely isotropic stiffness tensor. A generalization for transversely isotropic materials is presented in Section 6.1.4, next.

6.1.4 Transversely Isotropic Averaging

In order to obtain a transversely isotropic stiffness tensor (Section 1.12.4), equivalent in the average sense to the stiffness tensor with square symmetry, the following averaging procedure is used. A rotation θ of the tensor $\mathbf{C}^*$ about the x_1-axis produces

$$\mathbf{B}(\theta) = \overline{T}^T(\theta)\mathbf{C}^*\overline{T}(\theta) \tag{6.6}$$

where $\overline{T}(\theta)$ is the coordinate transformation matrix (see (1.49)). Then the equivalent transversely isotropic tensor is obtained by averaging as follows

$$\overline{\mathbf{B}} = \frac{1}{\pi} \int_0^\pi \mathbf{B}(\theta) d\theta \qquad (6.7)$$

Then, using the relations between the engineering constants and the components of the $\overline{\mathbf{B}}$ tensor, the following expressions are obtained explicitly in terms of the coefficients (6.2-6.5) of the tensor $\mathbf{C}^*$

$$E_1 = C_{11}^* - \frac{2\,C_{12}^{*2}}{C_{22}^* + C_{23}^*}$$

$$E_2 = \frac{\left(2\,C_{11}^*\,C_{22}^* + 2\,C_{11}^*\,C_{23}^* - 4\,C_{12}^{*2}\right)\left(C_{22}^* - C_{23}^* + 2\,C_{44}^*\right)}{3\,C_{11}^*\,C_{22}^* + C_{11}^*\,C_{23}^* + 2\,C_{11}^*\,C_{44}^* - 4\,C_{12}^{*2}}$$

$$G_{12} = G_{13} = C_{66}^* \qquad (6.8)$$

$$\nu_{12} = \nu_{13} = \frac{C_{12}^*}{C_{22}^* + C_{23}^*}$$

$$\nu_{23} = \frac{C_{11}^*\,C_{22}^* + 3\,C_{11}^*\,C_{23}^* - 2\,C_{11}^*\,C_{44}^* - 4\,C_{12}^{*2}}{3\,C_{11}^*\,C_{22}^* + C_{11}^*\,C_{23}^* + 2\,C_{11}^*\,C_{44}^* - 4\,C_{12}^{*2}}$$

Note that the transverse shear modulus G_{23} can be written in terms of the other engineering constants as

$$G_{23} = \frac{C_{22}^*}{4} - \frac{C_{23}^*}{4} + \frac{C_{44}^*}{2} = \frac{E_2}{2(1 + \nu_{23})}$$

or directly in terms of μ_m, μ_f as

$$G_{23} = \mu_m - \frac{f}{4D}\left[\left(-\frac{aS_3}{2\,\mu_m\,c} + \frac{a(S_7 + S_6)}{2\,\mu_m\,gc} - \frac{ba + 2b^2 - a^2}{4\,c^2}\right)\right.$$

$$\left. + 2\left(-\frac{2\,S_3}{\mu_m} + (\mu_m - \mu_f)^{-1} + \frac{4\,S_7}{\mu_m(2 - 2\nu_m)}\right)^{-1}\right] \qquad (6.9)$$

where D is given by (6.3), a, b, c and g are given by (6.4) and S_3, S_6 and S_7 can be evaluated by (6.5). These equations are implemented in CADEC [2], PMMIE.m and PMMIE.xls [15].

Example 6.1 *Compute the elastic properties of a composite material reinforced with parallel cylindrical fibers randomly distributed in the cross section. The constituent properties are $E_f = 241$ GPa, $\nu_f = 0.2$, $E_m = 3.12$ GPa, $\nu_m = 0.38$, fiber volume fraction $V_f = 0.6$*

Solution to Example 6.1 *Using MATLAB program PPMIE.m or PMMIE.xls in [15], the results shown in Table 6.1 are obtained.*

Table 6.1: Lamina elastic properties

Young's Moduli	Shear Moduli	Poisson's Ratio
$E_1 = 145880$ MPa	$G_{12} = G_{13} = 4386$ MPa	$\nu_{12} = \nu_{13} = 0.263$
$E_2 = E_3 = 13312$ MPa	$G_{23} = 4529$ MPa	$\nu_{23} = 0.470$

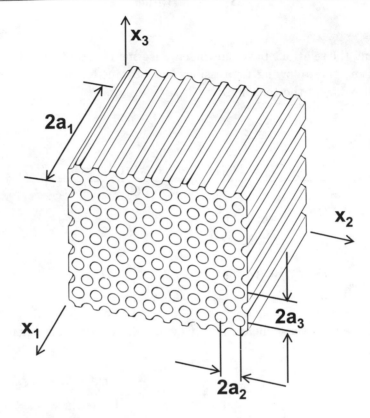

Fig. 6.2: Composite material with hexagonal array.

6.2   Numerical Homogenization

The composite material considered in this section has cylindrical fibers of infinite length, embedded in an elastic matrix, as shown in Figure 6.2. The cross section of the composite obtained by intersecting with a plane orthogonal to the fiber axis is shown in Figure 6.3, which clearly shows a periodic microstructure. Because of the periodicity, the three-dimensional representative volume element (RVE) shown in Figure 6.4 can be used for FE analysis.

In general, composites reinforced with parallel fibers display orthotropic material properties (Section 1.12.3) at the meso-scale (lamina level). In special cases, such as the hexagonal array shown in Figures 6.2 and 6.3, the properties become transversely isotropic (Section 1.12.4). In most commercially fabricated composites, it is impossible to control the placement of the fibers so precisely and most of the time the resulting microstructure is random, as shown in Figure 1.12. A random microstructure results in transversely isotropic properties at the meso-scale. The

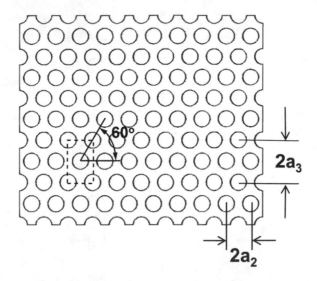

Fig. 6.3: Cross section of the composite material.

analysis of composites with random microstructure still can be done using a fictitious periodic microstructure, such as that shown in Figure 6.1, then averaging the
stiffness tensor $\mathbf{C}$ as in Section 6.1.4 to obtain the stiffness tensor of a transversely
isotropic material. A simpler alternative is to assume that the random microstructure is well approximated by the hexagonal microstructure displayed in Figure 6.3.
Analysis of such microstructure directly yields a transversely isotropic stiffness tensor, represented by (1.74), which is reproduced here for convenience

$$
\left\{
\begin{array}{c}
\bar{\sigma}_1 \\
\bar{\sigma}_2 \\
\bar{\sigma}_3 \\
\bar{\sigma}_4 \\
\bar{\sigma}_5 \\
\bar{\sigma}_6
\end{array}
\right\}
=
\left[
\begin{array}{cccccc}
C_{11} & C_{12} & C_{12} & 0 & 0 & 0 \\
C_{12} & C_{22} & C_{23} & 0 & 0 & 0 \\
C_{12} & C_{23} & C_{22} & 0 & 0 & 0 \\
0 & 0 & 0 & \frac{1}{2}(C_{22}-C_{23}) & 0 & 0 \\
0 & 0 & 0 & 0 & C_{66} & 0 \\
0 & 0 & 0 & 0 & 0 & C_{66}
\end{array}
\right]
\left\{
\begin{array}{c}
\bar{\epsilon}_1 \\
\bar{\epsilon}_2 \\
\bar{\epsilon}_3 \\
\bar{\gamma}_4 \\
\bar{\gamma}_5 \\
\bar{\gamma}_6
\end{array}
\right\}
\tag{6.10}
$$

where the 1-axis aligned with the fiber direction and an over-bar indicates the average computed over the volume of the RVE. Once the components of the transversely
isotropic tensor $\mathbf{C}$ are known, the five elastic properties of the homogenized material can be computed by (6.11), i.e. the longitudinal and transversal Young's moduli
E_1 and E_2, the longitudinal and transversal Poisson's ratios ν_{12} and ν_{23}, and the
longitudinal shear modulus G_{12}, as follows

$$
\begin{aligned}
E_1 &= C_{11} - 2C_{12}^2/(C_{22}+C_{23}) \\
\nu_{12} &= C_{12}/(C_{22}+C_{23}) \\
E_2 &= \left[C_{11}(C_{22}+C_{23}) - 2C_{12}^2\right](C_{22}-C_{23})/\left(C_{11}C_{22}-C_{12}^2\right) \\
\nu_{23} &= \left[C_{11}C_{23} - C_{12}^2\right]/\left(C_{11}C_{22}-C_{12}^2\right) \\
G_{12} &= C_{66}
\end{aligned}
\tag{6.11}
$$

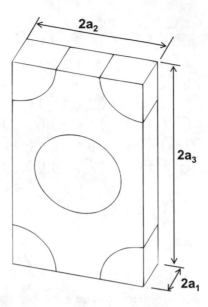

Fig. 6.4: Representative volume element (RVE).

The shear modulus G_{23} in the transversal plane can be obtained by the classical relation (1.78) or directly as follows

$$G_{23} = C_{44} = \frac{1}{2}\left(C_{22} - C_{23}\right) = \frac{E_T}{2\left(1 + \nu_T\right)} \tag{6.12}$$

In order to evaluate the overall elastic matrix $\mathbf{C}$ of the composite, the RVE is subjected to an average strain $\bar{\epsilon}_\beta$ [16]. The six components of strain ε_{ij}^0 are applied by enforcing the following boundary conditions on the displacement components

$$u_i\left(a_1, x_2, x_3\right) - u_i\left(-a_1, x_2, x_3\right) = 2a_1\varepsilon_{i1}^0 \qquad \begin{array}{l} -a_2 \le x_2 \le a_2 \\ -a_3 \le x_3 \le a_3 \end{array} \tag{6.13}$$

$$u_i\left(x_1, a_2, x_3\right) - u_i\left(x_1, -a_2, x_3\right) = 2a_2\varepsilon_{i2}^0 \qquad \begin{array}{l} -a_1 \le x_1 \le a_1 \\ -a_3 \le x_3 \le a_3 \end{array} \tag{6.14}$$

$$u_i\left(x_1, x_2, a_3\right) - u_i\left(x_1, x_2, -a_3\right) = 2a_3\varepsilon_{i3}^0 \qquad \begin{array}{l} -a_1 \le x_1 \le a_1 \\ -a_2 \le x_2 \le a_2 \end{array} \tag{6.15}$$

Note that tensor components of strain, defined in (1.5) are used in (6.13-6.15). Also, note that a superscript $()^0$ indicates an *applied* strain, while an over-line indicates a volume average. Furthermore, $2a_j\,\varepsilon_{ij}^0$ is the displacement necessary to enforce a strain ε_{ij}^0 over a distance $2a_j$ (Figure 6.4).

The strain ε_{ij}^0 applied on the boundary by using (6.13-6.15) results in a complex state of strain inside the RVE. However, the volume average of the strain in the RVE equals the applied strain, i.e.,

$$\bar{\varepsilon}_{ij} = \frac{1}{V}\int_V \varepsilon_{ij}dV = \varepsilon_{ij}^0 \tag{6.16}$$

For the homogeneous composite material, the relationship between average stress and strain is

$$\overline{\sigma}_\alpha = C_{\alpha\beta}\,\overline{\epsilon}_\beta \qquad (6.17)$$

where the relationship between $i, j = 1..3$ and $\beta = 1..6$ is given by the definition of contracted notation in (1.9). Thus, the components of the tensor $\mathbf{C}$ are determined solving six elastic models of the RVE subjected to the boundary conditions (6.13-6.15), where only one component of the strain ϵ_β^0 is different from zero for each of the six problems.

By choosing a unit value of applied strain, and once the problem defined by the boundary conditions (6.13-6.15) is solved, it is possible to compute the stress field σ_α, whose average gives the required components of the elastic matrix, one column at a time, as

$$C_{\alpha\beta} = \overline{\sigma}_\alpha = \frac{1}{V}\int_V \sigma_\alpha\,(x_1, x_2, x_3)\ dV \quad \text{with } \epsilon_\beta^0 = 1 \qquad (6.18)$$

where $\alpha, \beta = 1\ldots 6$ (see Section 1.5). The integrals (6.18) are evaluated within each finite element using the Gauss-Legendre quadrature. Commercial programs such as ANSYS have the capability to compute the average stress and volume, element by element. Therefore, computation of the integral (6.18) is a trivial matter. For more details see Example 6.1.

The coefficients in $\mathbf{C}$ are found by setting a different problem for each column in (6.10), as follows.

First Column of C

In order to determine the components C_{i1}, with $i = 1, 2, 3$, the following strain is applied to stretch the RVE in the fiber direction (x_1-direction)

$$\epsilon_1^o = 1 \qquad \epsilon_2^o = \epsilon_3^o = \gamma_4^o = \gamma_5^o = \gamma_6^o = 0 \qquad (6.19)$$

Thus, the displacement boundary conditions (6.13-6.15) for the RVE in Figure 6.4 become

$$
\begin{aligned}
u_1\,(+a_1, x_2, x_3) - u_1\,(-a_1, x_2, x_3) &= 2a_1 \\
u_2\,(+a_1, x_2, x_3) - u_2\,(-a_1, x_2, x_3) &= 0 \\
u_3\,(+a_1, x_2, x_3) - u_3\,(-a_1, x_2, x_3) &= 0
\end{aligned}
\qquad
\begin{aligned}
-a_2 \le x_2 \le a_2 \\
-a_3 \le x_3 \le a_3
\end{aligned}
$$

$$
u_i\,(x_1, +a_2, x_3) - u_i\,(x_1, -a_2, x_3) = 0
\qquad
\begin{aligned}
-a_1 \le x_1 \le a_1 \\
-a_3 \le x_3 \le a_3
\end{aligned}
\qquad (6.20)
$$

$$
u_i\,(x_1, x_2, +a_3) - u_i\,(x_1, x_2, -a_3) = 0
\qquad
\begin{aligned}
-a_1 \le x_1 \le a_1 \\
-a_2 \le x_2 \le a_2
\end{aligned}
$$

The conditions (6.20) are constraints on the relative displacements between opposite faces of the RVE. Because of the symmetries of the RVE and symmetry of

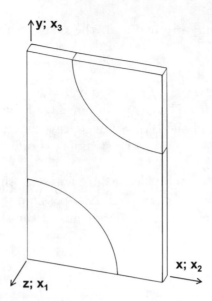

Fig. 6.5: One-eighth model of the RVE. Note that the model is set up with the fiber along the z-axis, which corresponds to the x_1-direction in the equations.

the constraints (6.20), only one-eighth of the RVE needs to be modeled in FEA. Assuming the top-right-front portion is modeled (Figure 6.5), the following equivalent external boundary conditions, i.e. boundary conditions on components of displacements and stresses, can be used

$$
\begin{array}{rcl}
u_1\left(a_1, x_2, x_3\right) & = & a_1 \\
u_1\left(0, x_2, x_3\right) & = & 0 \\
\sigma_{12}\left(a_1, x_2, x_3\right) & = & 0 \qquad 0 \le x_2 \le a_2 \\
\sigma_{12}\left(0, x_2, x_3\right) & = & 0 \qquad 0 \le x_3 \le a_3 \\
\sigma_{13}\left(a_1, x_2, x_3\right) & = & 0 \\
\sigma_{13}\left(0, x_2, x_3\right) & = & 0 \\
\\
u_2\left(x_1, a_2, x_3\right) & = & 0 \\
u_2\left(x_1, 0, x_3\right) & = & 0 \\
\sigma_{21}\left(x_1, a_2, x_3\right) & = & 0 \qquad 0 \le x_1 \le a_1 \\
\sigma_{21}\left(x_1, 0, x_3\right) & = & 0 \qquad 0 \le x_3 \le a_3 \\
\sigma_{23}\left(x_1, a_2, x_3\right) & = & 0 \\
\sigma_{23}\left(x_1, 0, x_3\right) & = & 0 \\
\\
u_3\left(x_1, x_2, a_3\right) & = & 0 \\
u_3\left(x_1, x_2, 0\right) & = & 0 \\
\sigma_{31}\left(x_1, x_2, a_3\right) & = & 0 \qquad 0 \le x_1 \le a_1 \\
\sigma_{31}\left(x_1, x_2, 0\right) & = & 0 \qquad 0 \le x_2 \le a_2 \\
\sigma_{32}\left(x_1, x_2, a_3\right) & = & 0 \\
\sigma_{32}\left(x_1, x_2, 0\right) & = & 0
\end{array}
\qquad (6.21)
$$

These boundary conditions are very easy to apply. Symmetry boundary conditions are applied on the planes $x_1 = 0$, $x_2 = 0$, $x_3 = 0$. Then, a uniform displacement is applied on the plane $x_1 = a_1$. The stress boundary conditions do not need to be applied explicitly in a displacement-based formulation. The displacement components in (6.21) represent strains that are non-zero along the x_1-direction and zero along the other two directions. The stress boundary conditions listed in (6.21) reflect the fact that, in the coordinate system used, the composite material is macroscopically orthotropic and that the constituent materials are orthotropic too. Therefore, there is no coupling between extension and shear strains. This is evidenced by the zero coefficients above the diagonal in columns 4 to 6 in (6.10).

The coefficients in column one of (6.10) are found by using (6.18), as follows

$$C_{\alpha 1} = \bar{\sigma}_\alpha = \frac{1}{V} \int_V \sigma_\alpha (x_1, x_2, x_3) \ dV \qquad (6.22)$$

Second Column of C

The components $C_{\alpha 2}$, with $\alpha = 1, 2, 3$, are determined by setting

$$\epsilon_2^o = 1 \qquad \epsilon_1^o = \epsilon_3^o = \gamma_4^o = \gamma_5^o = \gamma_6^o = 0 \qquad (6.23)$$

Thus, the following boundary conditions on displacements can be used

$$
\begin{aligned}
u_1 (a_1, x_2, x_3) &= 0 \\
u_1 (0, x_2, x_3) &= 0 \\
u_2 (x_1, a_2, x_3) &= a_2 \\
u_2 (x_1, 0, x_3) &= 0 \\
u_3 (x_1, x_2, a_3) &= 0 \\
u_3 (x_1, x_2, 0) &= 0
\end{aligned}
\qquad (6.24)
$$

The trivial stress boundary conditions have not been listed because they are automatically enforced by the displacement-based FEA formulation. Using (6.18), the stiffness terms in the second column of **C** are computed as

$$C_{\alpha 2} = \bar{\sigma}_\alpha = \frac{1}{V} \int_V \sigma_{\alpha 2} (x_1, x_2, x_3) \ dV \qquad (6.25)$$

Third Column of C

Because of transverse isotropy of the material (6.10), the components of the third column of the matrix **C** can be determined from the first and the second column, so no further computation is required. However, if desired, the components $C_{\alpha 3}$, with $\alpha = 1, 2, 3$, can be found by applying the following strain

$$\epsilon_3^o = 1 \qquad \epsilon_1^o = \epsilon_2^o = \gamma_4^o = \gamma_4^o = \gamma_5^o = 0 \qquad (6.26)$$

Thus, the following boundary conditions on displacement can be used

$$
\begin{aligned}
u_1\left(a_1, x_2, x_3\right) &= 0 \\
u_1\left(0, x_2, x_3\right) &= 0 \\
u_2\left(x_1, a_2, x_3\right) &= 0 \\
u_2\left(x_1, 0, x_3\right) &= 0 \\
u_3\left(x_1, x_2, a_3\right) &= a_3 \\
u_3\left(x_1, x_2, 0\right) &= 0
\end{aligned}
\tag{6.27}
$$

The required components of $\mathbf{C}$ are determined by averaging the stress field as in (6.18).

Example 6.2 *Compute E_1, E_2, ν_{12}, and ν_{23} for a unidirectional composite with isotropic fibers $E_f = 241$ GPa, $\nu_f = 0.2$, and isotropic matrix $E_m = 3.12$ GPa, $\nu_m = 0.38$ with fiber volume fraction $V_f = 0.4$. The fiber diameter is $d_f = 7$ μm, placed in an hexagonal array as shown in Figure 6.3.*

Solution to Example 6.2 *The dimensions a_2 and a_3 of the RVE, as shown in Figure 6.4, are chosen to obtain $V_f = 0.4$ with an hexagonal array microstructure. The fiber volume and the total volume of the RVE are*

$$
v_f = 4a_1\pi \left(\frac{d_f}{2}\right)^2 \quad ; \quad v_t = 2a_1\,2a_2\,2a_3
$$

The ratio between both is the volume fraction. Therefore,

$$
V_f = \pi \frac{(d_f/2)^2}{2\,a_2\,a_3} = 0.4
$$

Additionally, the relation between a_2 and a_3 is established by the hexagonal array pattern

$$
a_3 = a_2 \tan(60°)
$$

These two relations yield a_2 and a_3, while the a_1 dimension can be chosen arbitrarily. In this case, the RVE dimensions are

$$
a_1 = a_2/4 \quad ; \quad a_2 = 5.2701 \ \mu m \quad ; \quad a_3 = 9.1281 \ \mu m
$$

Since this RVE is symmetric, it is possible to model one-eighth of the RVE, as shown in Figure 6.5. The ANSYS command list below is used to model one-eighth of the RVE.

```
/TITLE,ONE-EIGHT Symmetric Model of RVE with hexagonal array fibers

rf=3.5       ! Radius fiber in microns
a2=5.2701    ! x2 length in microns
a3=9.1281    ! x3 length in microns
a1=a2/4      ! x1 length in microns

/PREP7                    ! Pre-processor module

MP,EX,1,0.241             ! Fiber material properties in TeraPascals [TPa]
MP,PRXY,1,0.2
```

```
MP,EX,2,3.12e-3        ! Matrix material properties in TeraPascals [TPa]
MP,PRXY,2,0.38

ET,1,SOLID186          ! Choose SOLID186 element type

BLOCK,0,a2,0,a3,0,a1 ! Geometry definition
CYLIND,rf,,0,a1,0,90
WPOFF,a2,a3
CYLIND,rf,,0,a1,180,270
VOVLAP,all             ! Overlap volumes
NUMCMP,all             ! Renumbering volumes

LSEL,u,loc,z,a1        ! Meshing Control
LSEL,u,loc,z,0
LESIZE,all,,,2         ! Number of divisions through the thickness
VSEL,s,,,1,2
ASLV,s
LSLA,s
LESIZE,all,,,6         ! Number of divisions on the fiber
LSEL,s,loc,y,a3
LSEL,a,loc,y,0
LESIZE,all,,,3         ! Number of divisions on the matrix
ALLSEL,all
LESIZE,all,,,8         ! Number of divisions on the matrix

MAT,1                  ! Associate material #1 with volumes 1 and 2
VMESH,1,2              ! Mesh area 1 and 2
MAT,2                  ! Associate material #2 with volume 3
VSWEEP,3              ! Mesh by sweep procedure area 3

FINISH                 ! Exit pre-processor module
```

The boundary conditions are defined in three load steps, which are then used to obtain the coefficients $C_{\alpha\beta}$ in columns one, two, and three. A unit strain is applied along each direction, each time. Equation (6.18) is then used to obtain the stiffness coefficients.

```
/SOLU                  ! Solution module
ANTYPE,STATIC          ! Set static analysis

lsclear,all            ! Boundary conditions Column 1
asel,s,loc,x,0         ! Model X direction = 2 material direction
asel,a,loc,x,a2
da,all,ux,0
asel,s,loc,y,0         ! Model Y direction = 3 material direction
asel,a,loc,y,a3
da,all,uy,0
asel,s,loc,z,0         ! Model Z direction = 1 material direction
da,all,uz,0
asel,s,loc,z,a1
da,all,uz,a1
asel,all
```

```
lswrite,1

lsclear,all            ! Boundary conditions Column 2
asel,s,loc,x,0         ! Model X direction = 2 material direction
DA,all,ux,0
asel,s,loc,x,a2
da,all,ux,a2
asel,s,loc,y,0         ! Model Y direction = 3 material direction
asel,a,loc,y,a3
da,all,uy,0
asel,s,loc,z,0         ! Model Z direction = 1 material direction
asel,a,loc,z,a1
da,all,uz,0
asel,all
lswrite,2

lsclear,all            ! Boundary conditions Column 3
asel,s,loc,x,0         ! Model X direction = 2 material direction
asel,a,loc,x,a2
da,all,ux,0
asel,s,loc,y,0         ! Model Y direction = 3 material direction
DA,all,uy,0
asel,s,loc,y,a3
da,all,uy,a3
asel,s,loc,z,0         ! Model Z direction = 1 material direction
asel,a,loc,z,a1
da,all,uz,0
asel,all
lswrite,3

LSSOLVE,1,3            ! Solve all load sets
FINISH                ! Exit solution module
```

The APDL language macro **srecover**, shown below, is defined in order to compute the average stress in the RVE.

```
*create,srecover !,mac   ! Create macro to calculate average stress
/nopr
ETABLE, ,VOLU,             ! Get element volume
ETABLE, ,S,X               ! Get element stress
ETABLE, ,S,Y
ETABLE, ,S,Z
ETABLE, ,S,XY
ETABLE, ,S,XZ
ETABLE, ,S,YZ
SMULT,SXV,VOLU,SX,1,1,      ! Stress by element volume
SMULT,SYV,VOLU,SY,1,1,
SMULT,SZV,VOLU,SZ,1,1,
SMULT,SXYV,VOLU,SXY,1,1,
SMULT,SXZV,VOLU,SXZ,1,1,
SMULT,SYZV,VOLU,SYZ,1,1,
```

```
SSUM
*get,totvol,ssum,,item,volu ! Integer stress along total volume
*get,totsx ,ssum,,item,sxv
*get,totsy ,ssum,,item,syv
*get,totsz ,ssum,,item,szv
*get,totsxy ,ssum,,item,sxyv
*get,totsxz ,ssum,,item,sxzv
*get,totsyz ,ssum,,item,syzv
Sxx0 = totsx/totvol     ! Compute average RVE stress
Syy0 = totsy/totvol
Szz0 = totsz/totvol
Sxy0 = totsxy/totvol
Sxz0 = totsxz/totvol
Syz0 = totsyz/totvol
/gopr
*end                    ! End of srecover macro
```

The coefficients $C_{\alpha\beta}$ and the equivalent engineering elastic constants are computed using the previous macro, as follows.

```
/POST1              ! Post-processor module

SET,1               ! First column coefficients
*use,srecover
C11 = Szz0
C21 = Sxx0
C31 = Syy0
SET,2               ! Second column coefficients
*use,srecover
C12 = Szz0
C22 = Sxx0
C32 = Syy0
SET,3               ! Third column coefficients
*use,srecover
C13 = Szz0
C23 = Sxx0
C33 = Syy0

EL=C11-2*C12*C21/(C22+C23)            ! Longitudinal E1 modulus
nuL=C12/(C22+C23)                     ! 12 Poisson coefficient
ET=(C11*(C22+C23)-2*C12*C12)*(C22-C23)/(C11*C22-C12*C21)
                                      ! Transversal E2 modulus
nuT=(C11*C23-C12*C21)/(C11*C22-C12*C21) ! 23 Poisson coefficient
GT=(C22-C23)/2 ! or GT=ET/2/(1+nuT)   ! 23 Shear stiffness

FINISH              ! Exit post-processor module
```

The results are shown in Table 6.2.

Table 6.2: Equivalent elastic properties of the unidirectional lamina

Young's Moduli	Poisson's Ratio
$E_1 = 98300$ MPa	$\nu_{12} = \nu_{13} = 0.299$
$E_2 = E_3 = 7482$ MPa	$\nu_{23} = 0.540$

Fourth Column of C

For a transversally isotropic material, according to (6.10), only the term C_{44} is expected to be different from zero and it can be determined as a function of the other components, so no further computation is needed. Therefore, it can be determined as

$$C_{44} = \frac{1}{2}(C_{22} - C_{23}) \tag{6.28}$$

If the material is orthotropic, a procedure similar to that used for column number six must be used.

Fifth Column of C

For a transversally isotropic material, according to (6.10), only the term C_{55} is different from zero and it is equal to C_{66}, which can be found from column number six. If the material is orthotropic, a procedure similar to that used for column number six must be used.

Sixth Column of C

Because of the lack of symmetry of the loads, in this case it is not possible to use boundary conditions as was done for the first three columns. Thus, the boundary conditions must be enforced by using coupling constraint equations (called CE in most FEA commercial packages).

According to (6.10), only the term C_{66} is different from zero. The components $C_{\alpha 6}$ are determined by setting

$$\gamma_6^0 = \varepsilon_{12}^0 + \varepsilon_{21}^0 = 1.0 \qquad \epsilon_1^0 = \epsilon_2^0 = \epsilon_3^0 = \gamma_4^0 = \gamma_5^0 = 0 \tag{6.29}$$

Note that $\varepsilon_{12}^0 = 1/2$ is applied between $x_1 = \pm a_1$ and another one-half is applied between $x_2 = \pm a_2$. In this case, the CE applied between two periodic faces (except points in the edges and vertices) are given as a particular case of (6.13-6.15) as follows

$$u_1\,(a_1, x_2, x_3) - u_1\,(-a_1, x_2, x_3) = 0$$
$$u_2\,(a_1, x_2, x_3) - u_2\,(-a_1, x_2, x_3) = a_1$$
$$u_3\,(a_1, x_2, x_3) - u_3\,(-a_1, x_2, x_3) = 0$$
$$-a_2 < x_2 < a_2$$
$$-a_3 < x_3 < a_3$$

$$u_1\,(x_1, a_2, x_3) - u_1\,(x_1, -a_2, x_3) = a_2$$
$$u_2\,(x_1, a_2, x_3) - u_2\,(x_1, -a_2, x_3) = 0$$
$$u_3\,(x_1, a_2, x_3) - u_3\,(x_1, -a_2, x_3) = 0$$
$$-a_1 < x_1 < a_1$$
$$-a_3 < x_3 < a_3$$

$$(6.30)$$

$$u_1\,(x_1, x_2, a_3) - u_1\,(x_1, x_2, -a_3) = 0$$
$$u_2\,(x_1, x_2, a_3) - u_2\,(x_1, x_2, -a_3) = 0$$
$$u_3\,(x_1, x_2, a_3) - u_3\,(x_1, x_2, -a_3) = 0$$
$$-a_1 < x_1 < a_1$$
$$-a_2 < x_2 < a_2$$

Note that (6.30) are applied between opposite points on the faces of the RVE but not on edges and vertices. In FEA, CE are applied between degrees of freedom (DOF). Once a DOF has been used in a CE, it cannot be used in another CE. For example, the first of (6.30) for $x_2 = a_2$ becomes

$$u_1(a_1, a_2, x_3) - u_1(-a_1, a_2, x_3) = 0 \qquad (6.31)$$

The DOF associated to $u_1(a_1, a_2, x_3)$ (for all $-a_3 < x_3 < a_3$) are eliminated because they are identical to $u_1(-a_1, a_2, x_3)$, as required by (6.31) and enforced by a CE based on the same. Once the DOF are eliminated, they cannot be used in another CE. For example, the fourth of (6.30) at $x_1 = a_1$ is

$$u_1(a_1, a_2, x_3) - u_1(a_1, -a_2, x_3) = 0 \qquad (6.32)$$

but this CE cannot be enforced because the DOF associated to $u_1(a_1, a_2, x_3)$ have been eliminated by the CE associated to (6.31). As a corollary, constraint equations on the edges and vertices of the RVE must be written separately from (6.30). Furthermore, only three equations, one for each component of displacement u_i can be written between a pair of edges or pair of vertices. Simply put, there are only three displacements that can be used to enforce periodicity conditions.

For pairs of edges, the task at hand is to reduce the first six equations of (6.30) to three equations that can be applied between pairs of edges for the interval $-a_3 < x_3 < a_3$. Note that the new equations will not be applied at $x_3 = \pm a_3$ because those are vertices, which will be dealt with separately. Therefore, the last three equations of (6.30) are inconsequential at this point.

The only way to reduce six equations to three, in terms of six unique DOF, is to add the equations for diagonally opposite edges. Figure 6.6 is a top view of the RVE looking from the positive x_3 axis. Point A in Figure 6.6 represents the edge formed by the planes $x_1 = a_1$ and $x_2 = a_2$. This location is constrained by the first of (6.30) at that location, which is precisely (6.31). Point C in Figure 6.6 represents the edge formed by the planes $x_1 = -a_1$ and $x_2 = -a_2$. This location is constrained by the fourth of (6.30), which at that location reduces to

$$u_1(-a_1, a_2, x_3) - u_1(-a_1, -a_2, x_3) = a_2 \qquad (6.33)$$

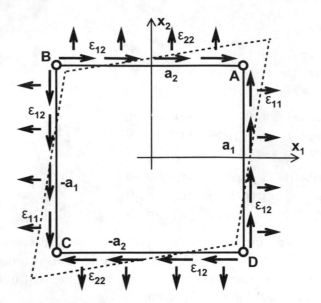

Fig. 6.6: Top view of the RVE showing that two displacements (vertical and horizontal) must be applied at edges to impose shear strain (shown as points A, B, C and D in the figure).

Adding (6.31) and (6.33) yields a single equation as follows

$$u_1(a_1, a_2, x_3) - u_1(-a_1, -a_2, x_3) = a_2 \qquad (6.34)$$

Repeating the procedure for the components u_2 and u_3, and grouping the resulting equations with (6.34) results in

$$
\begin{aligned}
u_1\,(a_1, a_2, x_3) - u_1\,(-a_1, -a_2, x_3) &= a_2 \\
u_2\,(a_1, a_2, x_3) - u_2\,(-a_1, -a_2, x_3) &= a_1 \qquad -a_3 < x_3 < a_3 \qquad (6.35) \\
u_3\,(a_1, a_2, x_3) - u_3\,(-a_1, -a_2, x_3) &= 0
\end{aligned}
$$

Considering (6.30) between edges B and D in Figure 6.6 results in

$$
\begin{aligned}
u_1\,(a_1, -a_2, x_3) - u_1\,(-a_1, a_2, x_3) &= -a_2 \\
u_2\,(a_1, -a_2, x_3) - u_2\,(-a_1, a_2, x_3) &= a_1 \qquad -a_3 < x_3 < a_3 \qquad (6.36) \\
u_3\,(a_1, -a_2, x_3) - u_3\,(-a_1, a_2, x_3) &= 0
\end{aligned}
$$

The planes $x_1 = \pm a_1$ and $x_3 = \pm a_3$ define two pairs of edges restrained by the following six CE

$$
\begin{aligned}
u_1\left(+a_1, x_2, +a_3\right) - u_1\left(-a_1, x_2, -a_3\right) &= 0 \\
u_2\left(+a_1, x_2, +a_3\right) - u_2\left(-a_1, x_2, -a_3\right) &= a_1 \qquad -a_2 < x_2 < a_2 \\
u_3\left(+a_1, x_2, +a_3\right) - u_3\left(-a_1, x_2, -a_3\right) &= 0 \\[6pt]
u_1\left(+a_1, x_2, -a_3\right) - u_1\left(-a_1, x_2, +a_3\right) &= 0 \\
u_2\left(+a_1, x_2, -a_3\right) - u_2\left(-a_1, x_2, +a_3\right) &= a_1 \qquad -a_2 < x_2 < a_2 \\
u_3\left(+a_1, x_2, -a_3\right) - u_3\left(-a_1, x_2, +a_3\right) &= 0
\end{aligned}
\tag{6.37}
$$

The six CE for the two pairs of edges defined by the planes $x_2 = \pm a_2$ and $x_3 = \pm a_3$ are

$$
\begin{aligned}
u_1\left(x_1, +a_2, +a_3\right) - u_1\left(x_1, -a_2, -a_3\right) &= a_2 \\
u_2\left(x_1, +a_2, +a_3\right) - u_2\left(x_1, -a_2, -a_3\right) &= 0 \qquad -a_1 < x_1 < a_1 \\
u_3\left(x_1, +a_2, +a_3\right) - u_3\left(x_1, -a_2, -a_3\right) &= 0 \\[6pt]
u_1\left(x_1, +a_2, -a_3\right) - u_1\left(x_1, -a_2, +a_3\right) &= -a_2 \\
u_2\left(x_1, +a_2, -a_3\right) - u_2\left(x_1, -a_2, +a_3\right) &= 0 \qquad -a_1 < x_1 < a_1 \\
u_3\left(x_1, +a_2, -a_3\right) - u_3\left(x_1, -a_2, +a_3\right) &= 0
\end{aligned}
\tag{6.38}
$$

Note that (6.35-6.38) are not applied at the vertices because redundant CE would appear among pairs of vertices that are located symmetrically with respect to the center of the RVE's volume. Therefore, each of the four pairs of vertices need to be constrained one at a time. The resulting CE are as follows

$$
\begin{aligned}
u_1(+a_1, +a_2, +a_3) - u_1(-a_1, -a_2, -a_3) &= a_1 \\
u_2(+a_1, +a_2, +a_3) - u_2(-a_1, -a_2, -a_3) &= a_2 \\
u_3(+a_1, +a_2, +a_3) - u_3(-a_1, -a_2, -a_3) &= 0 \\[6pt]
u_1(+a_1, +a_2, -a_3) - u_1(-a_1, -a_2, +a_3) &= -a_1 \\
u_2(+a_1, +a_2, -a_3) - u_2(-a_1, -a_2, +a_3) &= -a_2 \\
u_3(+a_1, +a_2, -a_3) - u_3(-a_1, -a_2, +a_3) &= 0 \\[6pt]
u_1(-a_1, +a_2, +a_3) - u_1(+a_1, -a_2, -a_3) &= -a_1 \\
u_2(-a_1, +a_2, +a_3) - u_2(+a_1, -a_2, -a_3) &= a_2 \\
u_3(-a_1, +a_2, +a_3) - u_3(+a_1, -a_2, -a_3) &= 0 \\[6pt]
u_1(+a_1, -a_2, +a_3) - u_1(-a_1, +a_2, -a_3) &= a_1 \\
u_2(+a_1, -a_2, +a_3) - u_2(-a_1, +a_2, -a_3) &= -a_2 \\
u_3(+a_1, -a_2, +a_3) - u_3(-a_1, +a_2, -a_3) &= 0
\end{aligned}
\tag{6.39}
$$

Equations (6.30-6.39) constrain the volume of the RVE with a unit strain given by (6.29). The FEA of this model yields all the component of stress. As discussed previously, element by element averages of these components of stress are available from the FEA (see macro `srecover` in Example 6.1) or they can be easily computed by post-processing. Therefore, the coefficient C_{66}, for this case is found using (6.18) written as

$$
C_{66} = \overline{\sigma}_6 = \frac{1}{V} \int_V \sigma_6\left(x_1, x_2, x_3\right)\, dV \quad \text{with } \gamma_6^0 = 1
\tag{6.40}
$$

Finally, the elastic properties of the composite are determined using (6.11).

Example 6.3 *Compute G_{12} for the composite in Example 6.1.*

Solution to Example 6.3 *To compute $G_{12} = C_{66}$, the RVE shown in Figure 6.4 must be used along with the CE explained in Eqs. (6.30-6.39).*
The dimensions to define the RVE are the same used in Example 6.1. Therefore, the fiber diameter is $d_f = 7$ μm and the RVE dimensions are

$$a_1 = a_2/4 \quad ; \quad a_2 = 5.2701 \ \mu m \quad ; \quad a_3 = 9.1281 \ \mu m$$

See the ANSYS command list below to model the whole RVE.

```
/TITLE, Full Model of RVE with hexagonal array fibers

rf=3.5                 ! Radius fiber in microns
a2=5.2701              ! x2 length in microns
a3=9.1281              ! x3 length in microns
a1=a2/4                ! x1 length in microns

/PREP7                 ! Pre-processor module

MP,EX,1,0.241          ! Fiber material properties in TeraPascals [TPa]
MP,PRXY,1,0.2
MP,EX,2,3.12e-3        ! Matrix material properties in TeraPascals [TPa]
MP,PRXY,2,0.38

ET,1,SOLID186          ! Choose SOLID186 element type

BLOCK,-a2,a2,-a3,a3,-a1,a1,
CYLIND,rf, ,-a1,a1,  0, 90,
CYLIND,rf, ,-a1,a1, 90,180,
CYLIND,rf, ,-a1,a1,180,270,
CYLIND,rf, ,-a1,a1,270,360,
CYLIND,rf, ,-a1,a1,  0, 90,
CYLIND,rf, ,-a1,a1, 90,180,
CYLIND,rf, ,-a1,a1,180,270,
CYLIND,rf, ,-a1,a1,270,360,
vgen,1,6,,,-a2,-a3,,,,1
vgen,1,7,,, a2,-a3,,,,1
vgen,1,8,,, a2, a3,,,,1
vgen,1,9,,,-a2, a3,,,,1
allsel,all
VOVLAP,all       ! Overlap volumes
NUMCMP,all       ! Renumbering all volumes, volume 9 is the matrix

LSEL,u,loc,z,a1       ! Meshing Control
LSEL,u,loc,z,-a1
LESIZE,all,,,4        ! Number of divisions through the thickness
VSEL,s,,,1,8
ASLV,s
```

```
LSLA,s
LESIZE,all,,,6        ! Number of divisions on the fiber
LSEL,s,loc,y,a3
LSEL,a,loc,y,-a3
LESIZE,all,,,6        ! Number of divisions on the matrix
ALLSEL,all
LESIZE,all,,,16       ! Number of divisions on the matrix

MAT,1                 ! Associate material #1 with volumes 1 and 2
VMESH,1,8             ! Mesh area 1 and 2
MAT,2                 ! Associate material #2 with volume 3
VSWEEP,9              ! Mesh by sweep procedure area 3

FINISH                ! Exit pre-processor module
```

The APDL macro ceRVE.mac, *available in [15], is used to define the CE and to implement (6.30-6.39). The macro is made available on the Web site [15] because it is too long to be printed here. The RVE dimensions and the applied strain are input arguments to the macro. In this example, only a strain $\gamma_6 = 1$ is applied.*

```
/SOLU                 ! Solution module
! ceRVe arguments:
! a1,a2,a3,eps1,eps2,eps3,eps4,eps5,eps6
    *use,ceRVE,a1,a2,a3,0,0,0,0,0,1.
    SOLVE             ! Solve analysis
FINISH                ! Exit solution module
```

To compute the average stress in the RVE, it is possible to use the macro **srecover**, *shown in Example 6.1. On account of the applied strain being equal to unity, the computed average stress is equal to C_{66}. Therefore, $G_{12} = C_{66} = 2584$ MPa.*

```
/POST1 ! Post-processor module
    *use,srecover
    C66 = Sxz0
FINISH                     ! Exit post-processor module
```

6.3   Local-Global Analysis

In local-global analysis (Figure 6.7), an RVE is used to compute the stress for a given strain at each Gauss integration point in the *global model*. The global model is used to compute the displacements and resulting strains, assuming that the material is homogeneous. The *local model* takes the inhomogeneities into account by modeling them with an RVE.

Equations (6.13-6.15) are used in Section 6.2 to enforce one component of strain at a time, with the objective of finding the equivalent elastic properties of the material. Equations (6.13-6.15) are still valid for a general state of strain applied to the RVE but care must be taken with the specification of periodic boundary

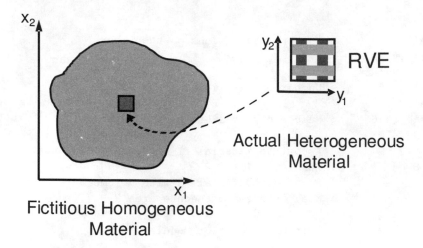

Fig. 6.7: Local-global analysis using RVE.

conditions at the edges and vertices, as discussed in page 155. Equations (6.13-6.15) are nine constraint equations that can be imposed between all the pairs of periodic points on the faces of the RVE except on the edges and vertices.

On the faces $x_1 = \pm a_1$, u_1 is used to impose ε_{11}^0, u_2 is used to impose $\varepsilon_{21}^0 = \gamma_6/2$, and u_3 is used to impose $\varepsilon_{31}^0 = \gamma_5/2$. To achieve this, (6.13) is expanded into its three components, using tensor notation for strains, as follows

$$
\begin{aligned}
u_1(a_1, x_2, x_3) - u_1(-a_1, x_2, x_3) &= 2a_1\varepsilon_{11}^0 \\
u_2(a_1, x_2, x_3) - u_2(-a_1, x_2, x_3) &= 2a_2\varepsilon_{21}^0 \\
u_3(a_1, x_2, x_3) - u_3(-a_1, x_2, x_3) &= 2a_3\varepsilon_{31}^0
\end{aligned}
\tag{6.41}
$$

On the faces $x_2 = \pm a_2$, u_1 is used to impose $\varepsilon_{12}^0 = \gamma_6/2$, u_2 is used to impose ε_{22}^0, and u_3 is used to impose $\varepsilon_{32}^0 = \gamma_4/2$. Therefore, (6.14) is expanded into its three components, using tensor notation for strains, as follows

$$
\begin{aligned}
u_1(x_1, a_2, x_3) - u_1(x_1, -a_2, x_3) &= 2a_1\varepsilon_{12} \\
u_2(x_1, a_2, x_3) - u_2(x_1, -a_2, x_3) &= 2a_2\varepsilon_{22} \\
u_3(x_1, a_2, x_3) - u_3(x_1, -a_2, x_3) &= 2a_3\varepsilon_{32}
\end{aligned}
\tag{6.42}
$$

On the faces $x_3 = \pm a_3$, u_1 is used to impose $\varepsilon_{13}^0 = \gamma_5/2$, u_2 is used to impose $\varepsilon_{23}^0 = \gamma_4/2$, and u_3 is used to impose ε_{33}^0. Therefore, (6.15) is expanded into its three components, using tensor notation for strains, as follows

$$
\begin{aligned}
u_1(x_1, x_2, a_3) - u_1(x_1, x_2, -a_3) &= 2a_1\varepsilon_{13} \\
u_2(x_1, x_2, a_3) - u_2(x_1, x_2, -a_3) &= 2a_2\varepsilon_{23} \\
u_3(x_1, x_2, a_3) - u_3(x_1, x_2, -a_3) &= 2a_3\varepsilon_{33}
\end{aligned}
\tag{6.43}
$$

Since each edge belongs to two faces, on every edge, it would seem that each component of displacement would be used to impose two CE, one from each face, as given by (6.41-6.43). However, as discussed in page 155, only one CE can be written for each component of displacement. Therefore, edges must be dealt with separately. Similarly, since three faces converge at a vertex, three periodic CE, one from each face, need to be imposed using a single component of displacement. Following a derivation similar to that presented in page 155, the following is obtained.

The planes $x_1 = \pm a_1$ and $x_2 = \pm a_2$ define two pairs of edges, for which (6.41-6.43) reduce to the following six equations (with $i = 1, 2, 3$), as follows

$$u_i(+a_1, +a_2, x_3) - u_i(-a_1, -a_2, x_3) - 2a_1\varepsilon_{i1} - 2a_2\varepsilon_{i2} = 0$$
$$u_i(+a_1, -a_2, x_3) - u_i(-a_1, +a_2, x_3) - 2a_1\varepsilon_{i1} + 2a_2\varepsilon_{i2} = 0 \qquad (6.44)$$

The planes $x_1 = \pm a_1$ and $x_3 = \pm a_3$ define two pairs of edges, for which (6.41-6.43) reduce to the following six equations (with $i = 1, 2, 3$), as follows

$$u_i(+a_1, x_2, +a_3) - u_i(-a_1, x_2, -a_3) - 2a_1\varepsilon_{i1} - 2a_3\varepsilon_{i3} = 0$$
$$u_i(+a_1, x_2, -a_3) - u_i(-a_1, x_2, +a_3) - 2a_1\varepsilon_{i1} + 2a_3\varepsilon_{i3} = 0 \qquad (6.45)$$

The planes $x_2 = \pm a_2$ and $x_3 = \pm a_3$ define two pairs of edges, for which (6.41-6.43) reduce to the following six equations (with $i = 1, 2, 3$), as follows

$$u_i(x_1, +a_2, +a_3) - u_i(x_1, -a_2, -a_3) - 2a_2\varepsilon_{i2} - 2a_3\varepsilon_{i3} = 0$$
$$u_i(x_1, +a_2, -a_3) - u_i(x_1, -a_2, +a_3) - 2a_2\varepsilon_{i2} + 2a_3\varepsilon_{i3} = 0 \qquad (6.46)$$

Four pairs of corners need to be analyzed one at a time. For each pair, the corners are located symmetrically with respect to the center of the RVE located at coordinates $(0, 0, 0)$. The resulting CE are as follows

$$u_i(+a_1, +a_2, +a_3) - u_i(-a_1, -a_2, -a_3) - 2a_1\varepsilon_{i2} - 2a_2\varepsilon_{i2} - 2a_2\varepsilon_{i3} = 0$$
$$u_i(+a_1, +a_2, -a_3) - u_i(-a_1, -a_2, +a_3) - 2a_1\varepsilon_{i2} - 2a_2\varepsilon_{i2} + 2a_2\varepsilon_{i3} = 0$$
$$u_i(-a_1, +a_2, +a_3) - u_i(+a_1, -a_2, -a_3) + 2a_1\varepsilon_{i2} - 2a_2\varepsilon_{i2} - 2a_2\varepsilon_{i3} = 0$$
$$u_i(+a_1, -a_2, +a_3) - u_i(-a_1, +a_2, -a_3) - 2a_1\varepsilon_{i2} + 2a_2\varepsilon_{i2} - 2a_2\varepsilon_{i3} = 0$$

$$(6.47)$$

Example 6.4 *Apply $\epsilon_2 = 0.2\ \%$ and $\gamma_4 = 0.1\ \%$ simultaneously to the composite in Example 6.1. Compute the maximum σ_2 and σ_{12} stresses on the matrix, and compute the average $\bar{\sigma}_2$ and $\bar{\sigma}_{12}$ in the RVE.*

Solution to Example 6.4 *The same procedure used in Example 6.3 is used to define the model. The APDL macro `ceRVE.mac` available in [15] is used to define the CE. The macro needs the RVE dimensions and the applied strain as input arguments. In this example, components of strain $\epsilon_2 = 0.2\ \%$ and $\gamma_4 = 0.1\ \%$ are applied, as follows*

```
/SOLU                ! Solution module
! units: TeraPascals, and microns, eps non-dimensional
! ceRVe arguments:
! a1,a2,a3,eps1,eps2,eps3,eps4,eps5,eps6
    *use,ceRVE,a1,a2,a3,0,2e-3,0,1e-3,0,0
    SOLVE                ! Solve analysis
FINISH               ! Exit solution module
```

The macro `srecover` *is used to compute the average stress of the RVE. The maximum stress in the RVE can be computed using the commands* PLESOL,S,1 *or* PRESOL,S,PRIN.

```
/POST1               ! Post-processor module
*use,srecover        ! Compute average properties
    S_1 = Szz0
    S_2 = Sxx0
    S_3 = Syy0
    S_4 = Sxy0
    S_5 = Syz0
    S_6 = Sxz0
VSEL,s,,,9
ESLV,S
plesol,s,x,1         ! Contour plot of S2 on matrix
plesol,s,xy,1        ! Contour plot of S23 on matrix
FINISH               ! Exit post-processor module
```

The results obtained are shown in Table 6.3.

Table 6.3: Maximum stress on the matrix and average stress in the RVE

Average Results	Maximum on the Matrix
$\sigma_2 = 10.0$ MPa	$\bar{\sigma}_2 = 29.5$ MPa
$\sigma_{12} = 2.42$ MPa	$\bar{\sigma}_{12} = 6.07$ MPa

6.4   Laminated RVE

A similar procedure to that used to obtain the RVE at the micro-scale can be used to analyze laminates on the meso-scale. In this case the RVE represents a laminate. Therefore, the through-thickness direction should remain free to expand along the thickness. For example, with layers parallel to the x-y plane, then $\sigma_z = 0$ and (6.15) is not enforced, so that the thickness coordinate is free to contract (see Figure 6.8). In general, the RVE must include the whole thickness. For symmetrical laminates subjected to in-plane loads, the RVE can be defined with half the thickness using symmetry boundary conditions (see Example 6.5).

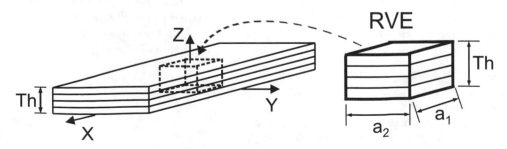

Fig. 6.8: Laminated RVE.

The CE for a laminated RVE are simpler. Only (6.13) and (6.14) must be enforced. In an hexahedral RVE, such as shown in Figure 6.8, only four faces ($x_1 = \pm a_1$ and $x_2 = \pm a_2$) and the four edges defined by these faces need to be considered.

Therefore, in a laminated RVE the constraint equations (6.13) and (6.14) become the following. On the periodic pair of faces $x_1 = \pm a_1$ and on the pair of faces $x_2 = \pm a_2$, the CE are

$$
\begin{aligned}
u_1(a_1, x_2, x_3) - u_1(-a_1, x_2, x_3) - 2a_1\varepsilon_{11} &= 0 \\
u_2(a_1, x_2, x_3) - u_2(-a_1, x_2, x_3) - 2a_2\varepsilon_{21} &= 0 \\
u_3(a_1, x_2, x_3) - u_3(-a_1, x_2, x_3) - 2a_3\varepsilon_{31} &= 0
\end{aligned}
\tag{6.48}
$$

$$
\begin{aligned}
u_1(x_1, a_2, x_3) - u_1(x_1, -a_2, x_3) - 2a_1\varepsilon_{12} &= 0 \\
u_2(x_1, a_2, x_3) - u_2(x_1, -a_2, x_3) - 2a_2\varepsilon_{22} &= 0 \\
u_3(x_1, a_2, x_3) - u_3(x_1, -a_2, x_3) - 2a_3\varepsilon_{32} &= 0
\end{aligned}
\tag{6.49}
$$

The planes $x_1 = \pm a_1$ and $x_2 = \pm a_2$ define two pairs of periodic edges, for which (6.13-6.14) reduce to the following equations,

$$
\begin{aligned}
u_1(+a_1, +a_2, x_3) - u_1(-a_1, -a_2, x_3) - 2a_1\varepsilon_{11} - 2a_2\varepsilon_{12} &= 0 \\
u_2(+a_1, +a_2, x_3) - u_2(-a_1, -a_2, x_3) - 2a_1\varepsilon_{21} - 2a_2\varepsilon_{22} &= 0 \\
u_3(+a_1, +a_2, x_3) - u_3(-a_1, -a_2, x_3) - 2a_3\varepsilon_{31} &= 0
\end{aligned}
\tag{6.50}
$$

$$
\begin{aligned}
u_1(+a_1, -a_2, x_3) - u_1(-a_1, +a_2, x_3) - 2a_1\varepsilon_{21} + 2a_2\varepsilon_{12} &= 0 \\
u_2(+a_1, -a_2, x_3) - u_2(-a_1, +a_2, x_3) - 2a_1\varepsilon_{21} + 2a_2\varepsilon_{22} &= 0 \\
u_3(+a_1, -a_2, x_3) - u_3(-a_1, +a_2, x_3) - 2a_3\varepsilon_{32} &= 0
\end{aligned}
\tag{6.51}
$$

For in-plane analysis, $\varepsilon_{31} = \varepsilon_{32} = 0$ and the third equation in (6.48)-(6.51) are automatically satisfied.

Example 6.5 *Compute G_{xy} for a $[0/90/-45/45]_S$ laminate with properties $E_1 = 139$ GPa, $E_2 = 14.5$ GPa, $G_{12} = G_{13} = 5.86$ GPa, $G_{23} = 2.93$ GPa, $\nu_{12} = \nu_{13} = 0.21$, $\nu_{23} = 0.38$ and layer thickness $t_k = 1.25mm$.*

Solution to Example 6.5 *A shear strain $\gamma^0_{xy} = 1$ is applied to the RVE. The laminate shear stiffness G_{xy} is obtained directly by computing the average stress in the RVE.*

As a result of laminate symmetry and in-plane load, an RVE of half thickness with symmetry boundary conditions in $z = 0$ can be used. The following ANSYS commands define the model and the laminate.

```
/TITLE,RVE of [0/90/-45/45]s laminate /PREP7
! Pre-processor module

th =1.25           ! Thickness of lamina in mm
a1 =1              ! Half length of RVE in x direction
a2 =1              ! Half length of RVE in y direction

! Equivalent Material properties
uimp,1,ex,ey,ez,139e3,14.5e3,14.5e3
```

```
uimp,1,gxy,gyz,gxz,5.86e3,2.93e3,5.86e3
uimp,1,prxy,pryz,prxz,0.21,0.38,0.21

ET,1,SOLID186                 ! Chooses SOLID186 element for analysis

! Define material orientation by local Coordinate
local,11,,0,0,0,45            ! defines 45 degree local cs
local,12,,0,0,0,-45           ! defines -45 degree local cs
local,13,,0,0,0,0             ! defines 0 degree local cs
local,14,,0,0,0,90            ! defines 90 degree local cs
CSYS,0                        ! set active cs to cart. system

! Generate Geometry
BLOCK,-a1,a1,-a2,a2,0,th      !  45 degrees layer
BLOCK,-a1,a1,-a2,a2,1*th,2*th ! -45 degrees layer
BLOCK,-a1,a1,-a2,a2,2*th,3*th !  90 degrees layer
BLOCK,-a1,a1,-a2,a2,3*th,4*th !   0 degrees layer
VGLUE,ALL                     ! Glue volumes

! Mesh Control and Mesh
NUMCMP,ALL
lesize,all,,,2
ESYS,11                       ! Selects 45 degrees material orientation
VMESH,1                       ! Meshes 45 degrees layer
ESYS,12                       ! Selects -45 degrees material orientation
VMESH,2                       ! Meshes -45 degree layer
ESYS,13                       ! Selects 0 degrees material orientation
VMESH,3                       ! Meshes 0 degree layer
ESYS,14                       ! Selects 90 degrees material orientation
VMESH,4                       ! Meshes 90 degree layer
FINISH                        ! Exit pre-processor module
```

The APDL macro **ceRVElaminate.mac** *available in [15] is used to define the CE, thus implementing (6.48-6.51). The macro needs the RVE dimensions and the applied strain as input arguments. In this example, only a strain $\gamma_{xy}^0 = 1$ is applied.*

```
/SOLU                 ! Solution module,
ANTYPE,STATIC         ! Set static analysis

NSEL,S,LOC,Z,0
D,all,UZ        ! Symmetry z=0
NSEL,R,LOC,Y,0
NSEL,R,LOC,X,0
D,all,all
NSEL,all
! ceRVElaminate arguments:
! a1,a2,epsX,epsY,epsXY
    *use,ceRVElaminate,a1,a2,0,0,1
SOLVE                 ! Solve analysis
FINISH                ! Exit solution module
```

To compute the average stress along the RVE, it is possible to use the macro `srecover`, used in Example 6.1. On account of the applied strain being equal to unity, the computed average stress is equal to C_{66}. Therefore, $G_{12} = 21,441$ MPa.

```
/POST1              ! Post-processor module
*use,srecover
G_xy =  Sxy0
PLESOL,s,xy,1
FINISH                    ! Exit post-processor module
```

Suggested Problems

Problem 6.1 *Compute the stiffness matrix components for a unidirectional composite with isotropic fibers $E_f = 241$ GPa, $\nu_f = 0.2$, and isotropic matrix $E_m = 3.12$ GPa, $\nu_m = 0.38$ with fiber volume fraction $V_f = 0.4$. The fiber diameter is $d_f = 7$ μm, placed in a square array as shown in Figure 6.1. Chose an RVE with vertical faces $x_2 = \pm a_2$ and horizontal faces $x_3 = \pm a_3$.*

Problem 6.2 *Compute the stiffness matrix components for the same material and fiber distribution used in Problem 6.1, but chose an RVE with faces rotated 45 degrees with respect to the horizontal and vertical direction in Figure 6.1. Therefore, the RVE size will be $\sqrt{2}a_2$ and horizontal faces $\sqrt{2}a_3$. Be careful to select a correct RVE that is periodic.*

Problem 6.3 *For both Problems 6.1 and 6.2 compute E_1, E_2, ν_{12}, ν_{23}, G_{12} and G_{23}. Compare and justify the results.*

References

[1] E. J. Barbero, Introduction to Composite Materials Design, Taylor & Francis, Philadelphia, PA, 1998.

[2] E. J. Barbero, Computer Aided Design for Composites
http://www.mae.wvu.edu/barbero/cadec.html

[3] J. D. Eshelby, The Determination of the Elastic Field of an Ellipsoidal Inclusion and Related Problems, Proceedings of the Royal Society, A241 (1957) 376-396.

[4] J. D. Eshelby, The Elastic Field Outside an Ellipsoidal Inclusion, Proceedings of the Royal Society, A252 (1959) 561-569.

[5] T. Mori, and K. Tanaka, Average Stress in Matrix and Average Elastic Energy of Materials with Misfitting Inclusions, Acta Metall., 21 (1973) 571-574.

[6] R. Hill, A self-consistent mechanics of composite materials, J. Mech. Phys. Solids, 13 (1965) 213-222.

[7] S. Nemat-Nasser, and M. Hori, Micromechanics: Overall Properties of Heterogeneous Materials, North-Holland, Amsterdam, 1993.

[8] R. Luciano, and E. J. Barbero, Formulas for the Stiffness of Composites with Periodic Microstructure, International Journal of Solids and Structures, 31(21) (1995) 2933-2944.

[9] R. Luciano, and E. J. Barbero, Analytical Expressions for the Relaxation Moduli of Linear Viscoelastic Composites with Periodic Microstructure, ASME J. Applied Mechanics, 62(3) (1995) 786-793.

[10] E. J. Barbero, and R. Luciano, R. Micromechanical Formulas for the Relaxation Tensor of Linear Viscoelastic Composites with Transversely Isotropic Fibers, International Journal of Solids and Structures, 32(13) (1995) 1859-1872.

[11] J. Aboudi, Mechanics of composite materials : A unified micromechanical approach, volume 29 of Studies in Applied Mechanics, Elsevier, New York, NY, 1991.

[12] Z. Hashin, and S. Shtrikman, A variational approach to the elastic behavior of multiphase materials, Journal of Mechanics and Physics of Solids, 11 (1963) 127-140.

[13] V. Tvergaard, Model studies of fibre breakage and debonding in a metal reinforced by short fibres, Journal of Mechanics and Physics of Solids, 41(8) (1993) 1309-1326.

[14] J. L. Teply, and G. J. Dvorak, Bound on overall instantaneous properties of elastic-plastic composites, Journal of Mechanics and Physics of Solids, 36(1) (1988) 29-58.

[15] E. J. Barbero, Web resource: http://www.mae.wvu.edu/barbero/feacm/

[16] R. Luciano, and E. Sacco, Variational Methods for the Homogenization of Periodic Media, European J. Mech. A/Solids, 17 (1998) 599-617.

Chapter 7

Viscoelasticity

Our interest in viscoelasticity is motivated by observed creep behavior of polymer matrix composites (PMC), which is a manifestation of viscoelasticity. The time dependent response of materials can be classified as elastic, viscous, and viscoelastic. On application of a sudden load, which is then held constant, an elastic material undergoes instantaneous deformation. In a *one-dimensional state of stress*, the elastic strain is $\varepsilon = D\sigma$, where $D = 1/E$ is the compliance or inverse of the modulus E. The deformation then remains constant. Upon unloading, the elastic strain reverses to its original value, thus all elastic deformation is recovered.

The viscous material flows at a constant rate $\dot{\varepsilon} = \sigma/\eta$ where $\eta = \tau E_0$ is the Newton viscosity, E_0 is the initial modulus, and τ is the time constant of the material. The accumulated strain $\varepsilon = \int \dot{\varepsilon} dt$ cannot be recovered by unloading.

A viscoelastic material combines the behavior of the elastic and viscous material in one, but the response is more complex than just adding the viscous strain to the elastic strain. Let H be the Heaviside function defined as

$$H(t - t_0) = 0 \text{ when } t < t_0$$
$$H(t - t_0) = 1 \text{ when } t \geq t_0 \tag{7.1}$$

Upon step loading $\sigma = H(t - t_0)\,\sigma_0$, with a constant load σ_0, the viscoelastic material experiences a sudden elastic deformation, just like the elastic material. After that, the deformation grows by a combination of recoverable and unrecoverable viscous flow.

A simple series addition of viscous flow and elastic strain (*Maxwell model*, Figure 7.1(a), with $\eta = \tau E_0$) yields totally unrecoverable viscous flow plus recoverable elastic deformation as

$$\dot{\varepsilon}(t) = \frac{\sigma(t)}{\tau E_0} + \frac{\dot{\sigma}(t)}{E_0} \tag{7.2}$$

A simple parallel combination of elastic and viscous flow (*Kelvin model*, Figure 7.1(b), with $\eta = \tau E$) yields totally recoverable deformation with no viscous flow as

$$\sigma(t) = \tau E \dot{\varepsilon}(t) + E \varepsilon(t) \tag{7.3}$$

but the deformation does not recover instantaneously.

Materials with unrecoverable viscous flow, such as (7.2), are called *liquids* even though the flow may occur very slowly. Glass is a liquid material over the time span of centuries; the thickness of window panes in medieval cathedrals is thicker at the bottom and thinner at the top, thus revealing the flow that took place over the centuries under the load imposed by gravity. Materials with fully recoverable viscous deformations, such as (7.3), are called *solids*. We shall see that structural design is much easier with solid materials than with liquid materials.

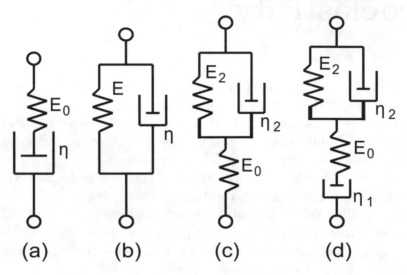

Fig. 7.1: Viscoelastic modes: (a) Maxwell, (b) Kelvin, (c) standard solid, (d) Maxwell-Kelvin.

Please take heed of the common misconception introduced in early strength of materials courses that most structural materials are elastic. Only perfectly crystalline materials are elastic. Most materials are viscoelastic if observed for sufficiently long periods of time, or at sufficiently high temperature. In other words, most real materials are viscoelastic.

For elastic materials, the compliance D is the inverse of the modulus E, both of which are constants, and they are related by

$$DE = 1 \qquad (7.4)$$

For viscoelastic materials in the time domain, the compliance is called $D(t)$ and it is related to the time varying relaxation $E(t)$ in a similar but not so simple way, as will be shown in Section 7.3. Note that the relaxation $E(t)$ takes the place of the modulus E. A brief derivation of the relationship between compliance and relaxation is presented next, in order to facilitate the presentation of viscoelastic models in Section 7.1. When both the compliance D and the relaxation E are functions of time, (7.4) simply becomes

$$D(t)E(t) = 1 \qquad (7.5)$$

Both $D(t)$ and $E(t)$ are functions of time and thus it is not possible to operate algebraically on (7.5) to get either function explicitly in terms of the other. To find one from the other, take the Laplace transform (see Section 7.3) to get

$$s^2 D(s) E(s) = 1 \qquad (7.6)$$

Since both $D(s)$ and $E(s)$ are algebraic functions of s, and the time t is not involved, it is possible to operate algebraically to get

$$E(s) = \frac{1}{s^2 D(s)} \qquad (7.7)$$

Finally, the relaxation in the time domain is the inverse Laplace of (7.7) or

$$E(t) = L^{-1}[E(s)] \qquad (7.8)$$

Similarly, the compliance $D(t)$ can be obtained from the relaxation $E(t)$ as

$$D(t) = L^{-1} \left[\frac{1}{s^2 L[E(t)]} \right] \qquad (7.9)$$

where $L[]$ indicates the Laplace transform and $L^{-1}[]$ indicates the inverse Laplace transform.

7.1 Viscoelastic Models

The viscoelastic material models presented in this section are convenient curve fits of experimental data. In the time domain, the usual experiments are the creep and relaxation tests. In the *creep test*, a constant stress σ_0 is applied and the ensuing strain is measured. The ratio of measured strain to applied stress is the compliance $D(t) = \varepsilon(t)/\sigma_0$. In the *relaxation test*, a constant strain ε_0 is applied and the stress needed to maintain that strain is measured. The ratio of measured stress to applied strain is the relaxation $E(t) = \sigma(t)/\varepsilon_0$.

7.1.1 Maxwell Model

To derive the compliance of the *Maxwell* model [1], a creep test is performed under constant stress σ_0 applied at the ends of the model shown in Figure 7.1(a). The rate of strain is given by (7.2). Integrating with respect to time we get

$$\varepsilon(t) = \frac{1}{\tau E_0} \int_0^t \sigma_0 dt + \frac{\sigma_0}{E_0} \qquad (7.10)$$

where E_0 is the elastic constant of the spring, τ is the time constant of the material, and $\eta = \tau E_0$ in Figure 7.1(a). The spring and dashpot are subject to the same load and to the same constant stress σ_0, so evaluating the integral yields

$$\varepsilon(t) = \frac{\sigma_0 \, t}{\tau E_0} + \frac{\sigma_0}{E} \qquad (7.11)$$

Then, the compliance is

$$D(t) = \frac{1}{E_0} + \frac{t}{\tau E_0} \tag{7.12}$$

To derive the relaxation of the Maxwell model, take the Laplace transform of (7.12) (using Table 7.1 or [2]) to get

$$D(s) = \frac{1}{sE_0} + \frac{1}{s^2\tau E_0} = \frac{s\tau + 1}{s^2\tau E_0} \tag{7.13}$$

At $t = 0$, the dashpot does not move, so E_0 is also the initial elastic modulus of the material. Now, the relaxation in the Laplace domain is

$$E(s) = \frac{1}{s^2 D(s)} = \frac{\tau E_0}{s\tau + 1} \tag{7.14}$$

and the relaxation in the time domain is obtained by taking the inverse Laplace transform (using Table 7.1 or [2]) to get

$$E(t) = E_0 \exp(-t/\tau) \tag{7.15}$$

Note that at $t = \tau$, the relaxation decays to 36.8% of its initial value, and thus τ is called the *time constant* of the material.

Table 7.1: Some common Laplace transforms

Function	$f(t) = L^{-1}\{f(s)\}$	$f(s) = L\{f(t)\}$
constant	a	a/s
linear	at	a/s^2
derivative	df/dt	$sf(s)-f(0)$
exponential	$\exp(at)$	$1/(s-a)$
convolution integral	$\int_0^t f(t-\tau)g(\tau)dr$	$L\{f\}L\{g\}$

7.1.2 Kelvin Model

For the *Kelvin* model, only the creep test is possible, since a relaxation test would require an infinitely large stress to stretch the dashpot in Figure 7.1(b) to a constant value in no time. For a creep test, a constant stress $\sigma = \sigma_0$ is applied. Then, (7.3) is an ordinary differential equation (ODE) in $\varepsilon(t)$, which is satisfied by $\varepsilon(t) = (\sigma_0/E)\left[1 - exp(-t/\tau)\right]$. Therefore, the compliance $D(t) = \varepsilon(t)/\sigma_0$ is

$$D(t) = 1/E[1 - \exp(-t/\tau)] \tag{7.16}$$

Using (7.8), the relaxation function can be written with the aid of the Heaviside step function $H(t)$ and the Dirac delta function $\delta(t)$ as follows

$$E(t) = EH(t) + E\tau\delta(t) \tag{7.17}$$

where $\delta(t-t_0) = \infty$ *if* $t = t_0$ and zero for any other time. The following MATLAB code yields (7.17):

```
syms s complex; syms Dt Et t E tau real;
Dt=expand((1-exp(-t/tau))/E)
Ds=laplace(Dt)
Es=1/Ds/s^2
Et=ilaplace(Es)
```

7.1.3 Maxwell-Kelvin Model

A crude approximation of a liquid material is the *Maxwell-Kelvin* model, also called the four-parameter model, described by Figure 7.1(d). Since the Maxwell and Kelvin elements are placed in series, the compliance is found by adding the compliances of the two individual modes, as

$$D(t) = \frac{1}{E_0} + \frac{t}{\tau_1 E_0} + \frac{1}{E_2}\left[1 - exp\left(-t/\tau2\right)\right] \tag{7.18}$$

where E_0 is the elastic modulus, τ_1 takes the place of τ in (7.12), and E_2, τ_2, take the place of E, τ, in (7.16). The relaxation modulus is given by ([1, page 28])

$$E(t) = \left(P_1^2 - 4P_2\right)^{-1/2}\left[\left(q_1 - \frac{q_2}{T_1}\right)exp\left(-t/T_1\right) - \left(q_1 - \frac{q_2}{T_2}\right)exp\left(-t/T_2\right)\right]$$

$$\eta_1 = E_0\tau_1 \quad ; \quad \eta_2 = E_0\tau_2$$

$$q_1 = \eta_1 \quad ; \quad q_2 = \frac{\eta_1\eta_2}{E_2}$$

$$T_1 = \frac{1}{2P_2}\left[P_1 + \sqrt{P_1^2 - 4P_2}\right] \quad ; \quad T_2 = \frac{1}{2P_2}\left[P_1 - \sqrt{P_1^2 - 4P_2}\right]$$

$$P_1 = \frac{\eta_1}{E_0} + \frac{\eta_1}{E_2} + \frac{\eta_2}{E_2} \quad ; \quad P_2 = \frac{\eta_1\eta_2}{E_0 E_2} \tag{7.19}$$

Another way to determine if a material is a liquid or a solid is to look at its long-term deformation. If the deformation is unbounded, then it is a liquid. If the deformation eventually stops, then it is a solid. A crude approximation of a solid material is the *standard solid model* (SLS) described by Figure 7.1(c) and

$$D(t) = 1/E_0 + 1/E_2\left[1 - \exp\left(\frac{-t}{\tau_2}\right)\right] \tag{7.20}$$

and

$$E(t) = E_\infty + (E_0 - E_\infty)\exp\left(\frac{-t(E_0 + E_2)}{\tau_2 E_2}\right) \tag{7.21}$$

where $E_\infty = (E_0^{-1} + E_2^{-1})^{-1}$ is the equilibrium modulus as time goes to infinity.

7.1.4 Power Law

Another model, which is popular to represent relatively short-term deformation of polymers is the *power law*

$$E(t) = At^{-n} \tag{7.22}$$

The parameters A and n are adjusted with experimental data. The power law is popular because it fits well the short-time behavior of polymers and because fitting the data is very easy; just take logarithm on both sides of (7.22) so that the equation becomes that of a line, then fit the parameters using linear regression. The compliance is obtained by using (7.9) as

$$D(t) = D_0 + D_c(t)$$
$$D_c(t) = [A\Gamma(1-n)\Gamma(1+n)]^{-1}t^n \tag{7.23}$$

where Γ is the Gamma function ([3, Section 2.6.4]), $D_0 = 1/E_0$ is the elastic compliance and the subscript $()_c$ indicates the creep component of the relaxation and compliance functions.

7.1.5 Prony Series

Although the short-term creep and relaxation of polymers can be described well by the power law, as the time range becomes longer, a more refined model becomes necessary. A general model is the *Prony series*, which consists of a number of decaying exponentials

$$E(t) = E_\infty + \sum E_i \exp(-t/\tau_i) \tag{7.24}$$

where τ_i are the time constants, E_i are the relaxation moduli, and E_∞ is the equilibrium modulus, if any exists. The larger the τ_i the slower the decay is.

7.1.6 Generalized Kelvin Model

While the Prony series can fit any material behavior if a large number of terms are used, other models are more efficient for fitting purposes, if harder to manipulate mathematically. For example the *generalized Kelvin* model

$$D(t) = D_0 + D_1'[1 - \exp(-t/\tau)^m] \tag{7.25}$$

can approximate well the long term compliance in the α-region of polymer creep [4]. At room temperature, this is the region of interest to structural engineers since it spans the range of time from seconds to years. In contrast the β-region [4], is of interest to sound and vibration experts, among others, since it spans the sub-second range of times. In other words, for long-term modeling, all compliance occurring in the β-region can be lumped in the term D_0, with D_1' representing all the compliance that could ever be accumulated in the α-region. Equation (7.25) has four parameters. When the data spans short times, it may be impossible to determine all four parameters because the material behavior cannot be distinguished from a 3-parameter power law (7.26). This can be easily understood if (7.25) is expanded in a power series, truncated after the first term as follows [4]

$$D(t) = D_0 + D_1'(t/\tau)^m[1 - (t/\tau)^m + ...] \approx D_0 + D_1 t^m \; ; \; D_1 = D_1'/\tau \tag{7.26}$$

For short times, all higher order powers of t can be neglected. What remains is a modified power law with only three parameters. Note that for short times, the parameter τ is combined with D_1' to form D_1. If the data cover a short time, the fitting algorithm will not be able to adjust both τ and D_1' in (7.26); virtually any combination of τ and D_1' will work. That means that short term data must be modeled by a smaller number of parameters, in this case three.

7.1.7 Nonlinear Power Law

All models described so far represent linear viscoelastic materials. In the context of viscoelasticity, linear means that the parameters in the model are not a function of stress (see Section (7.2.1)). That means that the deformation at any fixed time can be made proportionally larger by increasing the stress. If any of the parameters are a function of stress, the material is nonlinear viscoelastic. For example a *nonlinear power law* takes the form

$$\dot{\varepsilon} = At^B \sigma^D \tag{7.27}$$

Take logarithm to both sides of (7.27) to get a linear equation in two variables

$$y = \bar{A} + BX_1 + DX_2 \; ; \; \bar{A} = \log(A), X_1 = \log(t), X_2 = \log(\sigma) \tag{7.28}$$

that can be fitted with a multiple linear regression algorithm [2].

Although most materials are not linearly viscoelastic, they can be approximated as linear viscoelastic if the range of stress at which the structure operates is narrow.

Example 7.1 *Fit the creep data in Table 7.2 with (a) Maxwell (7.12), (b) Power Law (7.26), and (c) generalized Kelvin models (7.25).*

Table 7.2: Creep data

time [sec]	1	21	42	62	82	102	123	143	163	184	204
D(t) [GPa^{-1}]	1.49	1.99	2.21	2.35	2.56	2.66	2.75	2.85	2.92	2.96	3.01

Solution to Example 7.1 *To fit the Maxwell model, fit a line to the secondary creep data; that is, ignore the curvy portion for short times to get $E_0 = 0.460$ GPa, $\tau = 495$ s.*

To fit the Power Law, write (7.21) as

$$D(t) - D_0 = D_1 t^m$$

where $D_0 = 1.49$ GPa^{-1} is the first datum in Table 7.2 (see also (7.26)). Take logarithm to both sides of above equation and adjust a line using linear regression to get $D_0 = 1.49$ GPa^{-1}, $D_1 = 0.1117$ (GPa sec)$^{-1}$, and $m = 0.5$.

To fit the generalized Kelvin model you need to use a nonlinear solver [2] to minimize the error between the predicted (expected) values e_i and the experimental (observed) values o_i values. Such error is defined as the sum over all the available data points: $\chi^2 = \sum (e_i - o_i)^2/o_i^2$. In this way, the following are obtained: $D_0 = 1.657$ GPa^{-1}, $D_1' = 1.617$ GPa^{-1}, $\tau = 0.273$ sec, and $m = 0.0026$.

The experimental data and the fit functions are shown in Figure 7.2.

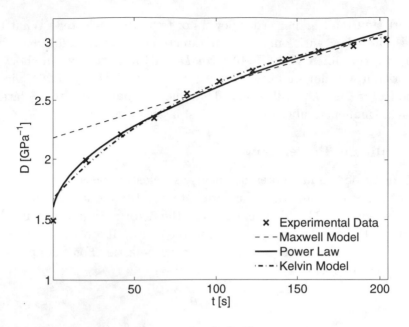

Fig. 7.2: Viscoelastic fit: Maxwell Model, Power Law, and Kelvin Model.

7.2   Boltzmann Superposition

7.2.1   Linear Viscoelastic Material

A viscoelastic material is linear if superposition applies. That is, given a stress history

$$\sigma(t) = \sigma_1(t) + \sigma_2(t) \tag{7.29}$$

the strain is given by

$$\varepsilon(t) = \varepsilon_1(t) + \varepsilon_2(t) \tag{7.30}$$

where $\varepsilon_1(t)$, $\varepsilon_2(t)$ are the strain histories corresponding to $\sigma_1(t)$ and $\sigma_2(t)$, respectively. For linear materials, the creep compliance and relaxation modulus are independent of stress

$$D(t) = \frac{\varepsilon(t)}{\sigma_0}$$

$$E(t) = \frac{\sigma(t)}{\varepsilon_0} \tag{7.31}$$

For nonlinear materials, $D(t,\sigma)$ is a function of stress and $E(t,\varepsilon)$ is a function of strain.

For a linear material subjected to a stress σ_0 applied at time $t = \theta_0$ (Figure 7.3) we have

$$\varepsilon(t) = \sigma_0 \, D(t,\theta_0) \quad ; \quad t > \theta_0 \tag{7.32}$$

Adding an infinitesimal load step $d\sigma$ at time $\theta_0 + d\theta$ results in

$$\varepsilon(t) = \sigma_0 \, D(t,\theta_0) + d\sigma \, D(t,\theta_0 + d\theta) \quad ; \quad t > \theta_0 + d\theta \tag{7.33}$$

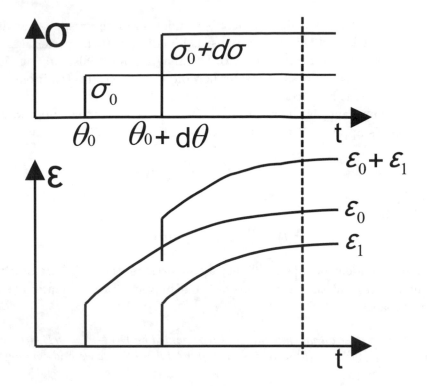

Fig. 7.3: Boltzmann superposition of strains.

If stress changes continuously by $d\sigma$ over intervals $d\theta$, the summation (7.33) can be replaced by an integral to yield the accumulated strain as

$$\varepsilon(t) = \sigma_0 \, D(t, \theta_0) + \int_{\theta_0}^{t} D(t, \theta) d\sigma = \sigma_0 \, D(t, \theta_0) + \int_{\theta_0}^{t} D(t, \theta) \frac{d\sigma}{d\theta} d\theta \qquad (7.34)$$

where the discrete times θ_0, $\theta_0 + d\theta$,, etc., are represented by the continuous function θ. Although aging effects are negligible over each infinitesimal $d\theta$, they are significant over time. Therefore, the compliance $D(t, \theta)$ is a function of the current time t and and all the time-history represented by θ in $D(t, \theta)$.

7.2.2 Unaging Viscoelastic Material

If the $\varepsilon_1(t, \theta)$ curve has the same shape as the $\varepsilon_1(t, \theta_0)$ curve, only translated horizontally, any curve can be shifted to the origin (Figure 7.3)

$$D(t, \theta) = D(t - \theta) \qquad (7.35)$$

Equation (7.35) is the definition of unaging material. For a discussion of aging materials see [1, 5]. Equation (7.35) means that all the curves have the same shape regardless of age θ, only shifted. Note θ in (7.35) is a continuous function $\theta < t$ that denotes the time of application of each load (σ_0, $d\sigma$, etc.)

The response $\varepsilon(t, \theta)$ at a fixed time t is a function of the response at all times $\theta < t$. Therefore, it is said that the response is hereditary. If the material is aging, t and θ are independent variables in $D(t, \theta)$. For unaging materials, only one variable, $t - \theta$, is independent, so it does not matter how old the material is (t), it only matters for how long $(t - \theta)$ it has been loaded with $d\sigma(\theta)$.

The creep compliance is the response of the material to stress and always starts when the stress is applied. If the change is gradual, from (7.34) we have

$$\varepsilon(t) = \int_0^t D(t - \theta)\, \dot{\sigma}(\theta)\, d\theta \tag{7.36}$$

The relaxation is

$$\sigma(t) = \int_0^t E(t - \theta)\, \dot{\varepsilon}(\theta)\, d\theta \tag{7.37}$$

The time dependent behavior of linear viscoelastic materials is hereditary, meaning that the behavior at time t depends on what happened to the material since the beginning of loading at $t = 0$.

Example 7.2 *Consider an unaging material represented by $D(t - \theta) = 1/E + (t - \theta)/\eta$ and loaded with (a) $\sigma_0 H(\theta)$ and (b) $\sigma_0 H(\theta - 1)$. Find $\varepsilon(t)$ in both cases and comment on the result.*

Solution to Example 7.2

(a) $\sigma = \sigma_0 H(\theta) \;\Rightarrow\; \dot{\sigma} = \sigma_0 \delta(0)$

$\varepsilon(t) = \int_0^t \left[\dfrac{1}{E} + \dfrac{(t - \theta)}{\eta} \right] \sigma_0 \delta(0) d\theta$

$\varepsilon(t) = \left[\dfrac{1}{E} + \dfrac{t}{\eta} \right] \sigma_0 \; ; \; t > 0$

(b) $\sigma = \sigma_0 H(\theta - 1) \;\Rightarrow\; \dot{\sigma} = \sigma_0 \delta(1)$

$\varepsilon(t) = \int_0^t \left[\dfrac{1}{E} + \dfrac{(t - \theta)}{\eta} \right] \sigma_0 \delta(1) d\theta$

$\varepsilon(t) = \left[\dfrac{1}{E} + \dfrac{(t - 1)}{\eta} \right] \sigma_0; \; t > 1$

It can be seen that (b) is identical to (a), only shifted; meaning that there is no aging.

7.3   Correspondence Principle

The Laplace transform of a function $f(t)$ in the time domain (t-domain) maps to the Laplace domain (s-domain) as $f(s)$. The Laplace transform is defined as

$$L[f(t)] = f(s) = \int_0^\infty \exp(-st) f(t) dt \tag{7.38}$$

Most of the time, the Laplace transform can be obtained analytically, just using a table of transforms, such as Table 7.1. Taking the Laplace transform of (7.36,

7.37), yields

$$\varepsilon(s) = L[D(t)] \; L[\dot{\sigma}(t)] = sD(s)\sigma(s) \tag{7.39}$$

$$\sigma(s) = L[E(t)] \; L[\dot{\varepsilon}(t)] = sE(s)\varepsilon(s) \tag{7.40}$$

Multiplying (7.39) times (7.40) yields

$$s^2 \; D(s)E(s) = 1 \tag{7.41}$$

or

$$s \; D(s) = [s \; E(s)]^{-1} \tag{7.42}$$

where it can be seen that $sD(s)$ is the inverse of $sE(s)$. This is analogous to (7.4) for elastic materials.

The correspondence principle states that all the equations of elasticity, available for elastic materials, are valid for linearly viscoelastic materials in the Laplace domain. This principle is the basis, for example, of the determination of creep and relaxation of polymer matrix composites in terms of fiber and matrix properties using standard micromechanics methods, as shown in Section 7.6.

The inverse mapping from the Laplace domain to the time domain

$$f(t) = L^{-1}(f(s)) \tag{7.43}$$

is more difficult to compute. Decomposition in partial fractions [6] is a useful technique to break up $f(s)$ into simpler component functions for which the inverse Laplace can be found analytically. Another useful technique is the convolution theorem defined in Table 7.1. Also, the limiting value theorems

$$f(0) = \lim_{s\to\infty} [sF(s)] \tag{7.44}$$

$$f(\infty) = \lim_{s\to 0} [sF(s)]$$

can be used to evaluate the initial and final response of a material in the time domain directly in the Laplace domain. As a last resort, the inverse Laplace can be found numerically using [7] or the collocation method described in Appendix D.

The Carson transform is defined as

$$\hat{f}(s) = sf(s) \tag{7.45}$$

In the Carson domain, the constitutive equations (7.36-7.37) become

$$\varepsilon(s) = \hat{D}(s)\sigma(s)$$

$$\sigma(t) = \hat{E}(s)\varepsilon(s) \tag{7.46}$$

which are analogous, in the Carson domain, to the stress-strain equations of elastic materials in the time domain. Furthermore, the relationship between compliance and relaxation becomes

$$\hat{D}(s) = 1/\hat{E}(s) \tag{7.47}$$

7.4   Frequency Domain

The Fourier transform maps the time domain into the frequency domain. It is defined as

$$F[f(t)] = f(\omega) = \int_{-\infty}^{\infty} \exp(-i\omega t) f(t)\, dt \tag{7.48}$$

and its inverse

$$f(t) = \frac{1}{\sqrt{2\pi}} \int_{-\infty}^{\infty} \exp(i\omega t) f(w)\, dw \tag{7.49}$$

Applying the Fourier transform to (7.36-7.37) yields

$$\varepsilon(\omega) = D(\omega)\sigma(\omega) \tag{7.50}$$
$$\sigma(\omega) = E(\omega)\varepsilon(\omega)$$

and

$$D(\omega) = 1/E(\omega) \tag{7.51}$$

where $D(\omega) = D' + iD''$ and $E(\omega) = E' + iE''$ are complex numbers. Here D', D'' are the storage and loss compliances, and E', E'' are the storage and loss moduli.

Using standard complex analysis we get

$$D' = \frac{E'}{E'^2 + E''^2}$$
$$D'' = \frac{E''}{E'^2 + E''^2} \tag{7.52}$$

The frequency domain has a clear physical meaning. If a sinusoidal stress $\sigma(\omega, t) = \sigma_0 \exp(-i\omega t)$ is applied to a viscoelastic material, it responds with an out-of-phase sinusoidal strain $\varepsilon(\omega, t) = \varepsilon_0 \exp(-i\omega t + \phi)$. Furthermore, the complex compliance is $D(\omega) = \varepsilon(\omega, t)/\sigma(\omega, t)$ and the complex relaxation is simply the inverse of the complex compliance, $E(\omega) = \sigma(\omega, t)/\varepsilon(\omega, t)$.

7.5   Spectrum Representation

The Prony series (7.24) provides a physical interpretation of polymer behavior as a series of Maxwell models, each with its own decay time. In the limit, a real polymer has an infinite number of such models [8], so that

$$E(t) - E_\infty = \int_{-\infty}^{\infty} H(\theta) \exp(-t/\theta) d\ln\theta = \int_{0}^{\infty} \frac{H(\theta)}{\theta} \exp(-t/\theta) d\theta \tag{7.53}$$

where $H(\theta)$ is the relaxation spectrum [9]. In terms of compliance, we have

$$D(t) - D_0 = \frac{t}{\eta} + \int_{-\infty}^{\infty} \frac{L(\theta)}{\theta} [1 - \exp(-t/\theta)] d\theta \tag{7.54}$$

where $L(\theta)$ is the retardation spectrum [9], D_0 is the elastic compliance, η is the asymptotic viscosity of liquids, with $\eta \to \infty$ for solids (see also [10]).

7.6 Micromechanics of Viscoelastic Composites

7.6.1 One-Dimensional Case

Recall the constitutive equations (7.46) in the Carson domain. By the correspondence principle, all equations of micromechanics for elastic materials are valid in the Carson domain for linear viscoelastic materials. For example, the Reuss micromechanical model assumes uniform identical strain in the matrix and fiber (see discussion on page 140). Therefore, the stiffness of the composite $\mathbf{C}$ is a linear combination of the stiffness of the constituents (fiber and matrix) weighed by their respective volume fractions V_m, V_f

$$\mathbf{C} = V_m \mathbf{C}^m + V_f \mathbf{C}^f \tag{7.55}$$

with $\mathbf{A}^m = \mathbf{A}^f = \mathbf{I}$ in (6.1). Taking into account the correspondence principle for a viscoelastic material (Section 7.3), it is possible to write the stiffness tensor in the Carson domain by analogy with (7.55) simply as

$$\hat{\mathbf{C}}(s) = V_m \hat{\mathbf{C}}^m + V_f \hat{\mathbf{C}}^f \tag{7.56}$$

From it, the stiffness tensor in the Laplace domain is (see (7.42))

$$\mathbf{C}(s) = \frac{1}{s}\hat{\mathbf{C}}(s) \tag{7.57}$$

Finally, the stiffness tensor in the time domain is obtained by finding the inverse Laplace transform (7.43) as

$$\mathbf{C}(t) = L^{-1}[\mathbf{C}(s)] \tag{7.58}$$

Example 7.3 *Derive the transverse compliance $D_2(t)$ in the time domain for a unidirectional composite with elastic fibers and a viscoelastic matrix represented by $D_m = 1/E_m + t/\eta_m$. Plot D_f, $D_m(t)$, and $D_2(t)$ for $0 < t < 0.1$, $E_f = 10$, $V_f = 0.5$, $E_m = 5$, $\eta_m = 0.05$. Use the Reuss model and discuss the results.*

Solution to Example 7.3 *The elastic behavior of the fiber and viscoelastic behavior of the matrix are defined as follows:*

Fiber (elastic): $E_f = constant \rightarrow D_f = \dfrac{1}{E_f}$

Matrix (Maxwell model (7.12) with $E_m = E_0$, $\eta_m = \tau E_0$): $\dfrac{1}{E_m} = \dfrac{1}{E_m} + \dfrac{t}{\eta_m}$

Take the Laplace transform,

$D_f(s) = \dfrac{1}{sE_f}$ *because* $\dfrac{1}{E_f}$ *is constant.*

$D_m(s) = \dfrac{1}{sE_m} + \dfrac{1}{s^2\eta_m}$

Then, the Carson transform is

$\hat{D}_f(s) = s\, D_f(s) = \dfrac{1}{E_f}$

$\hat{D}_m(s) = s\, D_m(s) = \dfrac{1}{E_m} + \dfrac{t}{s\eta_m}$

Using the Reuss model (page 140) to compute the composite behavior

$$\hat{D}_2 = V_f \hat{D}_f + V_m \hat{D}_m$$

$$\hat{D}_2 = V_f \frac{1}{E_f} + V_m \left(\frac{1}{E_m} + \frac{1}{s\eta_m} \right)$$

Back to Laplace domain

$$D_2(s) = \frac{V_f}{sE_f} + \frac{V_m}{sE_m} + \frac{V_m}{s^2\eta_m}$$

Back transform to the time domain (inverse Laplace)

$$D_2(t) = L^{-1}(D_2(s)) = \frac{V_f}{E_f} + \frac{V_m(E_m t + \eta_m)}{E_m \eta_m}$$

To make a plot, take $E_f = 10$, $V_f = 0.5$, $E_m = 5$, $\eta_m = 0.05$, which results in

$$D_f = 0.1 = 1/10$$
$$D_m(t) = 0.2 + 20t$$
$$D_2(t) = 0.15 + 10t$$

Since $V_f = 0.5$, the initial compliance is halfway between those of the fiber and the matrix. The elastic fiber has constant compliance. The creep rate of the composite $1/\eta_c$ is $1/2$ of the creep rate of the matrix $1/\eta_m$.

7.6.2   Three-Dimensional Case

The constitutive equation for an elastic, isotropic material (1.82) can be written in terms of just two material parameters λ and $\mu = G$ as

$$\boldsymbol{\sigma} = (\lambda \mathbf{I}^{(2)} \otimes \mathbf{I}^{(2)} + 2\mu \mathbf{I}^{(4)}) : \boldsymbol{\varepsilon} \qquad (7.59)$$

where $\mathbf{I}^{(2)}$ and $\mathbf{I}^{(4)}$ are the second- and fourth-order identity tensors[1] (Appendix A). The constitutive equation of isotropic viscoelastic materials can be written in terms of the viscoelastic Lame constants $\lambda(s)$ and $\mu(s)$ as follows [10]

$$\boldsymbol{\sigma}(t) = \int_0^t \lambda(t-\theta)\mathbf{I}^{(2)} \otimes \mathbf{I}^{(2)} : \dot{\boldsymbol{\varepsilon}}(\theta)d\theta + \int_0^t 2\mu(t-\theta)\mathbf{I}^{(4)} : \dot{\boldsymbol{\varepsilon}}(\theta)d\theta \qquad (7.60)$$

Using the convolution theorem (Table 7.1), the Laplace transform of (7.60) is

$$\boldsymbol{\sigma}(s) = s\lambda(s)\mathbf{I}^{(2)} \otimes \mathbf{I}^{(2)} : \boldsymbol{\varepsilon}(s) + s\, 2\mu(s)\mathbf{I}^{(4)} : \boldsymbol{\varepsilon}(s) \qquad (7.61)$$

or in terms of the Carson transform

$$\hat{\boldsymbol{\sigma}}(s) = \hat{\mathbf{C}}(s) : \hat{\boldsymbol{\varepsilon}}(s) \qquad (7.62)$$

Assuming that the fiber is elastic, and the matrix is viscoelastic; the latter represented with a Maxwell model

$$D_m(t) = 1/E_m + t/\eta_m \qquad (7.63)$$

the Carson transform is

$$\hat{D}_m = 1/E_m + 1/s\eta_m = \frac{E_m + s\eta_m}{s\eta_m E_m} \qquad (7.64)$$

---

[1]Tensors are indicated by boldface type, or by their components using index notation.

Using the correspondence principle yields

$$\hat{E}_m = 1/\hat{D}_m = \frac{s\eta_m E_m}{E_m + s\eta_m} = \frac{sE_m}{E_m/\eta_m + s} \tag{7.65}$$

Using (1.79) and assuming the Poisson ratio ν_m of the matrix to be constant, the Lame constant of the matrix in the Carson domain is

$$\hat{\lambda}_m = \frac{\hat{E}_m \nu_m}{(1 + \nu_m)(1 - 2\nu_m)} \tag{7.66}$$

and the shear modulus of the matrix is

$$\hat{\mu}_m = \frac{\hat{E}_m}{2(1 + \nu_m)} \tag{7.67}$$

Barbero and Luciano [11] used the the Fourier expansion method to get the components of the relaxation tensor in the Carson domain for a composite with cylindrical fibers arranged in a square array with fiber volume fraction V_f. The elastic, transversely isotropic fibers are represented by the transversely isotropic stiffness tensor $\mathbf{C}'$ defined by (1.74) in terms of fiber properties in the axial and transverse (radial) directions E_A, E_T, G_A, G_T, and ν_{AT}. Defining the matrix properties in the Laplace $\tilde{(\,)}$ and Carson domain $\hat{(\,)}$ as $\hat{\lambda}_m = s\tilde{\lambda}_m(s)$ and $\hat{\mu}_m = s\tilde{\mu}_m(s)$, the components of the relaxation tensor of the composite in the Carson domain $\widehat{\mathbf{L}}^*$ become [11]

$$\widehat{L}_{11}^*(s) = \widehat{\lambda}_m + 2\widehat{\mu}_m - V_f \left(-a_4^2 + a_3^2 \right)$$
$$\left(-\frac{\left(2\widehat{\mu}_m + 2\widehat{\lambda}_m - C'_{33} - C'_{23} \right)(a_4^2 - a_3^2)}{a_1} + \frac{2(a_4 - a_3)\left(\widehat{\lambda}_m - C'_{12} \right)^2}{a_1^2} \right)^{-1}$$

$$\widehat{L}_{12}^*(s) = \widehat{\lambda}_m + V_f \left(\frac{\left(\widehat{\lambda}_m - C'_{12} \right)(a_4 - a_3)}{a_1} \right)$$
$$\left(\frac{\left(2\widehat{\mu}_m + 2\widehat{\lambda}_m - C'_{33} - C'_{23} \right)(a_3^2 - a_4^2)}{a_1} + \frac{2(a_4 - a_3)\left(\widehat{\lambda}_m - C'_{12} \right)^2}{a_1^2} \right)^{-1}$$

$$\widehat{L}_{22}^*(s) = \widehat{\lambda}_m + 2\widehat{\mu}_m - V_f \left(\frac{\left(2\widehat{\mu}_m + 2\widehat{\lambda}_m - C'_{33} - C'_{23} \right)a_3}{a_1} - \frac{\left(\widehat{\lambda}_m - C'_{12} \right)^2}{a_1^2} \right)$$
$$\left(\frac{\left(2\widehat{\mu}_m + 2\widehat{\lambda}_m - C'_{33} - C'_{23} \right)(a_3^2 - a_4^2)}{a_1} + \frac{2(a_4 - a_3)\left(\widehat{\lambda}_m - C'_{12} \right)^2}{a_1^2} \right)^{-1}$$

$$\widehat{L}_{23}^{*}(s) = \widehat{\lambda}_m + V_f \left(\frac{\left(2\,\widehat{\mu}_m + 2\,\widehat{\lambda}_m - C_{33}' - C_{23}'\right) a_4}{a_1} - \frac{\left(\widehat{\lambda}_m - C_{12}'\right)^2}{a_1^2} \right)$$

$$\left(\frac{\left(2\,\widehat{\mu}_m + 2\,\widehat{\lambda}_m - C_{33}' - C_{23}'\right)\left(a_3^2 - a_4^2\right)}{a_1} + \frac{2\left(a_4 - a_3\right)\left(\widehat{\lambda}_m - C_{12}'\right)^2}{a_1^2} \right)^{-1}$$

$$\widehat{L}_{44}^{*}(s) = \widehat{\mu}_m - V_f \left(\frac{2}{2\,\widehat{\mu}_m - C_{22}' + C_{23}'} - \left(2\,S_3 - \frac{4\,S_7}{2 - 2\,\nu_m}\right)\widehat{\mu}_m^{-1} \right)^{-1}$$

$$\widehat{L}_{66}^{*}(s) = \widehat{\mu}_m - V_f \left(\left(\widehat{\mu}_m - C_{66}'\right)^{-1} - \frac{S_3}{\widehat{\mu}_m} \right)^{-1} \tag{7.68}$$

where

$$a_1 = 4\,\widehat{\mu}_m^2 - 2\,\widehat{\mu}_m\,C_{33}' + 6\,\widehat{\lambda}_m\,\widehat{\mu}_m - 2\,C_{11}'\,\widehat{\mu}_m - 2\,\widehat{\mu}_m\,C_{23}' + C_{23}'\,C_{11}' + 4\,\widehat{\lambda}_m\,C_{12}'$$
$$- 2\,C_{12}'^{\,2} - \widehat{\lambda}_m\,C_{33}' - 2\,C_{11}'\,\widehat{\lambda}_m + C_{11}'\,C_{33}' - \widehat{\lambda}_m\,C_{23}'$$

$$a_2 = 8\,\widehat{\mu}_m^3 - 8\,\widehat{\mu}_m^2 C_{33}' + 12\,\widehat{\mu}_m^2\widehat{\lambda}_m - 4\,\widehat{\mu}_m^2 C_{11}'$$
$$- 2\,\widehat{\mu}_m\,C_{23}'^{\,2} + 4\,\widehat{\mu}_m\,\widehat{\lambda}_m\,C_{23}' + 4\,\widehat{\mu}_m\,C_{11}'\,C_{33}'$$
$$- 8\,\widehat{\mu}_m\,\widehat{\lambda}_m\,C_{33}' - 4\,\widehat{\mu}_m\,C_{12}'^{\,2} + 2\,\widehat{\mu}_m\,C_{33}'^{\,2} - 4\,\widehat{\mu}_m\,C_{11}'\,\widehat{\lambda}_m + 8\,\widehat{\mu}_m\,\widehat{\lambda}_m\,C_{12}'$$
$$+ 2\,\widehat{\lambda}_m\,C_{11}'\,C_{33}' + 4\,C_{12}'\,C_{23}'\,\widehat{\lambda}_m - 4\,C_{12}'\,C_{33}'\,\widehat{\lambda}_m - 2\,\widehat{\lambda}_m\,C_{11}'\,C_{23}'$$
$$- 2\,C_{23}'\,C_{12}'^{\,2} + C_{23}'^{\,2}C_{11}' + 2\,C_{33}'\,C_{12}'^{\,2} - C_{11}'\,C_{33}'^{\,2} + \widehat{\lambda}_m\,C_{33}'^{\,2} - \widehat{\lambda}_m\,C_{23}'^{\,2}$$

$$a_3 = \frac{4\,\widehat{\mu}_m^2 + 4\,\widehat{\lambda}_m\,\widehat{\mu}_m - 2\,C_{11}'\,\widehat{\mu}_m - 2\,\widehat{\mu}_m\,C_{33}' - C_{11}'\,\widehat{\lambda}_m - \widehat{\lambda}_m\,C_{33}' - C_{12}'^{\,2}}{a_2}$$
$$+ \frac{C_{11}'\,C_{33}' + 2\,\widehat{\lambda}_m\,C_{12}'}{a_2} - \frac{S_3 - \dfrac{S_6}{2 - 2\nu_m}}{\widehat{\mu}_m}$$

$$a_4 = - \frac{-2\,\widehat{\mu}_m\,C_{23}' + 2\,\widehat{\lambda}_m\,\widehat{\mu}_m - \widehat{\lambda}_m\,C_{23}' - C_{11}'\,\widehat{\lambda}_m - C_{12}'^{\,2} + 2\,\widehat{\lambda}_m\,C_{12}' + C_{11}'\,C_{23}'}{a_2}$$
$$+ \frac{S_7}{\widehat{\mu}_m\,(2 - 2\,\nu_m)} \tag{7.69}$$

The coefficients S_3, S_6, S_7 account for the geometry of the microstructure, including the geometry of the inclusions and their geometrical arrangement [12]. For cylindrical fibers arranged in a square array [13] we have

$$
\begin{aligned}
S_3 &= 0.49247 - 0.47603V_f - 0.02748V_f^2 \\
S_6 &= 0.36844 - 0.14944V_f - 0.27152V_f^2 \\
S_7 &= 0.12346 - 0.32035V_f + 0.23517V_f^2
\end{aligned}
\tag{7.70}
$$

Note that (7.68) yield six independent components of the relaxation tensor. This is because (7.68) represent a composite with microstructure arranged in a square array. If the microstructure is random (Figure 1.12), the composite is transversely isotropic (Section 1.12.4) and only five components of the relaxation tensor are independent. When the axis x_1 is the axis of transverse isotropy for the composite, the averaging procedure (6.7) yields the relaxation tensor with transverse isotropy as

$$
\begin{aligned}
\hat{C}_{11} &= \hat{L}_{11} \\
\hat{C}_{12} &= \hat{L}_{12} \\
\hat{C}_{22} &= \frac{3}{4}\hat{L}_{22} + \frac{1}{4}\hat{L}_{23} + \frac{1}{2}\hat{L}_{44} \\
\hat{C}_{23} &= \frac{1}{4}\hat{L}_{22} + \frac{3}{4}\hat{L}_{23} - \frac{1}{2}\hat{L}_{44} \\
\hat{C}_{66} &= \hat{L}_{66}
\end{aligned}
\tag{7.71}
$$

where the remaining coefficients are found using (1.74) due to transverse isotropy of the material. This completes the derivation of the relaxation tensor $\hat{\mathbf{C}}_{ij} = s\mathbf{C}_{ij}(s)$ in the Carson domain. Next, the inverse Laplace transform of each coefficient yields the coefficients of the stiffness tensor in the time domain as

$$
\mathbf{C}_{ij}(t) = L^{-1}\left[\frac{1}{s}\hat{\mathbf{C}}_{ij}\right]
\tag{7.72}
$$

A MATLAB code based on [7] is available in [14] to perform the inverse Laplace numerically. Another algorithm is provided in Appendix D.

Example 7.4 *Consider a composite made with 60% by volume of transversally isotropic fibers with axial properties $E_A = 168.4\ GPa$, $G_A = 44.1\ GPa$, $\nu_A = 0.443$, and transverse properties $E_T = 24.8\ GPa$ and $\nu_T = 0.005$. The epoxy matrix is represented by a Maxwell model (7.12) with $E_0 = 4.08\ GPa$, $\tau = 39.17\ min$ and $\nu_m = 0.311$. Plot the relaxation $E_2(t)$ of the composite as a function of time for $0 < t < 100$ minutes, compared to the elastic value of the transverse modulus E_2.*

Solution to Example 7.4 *This example has been solved using MATLAB^{TM}. The elastic and viscoelastic values of the transverse modulus E_2 are shown in Figure 7.4. The calculation procedure is explained next:*

- *Program the equations of Section 7.6.2 and use them to calculate the elastic values of the composite's elastic properties such as E_2. These equations have been implemented in* PMMViscoMatrix.m.

- *Replace the elastic modulus of the matrix E_0 by the Maxwell model for the matrix Eq. (7.15) in the Carson domain $\hat{E}_0$ (See* PMMViscoMatrix.m), *as follows:*

    1. *The output from the portion of the code implementing (7.68-7.71) are equations for the relaxation moduli in terms of s in the Carson domain. Note that it is necessary to declare the variable s as symbolic.*

    2. *Divide them by s to go back to the Laplace domain.*

    3. *Back transform to the time domain using the function* invlapFEAcomp, *which is derived from [7].*

    4. *Finally, fit the numerical values of $E_2(t)$ with a viscoelastic model equation. Usually it is convenient to use the same model equation for the composite relaxation as that used for the matrix relaxation; in this case the Maxwell model. This step is implemented in* fitfunFEAcomp.m

The MATLAB codes PMMViscoMatrix.m, invlapFEAcomp.m, *and* fitfunFEAcomp.m *are available in [14]. The results are shown in Figure 7.4. The complete set of Maxwell parameters for the composite are calculated in Example 7.5.*

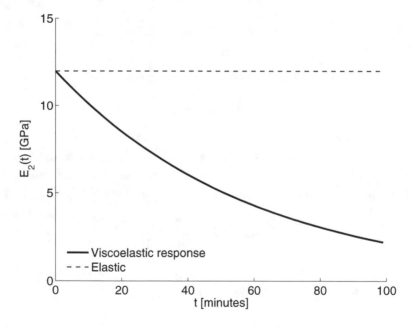

Fig. 7.4: Elastic and viscoelastic values of the transverse modulus E_2.

7.7   Macromechanics of Viscoelastic Composites

7.7.1   Balanced Symmetric Laminates

The in-plane viscoelastic behavior of a balanced symmetric laminate can be obtained using the procedure in Section 1.15 (Apparent Laminate Properties), but in the

Carson domain. Start with the stiffness of the laminae (7.71) in material coordinates, in the Carson domain. Rotate each matrix to global coordinates. Then, average them using (1.103). Using (1.106), find the laminate engineering properties in the Carson domain and divide by s to go back to the Laplace domain. Finally, take the inverse Laplace transform to find the laminate stiffness in the time domain. Then, fit them with a model equation as is done in Example 7.4.

7.7.2 General Laminates

Thanks to the correspondence principle, the stress-resultant vs. strain-curvature equations from classical lamination theory (CLT, see Chapter 3) are valid for linearly viscoelastic laminated composites in the Carson domain. The A, B, D, H matrices of a laminate in the Carson domain can be computed by using the equations from first-order shear deformation theory (FSDT, Section 3.1.1). This methodology was used in [15].

7.8 FEA of Viscoelastic Composites

Most commercial codes have implemented viscoelasticity (creep) for isotropic materials. This is a severe limitation for users interested in the analysis of viscoelastic behavior of polymer matrix composites.

However, it is possible to take advantage of the user programmable features of commercial software in order to implement the formulations presented in this chapter. This is relatively easy because the approach used in this chapter is not stress dependent, but a linear viscoelastic approach, and its implementation is not complicated. A USERMAT subroutine in ANSYS is used in Example 7.5 to implement the viscoelastic formulation.

Example 7.5 *Using the viscoelastic materials properties obtained in Example 7.4 compute the response of a $[0/90_8]_s$ laminate. The thickness of each layer is $t_k = 1.25$ mm. The laminate is $2b = 20$ mm width and $2L = 40$ mm length. Load the sample with a uniform strain $\varepsilon_x = 0.01$ by applying a uniform displacement at $x = L$. Use orthotropic solid elements on each layer and symmetry conditions. Plot the laminate and the 90-degree lamina stiffness. Also plot the stress in 0-degree and 90-degree lamina evolution for $0 > t > 300$ minutes.*

Solution to Example 7.5 *First, compute the viscoelastic engineering properties using the procedure described in Example 7.4. The resulting Maxwell parameters of the lamina are shown in Table 7.3.*

In ANSYS, using USERMAT subroutine for solid elements (`usermat3d.f`) it is possible to implement the constitutive equation of an orthotropic material with the following time-dependent properties:

$$E_1(t) = (E_1)_0 \exp(-t/\tau_1) \; ; \; E_2(t) = E_3(t) = (E_2)_0 \exp(-t/\tau_2)$$

$$G_{12}(t) = G_{13}(t) = (G_{12})_0 \exp(-t/\tau_{12}) \; ; \; G_{23}(t) = (G_{23})_0 \exp(-t/\tau_{23})$$

This subroutine is available in [14]. Next, the geometry can be modeled using a command file similar to that used in Example 5.2. The results are shown in Figures 7.5 and 7.6.

Table 7.3: Lamina viscoelastic properties

Young's Moduli	Shear Moduli	Poisson's Ratio
$(E_1)_0 = 102417$ MPa	$(G_{12})_0 = (G_{13})_0 = 5553.8$ MPa	$\nu_{12} = \nu_{13} = 0.4010$
$\tau_1 = 16551$ min	$\tau_{12} = \tau_{13} = 44.379$ min	
$(E_2)_0 = (E_3)_0 = 11975$ MPa	$(G_{23})_0 = 5037.3$ MPa	$\nu_{23} = 0.1886$
$\tau_2 = \tau_3 = 58.424$ min	$\tau_{23} = 54.445$ min	

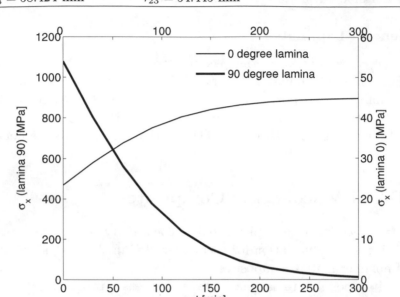

Fig. 7.5: Time stress evolution in 0°-lamina, and 90°-lamina.

Example 7.6 *Consider a composite made with 40% by volume of isotropic graphite fibers with properties $E_f = 168.4$ GPa, $\nu_f = 0.443$ and epoxy matrix represented by a Maxwell model with $E_0 = 4.082$ GPa, $\tau = 39.15$ min and $\nu_m = 0.311$ (independent of time). Construct a FE micromechanical model using hexagonal microstructure (see the mesh and macros used in Example 6.3), subject to shear strain $\varepsilon_{23} = 0.01$ ($\gamma_4 = 0.02$). Tabulate the average stress σ_4 over the RVE at times $t = 0, 20, 40, 60, 80,$ and 100 minutes.*

Solution to Example 7.6 *The fiber can be represented using standard elastic properties in ANSYS. The matrix should be modeled using the USERMAT provided in Example 7.5 (*usermat3d.f*). Therefore, the only part of the ANSYS model definition that changes with respect to Example 6.3 is the definition of the material, as follows*

```
MP,EX,1,168.4e-3    ! Fiber material properties in Tera Pascals [TPa]
MP,PRXY,1,0.443


TB,USER,2,1,12,     ! Material properties #2, Maxwell Model, 10 variables
TBTEMP,0            ! Matrix material properties in Tera Pascals [TPa]
!     Variable descriptions for the USERMAT subroutine
!     ------------------------------------------------
!       E1      E2      nu12    nu23    G12     G23
!       Tau1    Tau2    Tau12   Tau13
```

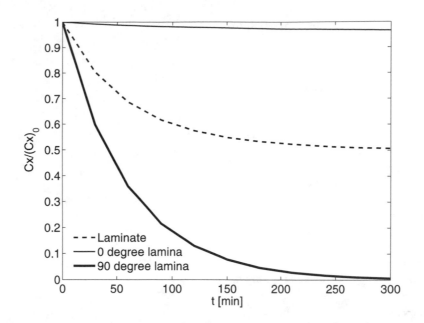

Fig. 7.6: Normalized stiffness $C_{22}(t)/(C_{22})_o$ for global laminate, $0°$-lamina, and $90°$-lamina.

```
TBDATA,,4.082e-3,4.082e-3,0.311,0.311,1.556e-3,1.556e-3
TBDATA,,39.15,39.15,39.15,39.15
```

The APDL macro ceRVE.mac *available in [14] is used to define the CE for the periodic model. The macro needs the RVE dimensions and the applied strain as input arguments. In this example, only a strain $\gamma_4 = 0.04$ is applied. The model is solved, using different substeps at times $t = 0, 20, 40, 60, 80,$ and 100 minutes.*

```
/SOLU                     ! Solution module,
ANTYPE,STATIC             ! Set static analysis
OUTRES,ALL,1              ! Store results for every SUBSTEP
! ceRVe arguments: a1,a2,a3,eps1,eps2,eps3,eps4,eps5,eps6
*use,ceRVE,a1,a2,a3,0,0,0,0.02,0,0
SOLVE                     ! Solve analysis
KBC,1                     ! Specifies stepped loading within a load step
NSUBST,1                  ! 1 = Number of substeps in this load step
TIME,1e-6                 ! Define time near to zero
NSUBST,5,10,5             ! 5 = Number of substeps in this load step
TIME,100                  ! Time at the end of load step
SOLVE                     ! Solve current load state
FINISH                    ! Exit solution module
```

To compute the average stress in the RVE, it is possible to use the macro srecover, *described in Example 6.2. The average stress obtained is shown in Table 7.4. The* SET,LIST *command listed all the load steps and substeps, which can be selected inside POST1 with the command* SET,#loadstep,#substep.

```
/POST1                  ! Post-processor module
SET,LIST
SET,1,1
*use,srecover
S_4 = Sxy0              ! Average stress t = 0
SET,2,1
*use,srecover
S_4 = Sxy0              ! Average stress t = 20
SET,2,2
*use,srecover
S_4 = Sxy0              ! Average stress t = 40
SET,2,3
*use,srecover
S_4 = Sxy0              ! Average stress t = 60
SET,2,4
*use,srecover
S_4 = Sxy0              ! Average stress t = 80
SET,2,5
*use,srecover
S_4 = Sxy0              ! Average stress t = 100
FINISH                        ! Exit post-processor module
```

Table 7.4: Average stress σ_4 along the time

Time [min]	0	20	40	60	80	100
Average σ_4 [MPa]	62.6	38.1	23.1	13.9	8.4	5.04

Using an exponential regression it is possible to calculate the values of $(G_0)_{23} = 3.13$ *GPa and* $\tau = 39.97$ *min that represent the relaxation of the composite in the 23-shear direction using a Maxwell model (see Figure 7.7).*

Suggested Problems

Problem 7.1 *Consider a composite made with 60% by volume of isotropic fibers with properties* $E_f = 168.4$ *GPa and* $\nu_f = 0.443$, *and epoxy matrix represented by a power law model (7.22) with* $D_0 = 0.222$ *GPa*$^{-1}$, $D_1 = 0.0135$ *(GPa min)*$^{-1}$, $m = 0.17$ *and* $\nu_m = 0.311$. *Plot the relaxation* $C_{22}(t)$ *of the composite as a function of time for* $0 < t < 100$ *minutes. Compare it to the elastic value of the stiffness* C_{22} *of the composite and the elastic stiffness* C_{22} *of the matrix.*

Problem 7.2 *Consider a composite made with 60% by volume of transversely isotropic graphite fibers with properties* $E_A = 168.4$ *GPa,* $E_T = 24.82$ *GPa,* $\nu_A = 0.443$, $\nu_T = 0.005$, $G_A = 44.13$ *GPa and epoxy matrix represented by a Maxwell model (7.15) with* $E_0 = 4.082$ *GPa,* $\tau = 39.15$ *min and* $\nu_m = 0.311$. *Plot the relaxation tensor stiffness components* $C(t)$ *of the composite as a function of time for* $0 < t < 100$ *minutes, compared to the elastic stiffness* C *of the composite and the elastic stiffness* C_m *of the matrix.*

Problem 7.3 *Compute the parameters in the Maxwell model for unidirectional lamina (see Section 1.14) of carbon/epoxy material used in Problem 7.2. Plot and compare the elastic and viscoelastic properties:* $E_1(t)$, $E_2(t)$ *and* $G_{12}(t)$. *Show all work in a report.*

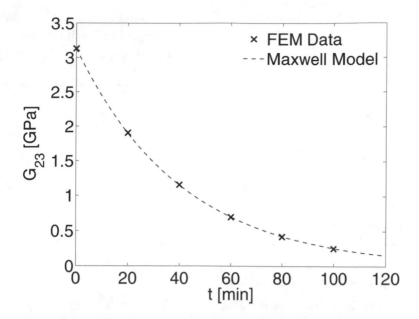

Fig. 7.7: Average $(G_0)_{23}$ evolution.

Problem 7.4 *In ANSYS, implement the Maxwell model constitutive equations for a transversally orthotropic lamina material under plane stress conditions. Use the user programmable feature USERMAT (for plane stress and shell elements, use subroutine* usermatps.f*). Using the viscoelastic materials properties obtained in Problem 7.3, compute the response of a* $[\pm45/90_2]_s$ *laminate. The thickness of each layer is* $t_k = 1.25$ *mm. Load the sample with uniform edge loads in the middle laminate surface* $N_x = N_y = 10$ *N/mm. Use LSS SHELL181 element type. Plot the laminate and the laminae relaxations, as well as the laminae stress* σ_x *as a function of time for* $0 > t > 300$ *minutes.*

Problem 7.5 *Compute the parameters in the Maxwell model for all the nine engineering properties of a* $[0/90]_S$ *laminate. Each layer is* 1.25 *mm thick. The material is carbon T300 and Epoxy 934(NR) with* $V_f = 0.62$ *and layer thickness 1.25 mm. Epoxy is represented by a Maxwell model (7.15) with* $E_0 = 4.082$ *GPa,* $\tau = 39.15$ *min and* $\nu_m = 0.311$*. Carbon T300 is transversely isotropic with axial modulus* $E_A = 202.8$ *GPa, transverse modulus* $E_T = 25.3$ *GPa,* $G_A = 44.1$ *GPa,* $\nu_A = 0.443$*, and* $\nu_T = 0.005$*, where the subscripts A and T indicate the axial and radial (transverse) directions of the fiber, respectively.*

References

[1] G. J. Creus, Viscoelasticity: Basic Theory and Applications to Concrete Structures, Springer-Verlag, Berlin, 1986.

[2] MATHLAB 7, Web resource: www.mathworks.com

[3] M. R. Spiegel, Mathematical Handbook, McGraw-Hill, New York, 1968.

[4] G. D. Dean, B. E. Read, and P. E. Tomlins, A Model For Long-Term Creep and the Effects of Physical Ageing In Poly (butylene terephthalate), Plastics and Rubber Processing and Applications, 13(1) (1990) 37-46.

[5] B. F. Oliveira and G. J. Creus, An Analytical-Numerical Framework for the Study of Ageing in Fibre Reinforced Polymer Composites, Composites B, 65(3-4) (2004) 443-457.

[6] K. Ogatha, Discrete-Time Control System, Prentice-Hall, Englewood Cliffs, NJ, 1987.

[7] K. J. Hollenbeck, INVLAP.M: A matlab function for numerical inversion of Laplace transforms by the de Hoog algorithm (1998)
http://www.mathworks.com/

[8] R. S. Lakes, Viscoelastic Solids, CRC Press, Boca Raton, FL, 1998.

[9] J. D. Ferry, Viscoelastic Properties of Polymers, 3rd. Ed. Wiley, New York, 1980.

[10] R. M. Christensen, Theory of Viscoelasticity, Accademic Press, New York, 1972.

[11] E. J. Barbero and R. Luciano, Micromechanical Formulas for the Relaxation Tensor of Linear Viscoelastic Composites with Transversely Isotropic Fibers, International Journal of Solids and Structures, 32(13) (1995) 1859-1872.

[12] S. Nemat-Nasser and M. Hori, Micromechanics: Overall Properties of Heterogeneous Materials, North-Holland, Amsterdam, 1993.

[13] R. Luciano and E. J. Barbero, Formulas for the Stiffness of Composites with Periodic Microstructure, International Journal of Solids and Structures, 31(21) (1995) 2933-2944.

[14] E. J. Barbero, Web resource: http://www.mae.wvu.edu/barbero/feacm/

[15] P. Qiao, E. J. Barbero and J. F. Davalos, On the Linear Viscoelasticity of Thin-Walled Laminated Composite Beams, Journal of Composite Materials, 34(1) (2000) 39-68.

Chapter 8

Damage Mechanics

Many modes of damage can be observed in composite materials, including matrix cracks, fiber breakage, fiber-matrix de-bonding, and many more. Much work has been done trying to quantify each of these damage modes, their evolution with respect to load, strain, time, number of cycles, etc., and their effect on stiffness, remaining life, etc. But it is difficult to relate all this information quickly to a practical design or operational situation. On the other hand, continuum damage mechanics (CDM) simply represents all these failure modes by the effect they have on the meso-scale (lamina level) behavior of the material. One notable effect of damage is a reduction of stiffness, which can be used to define damage [1]. One-dimensional models are used in Section 8.1 to introduce the concepts. The theoretical formulation for the general three-dimensional case is developed in Sections 8.2-8.4. A particular model for fiber reinforced composites is fully developed in Section 9.

8.1 One-Dimensional Damage Mechanics

The development of a one-dimensional damage mechanics solution involves the definition of three major entities: 1) a suitable damage variable, 2) an appropriate damage activation function, and 3) a convenient damage evolution, or kinetic equation.

8.1.1 Damage Variable

Consider a composite rod of nominal area $\overline{A}$, unloaded, and free of any damage (Figure 8.1.a). Upon application of a sufficiently large load P, damage appears (Figure 8.1.b). On a macroscopic level, damage can be detected by the loss of stiffness of the material. In CDM, damage is represented by a state variable D, called *damage variable*, which represents the loss of stiffness [1]

$$D = 1 - E/\overline{E} \tag{8.1}$$

where $\overline{E}$ is the initial (virgin) Young's modulus, and E is the modulus after damage.[1] Earlier work [2] conceptualized damage as the reduction of area due to accumulation of micro-cracks having the same effect of the actual damage

$$D = 1 - A/\overline{A} \tag{8.2}$$

where $\overline{A}$, A, are the initial and remaining cross-sectional areas, respectively. The complement to damage is the integrity [3]

$$\Omega = 1 - D = A/\overline{A} \tag{8.3}$$

which can be interpreted as the remaining cross-sectional area ratio, using the original area as basis. It is noted that, in principle, damage is a measurable parameter, which could be determined by measuring the damaged area, remaining area, or more practically measuring the initial and remaining moduli. Therefore, in thermodynamics terms, damage is a measurable state variable, in the same sense as the temperature is a measurable state variable that quantifies in macroscopic terms the random agitation of atoms, molecules, and other elementary particles. While it is possible, but extremely difficult, to track the agitation of atoms and molecules, it is very easy to measure the temperature with a thermometer or other device. The same holds true for damage in composite materials.

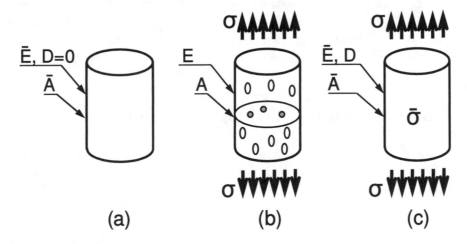

(a)   (b)   (c)

Fig. 8.1: (a) Unstressed material configuration, (b) stressed material configuration with distributed damage, (c) effective configuration.

The analysis of a structural component is done in terms of the nominal area $\overline{A}$, which is the only one known to the designer. The remaining area $A = (1-D)\overline{A}$ is not known a priori. The nominal stress is $\sigma = P/\overline{A}$. Neglecting stress concentrations[2] at the tips of the fictitious cracks representing damage in the damaged configuration

---

[1]See also (8.10).

[2]Even taking into account the stress concentrations, the volume average of the distribution of *effective stress* in the representative volume element (RVE, see Chapter 6) is still $\sigma = P/A$.

(Figure 8.1.b), the value of effective stress, acting on the remaining area A, is $\overline{\sigma} = P/A > P/\overline{A}$.

Therefore, we can envision a configuration (Figure 8.1.c) free of damage, with nominal area $\overline{A}$, loaded by nominal stress σ, but internally subjected to effective stress $\overline{\sigma}$. Thus, the effective configuration allows us to perform structural analysis using the nominal geometry but effectively taking into account the increase of effective stress and the decrement of stiffness caused by damage.

In the undamaged configuration (a), $D = 0$, $\sigma = \overline{\sigma}$, $\varepsilon = \overline{\varepsilon}$, and Hooke's law is

$$\overline{\sigma} = \overline{E}\,\overline{\varepsilon} \tag{8.4}$$

In the effective configuration (c)

$$\sigma = E(D)\varepsilon \tag{8.5}$$

The *principle of strain equivalence* assumes that the strain is the same in the configurations (b) and (c), or $\varepsilon = \overline{\varepsilon}$. Starting with the nominal stress $\sigma = P/\overline{A}$, multiplying by A/A and using (8.3), the relationship between effective stress $\overline{\sigma}$ and nominal stress σ (under strain equivalence) is

$$\sigma = \overline{\sigma}\,(1 - D) \tag{8.6}$$

Using (8.6), (8.4) and $\varepsilon = \overline{\varepsilon}$ in (8.5), the apparent modulus is

$$E(D) = \overline{E}(1 - D) \tag{8.7}$$

The *principle of energy equivalence* [4] states that the elastic strain energy is identical in the configurations (b) and (c). That is, $\sigma : \varepsilon = \overline{\sigma} : \overline{\varepsilon}$, which is satisfied by

$$\sigma = \overline{\sigma}(1 - D) \\ \overline{\varepsilon} = \varepsilon(1 - D) \tag{8.8}$$

Substituting (8.8) in (8.5) yields

$$E(D) = \overline{E}(1 - D)^2 \tag{8.9}$$

which redefines the damage variable as

$$D = 1 - \sqrt{E/\overline{E}} \tag{8.10}$$

Every state variable has a conjugate thermodynamic force driving its growth. In plasticity, the measurable state variable is the plastic strain tensor ε^p, which is driven to grow by its conjugate thermodynamic force, the stress tensor σ. The thermodynamic damage force Y is defined as conjugate to the state variable D.

A kinetic equation $\dot{D}(Y)$ governs the growth of the state variable D as a function of its conjugate thermodynamic force Y. In principle, any relevant variable can be chosen as independent variable Y to define the kinetic equation $\dot{D}(Y)$, as long as it is independent of its conjugate state variable. When the damage D is a scalar and it is used to analyze one-dimensional problems, various authors have chosen independent variables in the form of strain ε [5], effective stress $\bar{\sigma}$ [6, 7], excess energy release rate $G - 2\gamma_c$ [8], and so on. However, the choice is better based on the appropriate form of the thermodynamic principle governing the problem, as shown in Section 8.3.

8.1.2 Damage Threshold and Activation Function

The elastic domain is defined by a threshold value for the thermodynamic force below which no damage occurs. When the load state is in the elastic domain, damage does not grow. When the load state reaches the limit of the elastic domain, additional damage occurs. Furthermore, the elastic domain modifies its size or hardens. Typical one-dimensional responses of two materials are shown in Figure 8.2. Initially the elastic domain is defined by the initial threshold values, $\sigma \leq \sigma_0$ and $\varepsilon \leq \varepsilon_0$. While the load state is inside this domain, no damage occurs. When the load state is higher than the threshold, damage increases and the threshold changes. The elastic domain may evolve as hardening or softening. A stress threshold increases for materials with hardening (see Figure 8.2a), and it decreases for materials with softening (see Figure 8.2b). On the other hand, a strain or effective stress threshold always increases for hardening and softening behavior, as shown in Figure 8.2.

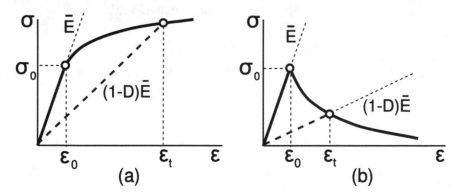

Fig. 8.2: (a) Hardening behavior and (b) softening behavior. No damage occurs until the strain reaches a threshold value ε_0, and no damage occurs during unloading.

The elastic domain can be defined by the *damage activation function g* as

$$g := \hat{g} - \hat{\gamma} \geq 0 \tag{8.11}$$

where $\hat{g}$ is a positive function (norm) that depends on the independent variable (in a one-dimensional case a scalar Y) and $\hat{\gamma}$ is the updated damage threshold for isotropic hardening. According to the positive dissipation principle (see Section 8.3), the updated damage threshold $\hat{\gamma}$ can be written as

$$\hat{\gamma} = \gamma(\delta) + \gamma_0 \tag{8.12}$$

where γ_0 denotes the virgin damage threshold, and γ is a positive monotonic function, called *hardening (or softening) function*, that depends on the internal variable δ, called *damage hardening variable*.

8.1.3 Kinetic Equation

The rate of damage accumulation is represented by a kinetic equation. The evolution of damage and hardening are defined by

$$\dot{D} = \dot{\lambda}\frac{\partial g}{\partial Y} \quad ; \quad \dot{\delta} = \dot{\lambda}\frac{\partial g}{\partial \gamma} \tag{8.13}$$

where Y is the independent variable and $\dot{\lambda} \geq 0$ is the damage multiplier that enforces consistency among the damage and hardening evolution as defined by (8.13). Furthermore, the values of $\dot{\lambda}$ and g allow us to distinguish among two possible situations, loading or unloading without damage growth, and loading with damage growth, according to the Kuhn-Tucker conditions [9]

$$\dot{\lambda} \geq 0 \quad ; \quad g \leq 0 \quad ; \quad \dot{\lambda}g = 0 \tag{8.14}$$

In other words, the Kuhn-Tucker conditions allow us to differentiate among two different cases:

1. Un-damaging loading or unloading, in the elastic domain. The damage activation function is $g \leq 0$ and by condition (8.14.c) $\dot{\lambda} = 0$, and by (8.13.a) $\dot{D} = 0$.

2. Damage loading. In this case $\dot{\lambda} > 0$ and condition (8.14.c) implies that $g = 0$. Then, the value of $\dot{\lambda}$ can be determined by the damage consistency condition

$$\begin{aligned} g &= 0 \\ \dot{g} &= 0 \end{aligned} \tag{8.15}$$

Example 8.1 *Compute $\dot{\lambda}$ for a one-dimensional model under tensile load where the independent variable is the effective stress $Y = \bar{\sigma}$, the activation function is defined by $\hat{g} = \bar{\sigma}$, and the hardening function is defined by*

$$\hat{\gamma} = (F_0 - F_R)\delta + F_0$$

where F_0 and F_R are the initial threshold and the strength of the strongest microscopic element in the material, respectively.

Solution to Example 8.1 *The damage activation function g is defined as*

$$g := \hat{g} - \hat{\gamma} = \bar{\sigma} - [(F_0 - F_R)\delta + F_0] \geq 0$$

Therefore,

$$\frac{\partial g}{\partial \bar{\sigma}} = +1 \quad ; \quad \frac{\partial g}{\partial \hat{\gamma}} = -1$$

Using (8.13), the kinetic equations can be written as

$$\dot{D} = \lambda \frac{\partial g}{\partial \overline{\sigma}} = \lambda \quad ; \qquad \dot{\delta} = \lambda \frac{\partial g}{\partial \hat{\gamma}} = -\lambda$$

When new damage appears, the consistency conditions (8.15) yield

$$g = 0 \quad \Rightarrow \quad \hat{\gamma} = \overline{\sigma}$$

and

$$\dot{g} = 0 \quad \Rightarrow \quad \dot{g} = \frac{\partial g}{\partial \overline{\sigma}}\dot{\overline{\sigma}} + \frac{\partial g}{\partial \hat{\gamma}}\dot{\hat{\gamma}} = \dot{\overline{\sigma}} - \dot{\hat{\gamma}} = 0$$

where

$$\dot{\hat{\gamma}} = \frac{\partial \hat{\gamma}}{\partial \delta}\dot{\delta} = (F_0 - F_R)(-\dot{\lambda}) = (F_R - F_0)\dot{\lambda}$$

Substituting into the second consistency condition (8.15) we obtain $\dot{\lambda}$ *as*

$$\dot{\lambda} = \frac{1}{F_R - F_0}\dot{\overline{\sigma}}$$

8.1.4   Statistical Interpretation of the Kinetic Equation

Let's assume that individual damaging events are caused by the failure of microscopic elements inside the material (e.g., fiber breaks, matrix cracks, fiber-matrix de-bond, etc.). Furthermore, assume each of these material points has a failure strength $\overline{\sigma}$ and that the collection of failure strengths for all these points, i.e., elements failing at a certain stress $\overline{\sigma}$ over the total number of elements available, is represented by a probability density $f(\overline{\sigma})$ (Figure 8.3.b). The fraction of elements broken during an effective stress excursion from zero to $\overline{\sigma}$ provides a measure of damage,

$$D(\overline{\sigma}) = \int_0^{\sigma} f(\overline{\sigma})d\overline{\sigma} = F(\overline{\sigma}) \tag{8.16}$$

where $F(\overline{\sigma})$ is the cumulative probability (Figure 8.3.b) corresponding to the probability density $f(\overline{\sigma})$. Then, the kinetic equation in terms of effective stress $\overline{\sigma}$ is

$$\dot{D} = \frac{dD}{d\overline{\sigma}}\dot{\overline{\sigma}} = f(\overline{\sigma})\dot{\overline{\sigma}} \tag{8.17}$$

8.1.5   One-Dimensional Random-Strength Model

As explained in Section 8.1.3, the rate of damage accumulation is represented by a kinetic equation. Equation (8.17) represents a generic kinetic equation, which becomes specific once a particular probability density of failure is adopted.

Consider a loose bundle of short fibers embedded in a matrix and subjected to a uniform stress. The fiber-matrix interfacial strength is assumed to be identical for all fibers but the embedment length is random. The fiber pull out strength is therefore random. Random means that the probability of finding a fiber pulling out at any value of stress $F_0 < \overline{\sigma} < F_R$ is constant. In other words, there is no stress level at which more fibers or less fibers pull out because the probability of pull out is random. This is represented in Figure 8.3 and given by the equation $f(\bar{\sigma}) = 1/(F_R - F_0)$. Substituting $\bar{\varepsilon}$ for $\bar{\sigma}$ as the independent variable in (8.17), and assuming strain equivalence $\varepsilon = \bar{\varepsilon}$, we have

$$f(\bar{\varepsilon}) = \frac{\overline{E}}{F_R - F_0} \quad ; \quad F_0 \leq \bar{\sigma} \leq F_R \tag{8.18}$$

Equation (8.18) yields the model proposed in [5], which represents well the damaging behavior of Haversian bone [10], concrete in tension [11], fiber composites when damage is controlled by fiber pull out [12], and transverse damage of unidirectional composites.

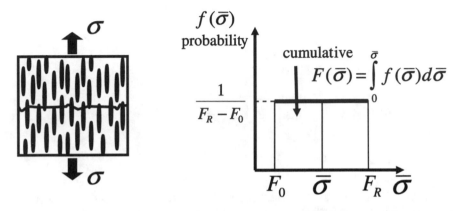

Fig. 8.3: One-dimensional random-strength model.

Damage Activation Function

For a one-dimensional problem, choosing strain as the independent variable, it is possible to write $\hat{g} = \varepsilon$. Therefore, the damage activation function can be written as

$$g := \varepsilon - \hat{\gamma} \geq 0 \tag{8.19}$$

where $\hat{\gamma}$ is the updated damage threshold. Assuming that no damage occurs until the strain reaches a threshold value $\varepsilon_0 = F_0 \overline{E}$, and applying the consistency conditions (8.15) and using (8.19), the updated damage threshold $\hat{\gamma}$ is given by the highest value of strain seen by the material, or

$$\hat{\gamma} = \max(\varepsilon_0, \varepsilon) \tag{8.20}$$

Kinetic Equation

The kinetic equation (8.17) for the case of random strength (8.18) in terms of strains $\bar{\varepsilon} = \varepsilon$ can be expressed as

$$\dot{D} = \frac{dD}{d\varepsilon}\dot{\varepsilon} = \begin{cases} \overline{E}/(F_R - F_0)\dot{\varepsilon} & \text{when} ; \ \varepsilon > \hat{\gamma} \\ 0 & \text{otherwise} \end{cases} \tag{8.21}$$

In this case, the independent variable is ε, and using (8.19), the kinetic equation (8.13) reduces to

$$\dot{D} = \dot{\lambda} \tag{8.22}$$

Using the Kuhn-Tucker conditions and (8.21), the consistency condition (8.15) reduces to

$$\dot{\lambda} = \overline{E}/(F_R - F_0)\dot{\varepsilon} \tag{8.23}$$

when damage occurs and $\dot{\lambda} = 0$ otherwise. In this particular case, the kinetic equation is known explicitly (8.22-8.23). Therefore, it is not necessary to evaluate the evolution of hardening (8.12) because hardening is computed explicitly by (8.20). Note that (8.23) is identical to the solution of Example 8.1 because the hardening function was chosen deliberately to yield this result.

Secant Constitutive Equation

In this particular case, the damage variable is active when tensile load appears, and it can be obtained by integrating (8.21) as

$$D_t = \overline{E}\,\frac{\hat{\gamma} - \varepsilon_0}{F_R - F_0} \quad \text{when } \varepsilon > 0 \tag{8.24}$$

Note that the damage state does not depend on the actual load state ε, it only depends on the history of the load state $\hat{\gamma}$. In this example, crack closure is assumed in compression, damage becomes passive, and $D_c = 0$. Mathematically, damage under unilateral contact conditions can be defined by the following equation

$$D = D_t \frac{\langle \varepsilon \rangle}{|\varepsilon|} + D_c \frac{\langle -\varepsilon \rangle}{|\varepsilon|} \tag{8.25}$$

where the McCauley operator $\langle x \rangle$ is defined as $\langle x \rangle := \frac{1}{2}(x + |x|)$.

Substituting (8.24) into (8.5), and using strain equivalence, yields the following constitutive equation

$$\sigma = E(D)\,\varepsilon = \begin{cases} \left(1 - \overline{E}\,\dfrac{\hat{\gamma} - \varepsilon_0}{F_R - F_0}\right)\overline{E}\,\varepsilon & \text{when } \varepsilon > 0 \\[2mm] \overline{E}\,\varepsilon & \text{when } \varepsilon < 0 \end{cases} \tag{8.26}$$

Tangent Constitutive Equation

In a finite element formulation, it is necessary to provide the constitutive equation in rate form, where the rates of stress $\dot{\sigma}$ and strain $\dot{\varepsilon}$ are expressed as functions of pseudo-time. In this particular case, the tangent constitutive equation can be obtained by differentiation of the secant constitutive equation as

$$\dot{\sigma} = E(D)\dot{\varepsilon} + \dot{E}(D)\varepsilon \tag{8.27}$$

The term $\dot{E}(D)$ is zero when new damage does not appear, i.e., when there is elastic loading or unloading. When damaging behavior occurs (8.20) yields $\hat{\gamma} = \varepsilon$, and differentiating $E(D)$ in (8.26) we obtain

$$\dot{E}(D) = -\frac{\overline{E}^2}{F_R - F_0} \dot{\varepsilon} \tag{8.28}$$

Substituting (8.28) into (8.27) if damage occurs, or $\dot{E}(D) = 0$ if no damage occurs, the tangent constitutive equation can be written as

$$\dot{\sigma} = \begin{cases} \left(1 - \overline{E}\, \dfrac{2\hat{\gamma} - \varepsilon_0}{F_R - F_0}\right) \overline{E}\dot{\varepsilon} & \text{when } \varepsilon > \hat{\gamma} \\ E(D)\,\dot{\varepsilon} & \text{when } \varepsilon < \hat{\gamma} \end{cases} \tag{8.29}$$

Model Identification

The initial damage threshold ε_0 represents the minimum strain to initiate damage and it is proportional to F_0 as follows

$$F_0 = \overline{E}\varepsilon_0 \tag{8.30}$$

Under load control, a tensile specimen breaks at $\varepsilon = \hat{\gamma} = \varepsilon_{cr}$ when $d\sigma/d\varepsilon = 0$. Then, using (8.29.a), the only unknown parameter in the model can be computed as

$$F_R = 2\overline{E}\varepsilon_{cr} \tag{8.31}$$

The material parameters F_0 and F_R can be calculated from the experimental data using (8.30) and (8.31) with $\overline{E}$ being the undamaged modulus of the material. The measurable values ε_0 and ε_{cr} can be obtained easily from material testing at the macroscopic level.

For the particular case $\varepsilon_0 = 0$, using (8.24) and (8.31) at $\varepsilon = \varepsilon_{cr}$, the critical damage at failure under tensile load is

$$D_{cr} = 0.5 \tag{8.32}$$

Therefore the critical effective stress is

$$\overline{\sigma}_{T\,cr} = \overline{E}\varepsilon_{cr} = 0.5F_R \tag{8.33}$$

and using (8.7) the critical applied stress is

$$\sigma_{T\,cr} = 0.25F_R \tag{8.34}$$

Therefore, in a material with initial threshold $\varepsilon_0 = 0$, a tensile specimen under load control fails when $D = 1/2$ and applied stress $F_R/4$.

A conservative estimate of transverse tensile strength of a fiber reinforced lamina can be obtained assuming that the fiber-matrix bond strength is negligible. In the limit, only the matrix between fibers carries the transverse load, with the fibers acting as holes. In this limit case, the matrix links can be assumed to have a random distribution of strength (8.18). Therefore, the random-strength model (8.29) applies, and the critical damage for transverse tensile loading of unidirectional fiber-reinforced lamina can be estimated by (8.32) as $D_{2t}^{cr} = 0.5$. At the present time, there is no model available to estimate the critical transverse-direction compression damage D_{2c}^{cr}.

Example 8.2 *A beam of rectangular cross section, width $b = 100$ mm and height $2h = 200$ mm is subjected to pure bending. The bending moment at failure is 25.1 MN mm. The beam is made of carbon/epoxy composite with randomly oriented short fibers with undamaged Young's modulus $\overline{E} = 46$ GPa. Find the bending moment at failure in terms of F_R in (8.21). Assume that the material does not damage in compression and it has a random distribution of strength in tension, with the strongest material element having unknown strength $F_R > 0$ and $F_0 = 0$. Determine F_R using the data given.*

Solution to Example 8.2 *This problem was solved in [12]. With reference to Figure 8.4, M is the applied bending moment, and y_c, y_t, are the distances from the neutral axis to the stress resultants N_c, N_t, on the tensile and compression portions of the beam.*

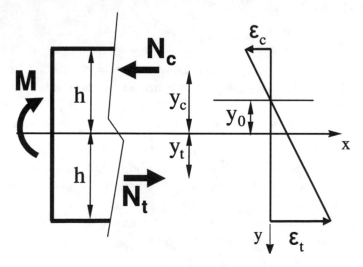

Fig. 8.4: One-dimensional random-strength model.[3]

Denoting by ε_t and ε_c the tension and compression strain on the outer surfaces of the beam, y_0 the distance from the mid-plane to the neutral surface, and assuming linear strain distribution through the thickness we have

$$\varepsilon(y) = \frac{y - y_0}{h - y_0}\varepsilon_t \quad or \quad \varepsilon(y) = \frac{-y + y_0}{h + y_0}\varepsilon_c$$

Since there is no damage in compression, the compression stress distribution is linear, and the resulting compression stress resultant is

$$N_c = \frac{1}{2}b(h + y_0)\overline{E}\varepsilon_c$$

---

[3]Reprinted from Mechanics of Materials, vol. 8 (1998), D. Kracjcinovic, Damage Mechanics, Fig. 2.11, pp. 134, copyright (1998), with permission from Elsevier.

and the distance y_c is

$$y_c = \frac{1}{3}(y_0 - 2h)$$

As the tensile side of the beam damages, the neutral axis moves away from the mid-surface. The tensile stress resultant is obtained using (8.26) and integrating the stress between y_0 and h as

$$N_t = \int_{y_0}^{h} dN_t = b \int_{y_0}^{h} E(D)\varepsilon(y)dy = \frac{1}{6}(h - y_0)b\left(3 - 2(\overline{E}/F_R)\varepsilon_t\right)\overline{E}\varepsilon_t$$

where $\overline{E}$ is the undamaged elastic moduli. The distance y_t is

$$y_t = \frac{1}{N_t}\int_{y_0}^{h} y\,dN_t = \frac{4h - 2y_0 - (\overline{E}/F_R)\varepsilon_t\,(3h + y_0)}{6 - 4(\overline{E}/F_R)\varepsilon_t}$$

The force and moment equations of equilibrium are

$$N_c + N_t = 0$$
$$N_c y_c + N_t y_t = M$$

Using the force equilibrium equation and assuming linear strain distribution through the thickness, it is possible to obtain the strains ε_t and ε_c in terms of y_0 as

$$\varepsilon_t = -\frac{6hy_0}{(h - y_0)^2}\frac{F_R}{\overline{E}} \quad ; \quad \varepsilon_c = \frac{6hy_0(h + y_0)}{(h - y_0)^3}\frac{F_R}{\overline{E}}$$

Using the above relation, it is possible to reduce the moment equilibrium equation to a single cubic equation in y_0

$$M = \frac{-y_0(4h^2 + 9hy_0 + 3y_0^2)}{(h - y_0)^3}\,bh^2 F_R$$

The ultimate bending moment can be determined by differentiation w.r.t. y_0

$$\frac{dM}{dy_0} = 0$$

that yields $y_{0\ cr} = -0.175\,h$ at beam failure. Therefore, the rupture bending moment is

$$M_{cr} = 0.2715\,bh^2\,F_R$$

A simple test (ASTM D790 or D6272) can be used to obtain the bending moment at failure; in this example $M_{cr} = 25.1\ 10^6 N\,mm$. Therefore, F_R can be estimated as $F_R = 92\ MPa$. As is customary in structural engineering, the equivalent bending strength is defined as

$$\sigma_{Bcr} = \frac{M_{cr}}{S} = 0.407\,F_R$$

where S is the section modulus (for a rectangular beam $S = \frac{2}{3}bh^2$). Note that according to (8.34), the tensile strength of the same material assuming the same kinetic equation (8.26) would be $\sigma_{T\ cr} = 0.25\,F_R$. This gives a ratio of equivalent bending to tensile strength equal to $\sigma_{B\ cr}/\sigma_{T\ cr} = 1.63$, which is in good agreement with experimental data $\sigma_{B\ cr}/\sigma_{T\ cr} = 1.6$ [13] obtained for unreinforced concrete and also with the value $\sigma_{B\ cr}/\sigma_{T\ cr} = 1.5$ recommended by the ACI Code [14].

8.1.6 Fiber-Direction, Tension Damage

If a lamina is subject to tensile stress in the fiber direction, it is reasonable to assume that the matrix carries only a small portion of the applied load and no damage is expected in the matrix during loading. Then, the ultimate tensile strength of the composite lamina can be accurately predicted by computing the strength of a bundle of fibers.

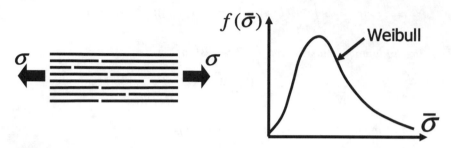

Fig. 8.5: One-dimensional random-strength model.

Fiber strength is a function of the gauge length used during fiber strength tests. The length scale that determines how much of the fiber strength is actually used in a composite is the ineffective length δ. Starting at a fiber break point, the ineffective length is that length over which a fiber recovers a large percentage of its load (say, 90%). Rosen [15] recognized this fact and proposed that the longitudinal ultimate strength of fibers embedded in a ductile-matrix can be accurately predicted by the strength of a dry bundle of fiber with length δ. A dry fiber is defined as a number of parallel fibers of some given length and diameter which, if unbroken, carry the same load. After a fiber within a dry bundle fails, the load is shared equally by the remaining unbroken fibers. A dry bundle typically refers to fibers which have not yet been combined with matrix. As tensile load is slowly applied to a dry bundle of fibers, the weaker fibers (with large flaw sizes) begin to fail and the stress on the remaining unbroken fibers increases accordingly. The Weibull expression [16]

$$F(\bar{\sigma}) = 1 - \exp\left(-\frac{\delta}{L_0}\left(\frac{\bar{\sigma}}{\bar{\sigma}_0}\right)^m\right) \tag{8.35}$$

is often used to describe the cumulative probability $F(\bar{\sigma})$ that a fiber of length δ will fail at or below an effective stress $\bar{\sigma}$. The values of $\bar{\sigma}_0$ and m, which represent the characteristic strength of the fiber, and the dispersion of fiber strength, respectively, can be determined from fiber strength experiments performed with a gauge length L_0. Equation (8.35) can be simplified as

$$F(\bar{\sigma}) = 1 - \exp\left(-\delta\alpha\bar{\sigma}^m\right) \tag{8.36}$$

where

$$\alpha = \frac{1}{L_0\bar{\sigma}_0^m} = \left[\frac{\Gamma(1+1/m)}{\bar{\sigma}_{av}}\right]^m \frac{1}{L} \tag{8.37}$$

where $\Gamma(x)$ is the Gamma function [17], $\bar{\sigma}_{av}$ is the average strength for a gauge length L. Equation (8.36) provides the percentage of fibers in a bundle which are broken as a function of the actual (or apparent) stress in the unbroken fibers. The percentage of fibers which are unbroken is $1 - F(\bar{\sigma})$. The apparent stress or bundle stress $\sigma = \sigma_b$ is equal to the applied load divided by the total fiber cross-sectional area. It is also equal to the product of the stress in unbroken fibers and the percentage of fibers which are unbroken

$$\sigma = \sigma_b = \bar{\sigma} \exp(-\delta\alpha\bar{\sigma}^m) \tag{8.38}$$

The value $\bar{\sigma}_{\max}$ which maximizes (8.38), can be easily determined and is given by

$$\bar{\sigma}_{\max} = (\delta\alpha m)^{-1/m} \tag{8.39}$$

The maximum (or critical) bundle strength σ_{cr} is determined by substituting (8.39) into (8.38)

$$\sigma_{cr} = (\delta\alpha m)^{-1/m} \exp(-1/m) \tag{8.40}$$

The composite longitudinal tensile strength is [18]

$$F_{1t} = \left[V_f + \frac{E_m}{E_f}(1 - V_f) \right] \sigma_{cr} \tag{8.41}$$

where V_f is the fiber volume fraction, and E_f and E_m are the fiber and matrix elastic Young's moduli, respectively.

Combining (8.36) and (8.39), we get

$$D_{1t}^{cr} = 1 - \exp(-1/m) \tag{8.42}$$

Therefore, the critical or maximum damage D_{1t}^{cr} for longitudinal tensile loading can be computed as the area fraction of broken fibers in the lamina prior to catastrophic failure, [7, 6], which turns out to be a function of the Weibull shape modulus m only.

Example 8.3 *The Data Sheet of carbon fiber T300 of Toray^{TM} Carbon Fibers, Inc. gives average tensile strength of the fiber $\sigma_{av} = 3.53\ GPa$, and tensile modulus $E_f = 230\ GPa$. Also, the same Data Sheet provides results of tensile tests of a UD composite specimen with epoxy $E_m = 4.5\ GPa$ and fiber volume fraction $V_f = 0.6$. The tensile strength reported is $F_{1t} = 1860\ MPa$. Using this experimental data, and assuming that a Weibull shape parameter $m = 8.9$, identify the damage model for a bundle of T300 fibers under tensile load. Then, formulate the damage model and implement it in ANSYS for a one-dimensional bar element. Finally, obtain the strain vs. stress response of a bundle of T300 fibers.*

Solution to Example 8.3

MODEL IDENTIFICATION

From (8.41) and using the experimental data available, it is possible to obtain σ_{cr} as

$$\sigma_{cr} = \frac{F_{1t}}{V_f + \frac{E_m}{E_f}(1 - V_f)} = 3060\ MPa$$

Then, the product $\delta\alpha$ can be obtained using (8.40) as

$$\delta\alpha = \frac{(\sigma_{cr})^{-m}}{me} = 3.92 \ 10^{-33}$$

The properties $E_f = 230 \ GPa$, $m = 8.9$, and $\delta\alpha = 3.92 \ 10^{-33}$ are sufficient for the identification of the model.

MODEL FORMULATION

Following a procedure similar to that shown in Section 8.1.5 to implement a damage model, the following items are needed:

Damage Activation Function

In this example, the effective stress is chosen as the independent variable. Therefore, the damage activation function can be written as

$$g := \overline{\sigma} - \hat{\gamma} \geq 0 \tag{8.43}$$

where $\hat{\gamma}$ is the updated damage threshold. Assuming an initial threshold value $\sigma_0 = 0$, from the consistency conditions (8.15) and (8.19), $\hat{\gamma}$ is given by the highest value of effective stress seen by the material

$$\hat{\gamma} = \max(0, \overline{\sigma}) \tag{8.44}$$

Secant Constitutive Equation

In this example, the kinetic equation (8.1.3) is available in integral form and given explicitly by (8.36) as

$$D = 1 - \exp\left(-\delta\alpha\hat{\gamma}^m\right) \quad \text{when } \overline{\sigma} > 0; \ \varepsilon > 0 \tag{8.45}$$

where the damage state does not depend on the actual load state $\overline{\sigma}$; it only depends on the load history state $\hat{\gamma}$.

Substituting (8.45) into (8.5) and (8.7), and using strain equivalence, yields the constitutive equation

$$\sigma = E(D) \ \varepsilon = \exp\left(-\delta\alpha\hat{\gamma}^m\right) \overline{E}\varepsilon \quad \text{when } \varepsilon > 0 \tag{8.46}$$

Tangent Constitutive Equation

The tangent constitutive equation can be obtained by differentiating the secant constitutive equation as

$$\dot{\sigma} = E(D)\dot{\varepsilon} + \dot{E}(D)\varepsilon \tag{8.47}$$

The factor $\dot{E}(D)$ is zero when no new damage appears, i.e., during elastic loading or unloading. When damage occurs, (8.44) yields $\hat{\gamma} = \overline{E}\varepsilon$, and differentiating $E(D)$ in (8.46) we obtain

$$\dot{E}(D) = -\delta\alpha m\hat{\gamma}^{m-1}\exp\left(-\delta\alpha\hat{\gamma}^m\right)\overline{E}^2\dot{\varepsilon} \tag{8.48}$$

The tangent constitutive equation is obtained by substituting (8.48) into (8.47) when damage occurs, or $\dot{E}(D) = 0$ when no new damage appears. Therefore, the tangent constitutive equation can be written as

$$\dot{\sigma} = \begin{cases} (1 - \delta\alpha m\hat{\gamma}^m)\exp\left(-\delta\alpha\hat{\gamma}^m\right)\overline{E}\dot{\varepsilon} & \text{when } \varepsilon > \hat{\gamma}/\overline{E} \\ E(D)\dot{\varepsilon} & \text{when } \varepsilon < \hat{\gamma}/\overline{E} \end{cases} \tag{8.49}$$

NUMERICAL ALGORITHM

The one-dimensional damage model is implemented in ANSYS using the USERMAT sub-routine usermat1d.f, available in [19]. The following items describe the procedure used to explicitly evaluate the damage constitutive equation.[4]

1. Read the strain at time t

$$\varepsilon_t$$

2. Compute the effective stress (assuming strain equivalence)

$$\overline{\sigma}_t = \overline{E}\varepsilon_t$$

3. Update the threshold value

$$\hat{\gamma}_t = \max(\hat{\gamma}_{t-1}, \overline{\sigma}_t)$$

4. Compute the damage variable

$$D_t = 1 - \exp\left(-\delta\alpha(\hat{\gamma}_t)^m\right)$$

5. Compute the nominal stress

$$\sigma_t = (1 - D_t)\,\overline{E}\,\varepsilon_t$$

6. Compute the tangent stiffness

$$E_t^{dam} = \begin{cases} (1 - \delta\alpha m(\hat{\gamma}_t)^m)\exp\left(-\delta\alpha(\hat{\gamma}_t)^m\right)\overline{E} & \text{when } \hat{\gamma}_t > \hat{\gamma}_{t-1} \\ (1 - D_t)\overline{E} & \text{when } \hat{\gamma}_t = \hat{\gamma}_{t-1} \end{cases}$$

MODEL RESPONSE

See the ANSYS command list below, and user material subroutine usermat1d.F in [19], which are used to model a one-dimensional bar representative of a carbon fiber bundle. The nominal stress-strain response is shown with a solid line in Figure 8.6. The bundle fails at $\varepsilon_{cr} = 1.5\%$, in good agreement with the strain to failure reported by Toray.

```
/TITLE, Tensile response bundle Carbon Fiber T300, FEAcomp Example 8.3
/PREP7                 ! Start pre-processor module

!=== USER MATERIAL DECLARATION ====================================
TB,USER,1,1,3,   ! DECLARES USAGE OF USERMAT 1, MAT 1, PROPERTIES 3
TBTEMP,0
TBDATA,,230000,8.9,3.92e-033     ! PROPERTIES: E, m, delta_alpha
TB,STAT,1,,2,                    ! NUMBER OF STATE VARIABLES 2
!=================================================================
ET,   1,   180       ! LINK180, link element for analysis
R,1,1                ! Real constant #1, Area = 1
N,1                  ! Define node 1, coordinates=0,0,0
NGEN,6,1,1,,,2       ! Generate 5 additional nodes, x-inc= 2mm
E,1,2                ! Generate element 1 by node 1 to 2
EGEN,5,1,1           ! Generate element 2,3,4 and 5
FINISH               ! Exit pre-processor module
```

---

[4]See Section 8.4.1 for those cases for which it is not possible to integrate the constitutive equation explicitly.

```
/SOLU                  ! Start Solution module
ANTYPE,STATIC
OUTRES,ALL,1           ! Store results for each sub-step
D,1,all                ! Define b.c. on node 1, totally fixed
D,6,UX,0.25            ! Define horizontal displacement on node 6.
NSUBST,50,75,50        ! 50 = Number of sub-steps in this load step
SOLVE                  ! Solve load step
FINISH                 ! Exit solution module

/POST26          ! Start time-historic post-process
NSOL,2,6,U,X, UXnode6     ! Load displacements node 6
RFORCE,3,6,F,X, FXnode6 ! Load reaction force node 6
XVAR,2                 ! displacement x-graph variable
PLVAR,3                ! plot reaction as y-graph variable
lines,1000             !
PRVAR,2,3              ! list displacements and reactions
FINISH                 ! Exit post-process module
```

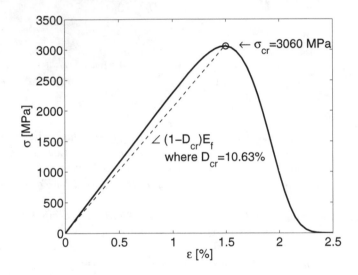

Fig. 8.6: Fiber tensile damage model response.

8.1.7 Fiber-Direction, Compression Damage

Many models have been proposed trying to improve the prediction of compression strength of composites first introduced by Rosen [20]. The literature encompasses fiber buckling modes [21, 22, 23, 24], kink-band models [25], and kink-bands induced by buckling [26]. In fiber buckling models, it is assumed that buckling of the fibers initiates a process that leads to the collapse of the material [20]. Rosen's model has been refined with the addition of initial fiber misalignment and nonlinear shear stiffness [21]. Experimental evidence suggests that fiber buckling of perfectly aligned fibers (Rosen's model) is an imperfection sensitive problem (see Section

4.1.1). Therefore, small amounts of imperfection (misalignment) cause large reductions in the buckling load, thus the reduction of the compression strength with respect to Rosen's prediction. Each fiber has a different value of fiber misalignment. The probability of finding a fiber with misalignment angle α is given by a Gaussian distribution [27, 23].

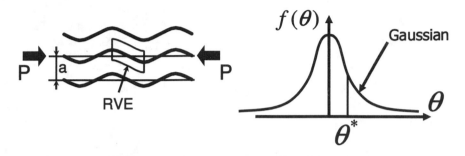

Fig. 8.7: One-dimensional random-strength model.

An optical technique [13] can be used to measure the misalignment angle of each fiber in the cross section. The resulting distribution of fiber misalignment was shown to be Gaussian by using the cumulative distribution function (CDF) plot and the probability plot [23]. Therefore, the probability density is

$$f(z) = \frac{\exp(-z^2)}{\Lambda\sqrt{2\pi}} \quad ; \quad z = \frac{\alpha}{\Lambda\sqrt{2}} \tag{8.50}$$

where Λ is the standard deviation and α is the continuous random variable, in this case being equal to the misalignment angle. The CDF gives the probability of obtaining a value smaller than or equal to some value of α, as follows

$$F(z) = \operatorname{erf}(z) = \frac{2}{\sqrt{\pi}} \int_0^z exp(-z'^2)dz' \tag{8.51}$$

where $erf(z)$ is the error function.

The relationship between the buckling stress and the imperfection (misalignment) is known in stability theory as the imperfection sensitivity curve. Several models from the literature can be used to develop this type of curve. The deterministic model, similar to the one presented by Wang [21] is developed in [24] but using the representation of the shear response given by Equation (8.52).

The shear stress-strain response of polymer-matrix composites can be represented [26, 23] by

$$\sigma_6 = F_6 \tanh\left(\frac{G_{12}}{F_6}\gamma_6\right) \tag{8.52}$$

where γ_6 is the in-plane shear strain. Furthermore, G_{12} is the initial shear stiffness and F_6 is the shear strength, which are obtained by fitting the stress-strain experimental data. Complete polynomial expansions [28] fit the experimental data well but they are not antisymmetric with respect to the origin. This introduces an artificial asymmetric bifurcation during the stability analysis [22]. Shear experimental data can be obtained by a variety of techniques including the ±45 coupon, 10° off-axis, rail shear, Iosipescu, Arcan, and torsion tests [29]. The nonlinear shear stress-strain curve should be measured for the actual composite being tested in compression.

Barbero [24] derived the equilibrium stress σ_{eq} as a function of the shear strain and the misalignment angle as

$$\bar{\sigma}_{eq}(\alpha, \gamma_6) = \frac{F_6}{2(\gamma_6 + \alpha)} \frac{(\sqrt{2}-1)a + (\sqrt{2}+1)(b-1)}{1-a+b}$$

$$a = \exp(\sqrt{2}g) - \exp(2g)$$
$$b = \exp(2g + \sqrt{2}g)$$

$$g = \frac{\gamma_6 G_{12}}{F_6}$$

(8.53)

with G_{12} and F_6 as parameters. Note that if the shear behavior is assumed to be linear $\bar{\sigma}_6 = G_{12}\gamma_6$ [30], then (8.53) does not have a maximum with respect to γ_6 and thus misaligned fibers embedded in a linear elastic matrix do not buckle. On the contrary, by using the hyperbolic tangent representation of shear (8.52), a maximum with respect to γ_6 is shown in (8.53). The maxima of the curves $\bar{\sigma}(\gamma_6)$ as a function of the misalignment angle α is the imperfection sensitivity curve, which represents the compression strength of a fiber (and surrounding matrix) as a function of its misalignment. For negative values of misalignment, it suffices to assume that the function is symmetric $\bar{\sigma}(-\alpha) = \bar{\sigma}(\alpha)$.

The stress carried by a fiber reduces rapidly after reaching its maximum because the load carrying capacity of a buckled fiber is much lower than the applied load. Several models can be constructed depending on the assumed load that a fiber carries after buckling. A lower bound can be found assuming that buckled fibers carry no more load because they have no post-buckling strength. According to the imperfection sensitivity equation (8.53), fibers with large misalignment buckle under low applied stress. If the post-buckling strength is assumed to be zero, the applied stress is redistributed onto the remaining, unbuckled fibers, which then carry a higher effective stress $\bar{\sigma}(\alpha)$. At any time during loading of the specimen, the applied load σ (applied stress times initial fiber area) is equal to the effective stress times the area of fibers that remain unbuckled

$$\sigma = \bar{\sigma}(\alpha)[1 - D(\alpha)]$$

(8.54)

where $0 \le D(\alpha) \le 1$ is the area of the buckled fibers per unit of initial fiber area. For any value of effective stress, all fibers having more than the corresponding value of misalignment have buckled. The area of buckled fibers $D(\alpha)$ is proportional to the area under the normal distribution located beyond the misalignment angle $\pm\alpha$.

Equation (8.54) has a maximum that corresponds to the maximum stress that can be applied to the composite. Therefore, the compression strength of the composite is found as

$$\sigma_c = \max\left[\bar{\sigma}(\alpha)\int_{-\alpha}^{\alpha} f(\alpha')d\alpha'\right]$$

(8.55)

where $\bar{\sigma}(\alpha)$ is given by Equation (8.53) and $f(\alpha')$ is given by (8.50). The maximum of (8.54), given by Equation (8.55) is a unique value for the compression strength of the composite that incorporates both the imperfection sensitivity and the distribution of fiber misalignment. Note that the standard deviation Λ is a parameter that describes the actual, measured, distribution of fiber misalignment, and it is not to be chosen arbitrarily as a representative value of fiber misalignment for all the fibers.

Since the distribution given in (8.50) cannot be integrated in closed form, (8.55) is evaluated numerically. However, it is advantageous to develop an explicit formula so that the compression strength can be easily predicted. Following the explicit formulation in [24], the compression strength of the unidirectional composite, explicitly in terms of the standard deviation of fiber misalignment Λ, the in-plane shear stiffness G_{12}, and the shear strength F_6 is

$$\frac{F_{1c}}{G_{12}} = \left(\frac{B_a}{a} + 1\right)^b \tag{8.56}$$

where $a = 0.21$ and $b = -0.69$ are two constants chosen to fit the numerical solution to the exact problem [24], with the dimensionless group B_a given by

$$B_a = \frac{G_{12}\Lambda}{F_6} \tag{8.57}$$

The misalignment angle of the fibers that buckle just prior to compression failure is given by

$$
\begin{aligned}
\alpha_{cr} &= a/b \\
a &= 1019.011G_{12}C_2^2\Lambda^3 - 375.3162C_2^3\Lambda^4 - 845.7457G_{12}^2C_2\Lambda^2 \\
&\quad + g\left(282.1113G_{12}C_2\Lambda^2 - 148.1863G_{12}^2\Lambda - 132.6943C_2^2\Lambda^3\right) \\
b &= 457.3229C_2^3\Lambda^3 - 660.77G_{12}C_2^2\Lambda^2 - 22.43143G_{12}^2C_2\Lambda \\
&\quad + g\left(161.6881C_2^2\Lambda^2 - 138.3753G_{12}C_2\Lambda - 61.38939G_{12}^2\right) \\
g &= \sqrt{C_2\Lambda\left(8.0C_2\Lambda - 9.424778G_{12}\right)} \\
C_2 &= -G_{12}^2/(4F_6)
\end{aligned}
\tag{8.58}
$$

Additionally, the shear strain at failure is

$$\gamma_{cr} = -\alpha_{cr} + \sqrt{\alpha_{cr}^2 + \frac{3}{2}\frac{\pi F_6 \alpha_{cr}}{G_{12}}} \tag{8.59}$$

In summary, when a fiber-reinforced lamina is compressed, the predominant damage mode is fiber buckling. However, the buckling load of the fibers is lower than that of the perfect system because of fiber misalignment, so much that a small amount of fiber misalignment could cause a large reduction in the buckling load. For each misalignment angle α, the composite area-fraction with buckled fibers $D(\alpha)$, corresponding to fibers with misalignment angle greater than α, can be taken as a measure of damage. If the fibers are assumed to have no post-buckling strength, then the applied stress is redistributed onto the remaining unbuckled fibers, which will be carrying a higher effective stress. The applied stress, which is lower than the effective stress by the factor $(1 - D)$, has a maximum, which corresponds to the compressive strength of the composite. Therefore, it is possible to compute the critical damage D_{1c} for longitudinal compressive loading as

$$D_{1c}^{cr} = 1 - \Omega_{1c} = 1 - erf\left(\frac{\alpha_{cr}}{\Lambda\sqrt{2}}\right) \tag{8.60}$$

where erf is the error function, Λ is the standard deviation of the actual Gaussian distribution of fiber misalignment (obtained experimentally [27]), and α_{cr} is the critical misalignment angle at failure. The three-dimensional theoretical formulation is developed in the next three sections.

8.2   Multidimensional Damage and Effective Spaces

The first step in the formulation of a general multidimensional damage model is to define the damage variable as well as the effective stress and strain spaces, as shown in this section. The second step is to define the form of either the Helmholtz free energy or the Gibbs energy and from them derive the thermodynamic forces conjugate to the state variables representing damage and hardening, as shown in Section 8.3. The third step is to derive the kinetic laws governing the rate of damage hardening, which are functions of the damage and hardening potentials, as shown in Section 8.4.

Experimental knowledge of the degradation and subsequent material response is used to guide the selection on the variable used to represent damage. A second-order damage tensor $\mathbf{D}$ can be used to represent damage of orthotropic fiber-reinforced composite materials, following Kachanov-Rabotnov's approach [3, 31]. For composite materials reinforced with stiff and strong fibers, damage can be accurately represented by a second-order tensor[5] with principal directions aligned with the material directions $(1, 2, 3)$ [32, 33, 34, 35, 36]. This is due to the fact that the dominant modes of damage are micro-cracks, fiber breaks, and fiber-matrix de-bond, all of which can be conceptualized as cracks either parallel or perpendicular to the fiber direction.[6] Therefore, the damage tensor can be written (see Appendix C) as

$$\mathbf{D} = D_{ij} = D_i \delta_{ij} \quad \text{no sum on i} \tag{8.61}$$

where D_i are the eigenvalues of $\mathbf{D}$, which represent the net stiffness reduction along the principal material directions n_i, and δ_{ij} is the Kronecker delta ($\delta_{ij} = 1$ if $i = j$, or zero otherwise). The integrity tensor is also diagonal, and using energy equivalence (8.8), we have

$$\mathbf{\Omega} = \Omega_{ij} = \sqrt{1 - D_i}\,\delta_{ij} \quad \text{no sum on i} \tag{8.62}$$

The integrity tensor is always symmetric and positive, because the net area reduction must be positive definite during damage evolution [37]. Both tensors

---

[5]Tensors are denoted by boldface type, or by their components with index notation.

[6]Strictly speaking, damage is transversely isotropic since cracks can also be aligned along any direction in the 2-3 plane.

are diagonal when represented in the principal system. Introducing a symmetric fourth-order tensor, $\mathbf{M}$, called the *damage effect tensor*, as

$$\mathbf{M} = M_{ijkl} = \frac{1}{2} \left(\Omega_{ik}\Omega_{jl} + \Omega_{il}\Omega_{jk} \right) \tag{8.63}$$

the transformation of stress and strain between the effective and damaged configurations is accomplished as follows

$$\overline{\boldsymbol{\sigma}} = \mathbf{M}^{-1} : \boldsymbol{\sigma} \quad ; \quad \overline{\boldsymbol{\varepsilon}}^e = \mathbf{M} : \boldsymbol{\varepsilon}^e \tag{8.64}$$

where an over-bar indicates that the quantity is evaluated in the effective configuration and the superscript e denotes quantities in the elastic domain.

By the energy equivalence hypothesis [4], it is possible to define the constitutive equation in the effective configuration (Figure 8.1.c) as

$$\overline{\boldsymbol{\sigma}} = \overline{\mathbf{C}} : \overline{\boldsymbol{\varepsilon}}^e \quad ; \quad \overline{\boldsymbol{\varepsilon}}^e = \overline{\mathbf{C}}^{-1} : \overline{\boldsymbol{\sigma}} = \overline{\mathbf{S}} : \overline{\boldsymbol{\sigma}} \tag{8.65}$$

where the fourth-order tensors $\mathbf{C}$ and $\mathbf{S}$ denote the secant stiffness tensor and compliance tensor, respectively. The stress-strain equations in the damaged configuration (Figure 8.1.b) are obtained by substituting (8.65) into (8.64),

$$\begin{aligned}
\boldsymbol{\sigma} &= \mathbf{M} : \overline{\boldsymbol{\sigma}} = \mathbf{M} : \overline{\mathbf{C}} : \overline{\boldsymbol{\varepsilon}}^e, & \boldsymbol{\varepsilon}^e &= \mathbf{M}^{-1} : \overline{\boldsymbol{\varepsilon}}^e = \mathbf{M}^{-1} : \overline{\mathbf{S}} : \overline{\boldsymbol{\sigma}}, \\
\boldsymbol{\sigma} &= \mathbf{M} : \overline{\mathbf{C}} : \mathbf{M} : \boldsymbol{\varepsilon}^e, & \boldsymbol{\varepsilon}^e &= \mathbf{M}^{-1} : \overline{\mathbf{S}} : \mathbf{M}^{-1} : \boldsymbol{\sigma}, \\
\boldsymbol{\sigma} &= \mathbf{C} : \boldsymbol{\varepsilon}^e & \boldsymbol{\varepsilon}^e &= \mathbf{S} : \boldsymbol{\sigma}
\end{aligned} \tag{8.66}$$

with

$$\mathbf{C} = \mathbf{M} : \overline{\mathbf{C}} : \mathbf{M} \qquad\qquad \mathbf{S} = \mathbf{M}^{-1} : \overline{\mathbf{S}} : \mathbf{M}^{-1} \tag{8.67}$$

Given that the tensor $\mathbf{M}$ is symmetric, the secant stiffness and compliance tensors are also symmetric.

8.3 Thermodynamics Formulation

The damage processes considered in this chapter can be described by a series of equilibrium states reached while the system traverses a nonequilibrium path due to the irreversibility of damage and plasticity. In general, the current state of a system (e.g., stress, stiffness, compliance) depends on the current state (e.g., strain) as well as on the history experienced by the system. This is the case for viscoelastic materials discussed in Chapter 7. However, for damaging and elastoplastic materials, the current state can be described in terms of the current strain and the *effects of history* on the material, which in this chapter are characterized by the damage tensor $\mathbf{D}$ and the plastic strain tensor ε^{p}.

8.3.1   First Law

The *first law* of thermodynamics states that any increment of internal energy of the system is equal to the heat added to the system minus the work done by the system on its surroundings

$$\delta U = \delta Q - \delta W \tag{8.68}$$

The system under consideration in this section is a representative volume element (RVE), which is the smallest volume element that contains sufficient features of the microstructure and irreversible processes, such as damage and plasticity, to be representative of the material as a whole. Further discussions about the RVE can be found in Chapter 6.

In rate form, (8.68) is

$$\dot{U} = \dot{Q} - \dot{W} \tag{8.69}$$

where

$$\dot{U} = \frac{d}{dt} \int_{\Omega} \rho u \, dV \tag{8.70}$$

Here ρ is the density, Ω is the volume of the RVE, and u is the *internal energy density*, which is an internal variable and a potential function.[7]

For a deformable solid, the rate of work done *by* the system is minus the product of the stress *applied* on the system times the rate of strain

$$\dot{W} = - \int_{\Omega} \boldsymbol{\sigma} : \dot{\boldsymbol{\varepsilon}} \, dV \tag{8.71}$$

where $\boldsymbol{\varepsilon}$ is the *total* strain (see (8.124)).

The heat flow into the RVE is given by

$$\dot{Q} = \int_{\Omega} \rho r \, dV - \int_{\partial\Omega} \mathbf{q} \cdot \mathbf{n} dA \tag{8.72}$$

where r is the radiation heat per unit mass, $\mathbf{q}$ is the heat flow vector per unit area, and $\mathbf{n}$ is the outward normal vector to the surface $\partial\Omega$ enclosing the volume Ω. Since the volume Ω of the RVE does not change with time, and using the divergence theorem,[8] the *first law at the local level* becomes

$$\rho\dot{u} = \boldsymbol{\sigma} : \dot{\boldsymbol{\varepsilon}} + \rho r - \nabla \cdot \mathbf{q} \tag{8.73}$$

The internal energy accounts for all the energy stored into the system. For example, a system undergoing elastic deformation $\delta\varepsilon^e$, raising temperature δT, and

---

[7]The values of the potential functions depend on the state and not on the path or process followed by the system to reach such state [38].

[8]$(\int_{\partial\Omega} \mathbf{q} \cdot \mathbf{n} dA = \int_{\Omega} \nabla \cdot \mathbf{q} dV)$; $div(\mathbf{q}) = \nabla \cdot \mathbf{q} = \partial \mathbf{q}_i / \partial x_i$

damage in the form of cracks of area growing by δA_c, undergoes a change of internal energy u given[9] by

$$\delta u = \boldsymbol{\sigma} : \delta \boldsymbol{\varepsilon}^e + C_p \delta T - (G - G_c) \delta A_c \qquad (8.74)$$

where G is the strain energy release rate, G_c is the surface energy needed to create the increment of the two surfaces of an advancing crack, and $C_p = C_v$ is the specific heat capacity of the solid.

In general $\boldsymbol{\varepsilon} = \boldsymbol{\varepsilon}(\boldsymbol{\sigma}, u, s_\alpha)$ where s_α are internal variables. Let's assume for the time being that the system is adiabatic, i.e., $\rho r - \nabla \cdot \mathbf{q} = 0$. Further, if there are no dissipative effects or heat transfer, then u is a function of $\boldsymbol{\varepsilon}$ only, $u = u(\boldsymbol{\varepsilon}^e)$, where $\boldsymbol{\varepsilon}^e$ is the elastic strain. In such case, the internal energy density reduces to the strain energy density, which in rate form is

$$\dot{\varphi}(\boldsymbol{\varepsilon}) = \boldsymbol{\sigma} : \dot{\boldsymbol{\varepsilon}}^e \qquad (8.75)$$

and the complementary strain energy density is

$$\dot{\varphi}^*(\boldsymbol{\varepsilon}) = \boldsymbol{\sigma} : \dot{\boldsymbol{\varepsilon}}^e - \dot{\varphi} = \dot{\boldsymbol{\sigma}} : \boldsymbol{\varepsilon}^e \qquad (8.76)$$

8.3.2   Second Law

The *second law* of thermodynamics formalizes the fact that heat flows from hot to cold. Mathematically, the heat flow $\mathbf{q}$ has opposite direction to the gradient[10] of temperature T, which is formally written as

$$\mathbf{q} \cdot \nabla T \leq 0 \qquad (8.77)$$

where the equal sign holds true only for adiabatic processes, i.e., when there is no heat exchange and thus no thermal irreversibility.

Let's visualize a process of heat transfer from a hot reservoir to a cold reservoir, happening in such a way that no heat is lost to, and no work is exchanged with the environment. Once heat has flowed to the cold reservoir, it is impossible to transfer it back to the hot reservoir without adding external work. That is, the process of heat transfer is irreversible even though, on account of the first law energy balance (8.73), no energy has been lost. For future use (8.77) can be written[11] as

$$\mathbf{q} \cdot \nabla T^{-1} \geq 0 \qquad (8.78)$$

The second law justifies the introduction of a new internal variable, the entropy density $s = s(u, \varepsilon)$, which is also a potential function [39]. According to the second law, the entropy density rate is $\dot{s} \geq 0$, where the equal sign holds true only for adiabatic processes.

---

[9]Thermodynamics custom and [41] are followed here in representing the internal energy with the letter u, not to be confused with the displacement vector $\mathbf{u}$ used elsewhere.

[10]The gradient of a scalar yields a vector, $\nabla T = \partial T / \partial x_i$

[11]$\nabla T^{-1} = -T^{-2} \nabla T$

Assume the specific entropy $s = s(u, \varepsilon)$ is a potential function such that for a reversible process [39]

$$ds = \left(\frac{\delta q}{T}\right)_{rev} \tag{8.79}$$

with $\delta Q = \int_{\Omega} \rho \, \delta q \, d\Omega$, where $\delta q = r - \rho^{-1}\nabla \cdot \mathbf{q}$ is the heat input per unit mass, and $S = \int_{\Omega} s \, \rho \, d\Omega$ is the entropy. We use δ, not d, to emphasize that δq is not the differential (perfect or total) of any (potential) function.

As a preamble to the definition of conjugate variables (see (8.86, 8.92, 8.99)), note that using (8.79), the first law can be rewritten for a *reversible* process on an ideal gas ($pv = RT$), as the *Gibbs* equation for an ideal gas,

$$du = T \, ds - p \, dv \tag{8.80}$$

where v is the specific volume (volume per unit mass). It can be seen in (8.80) that v is conjugate to $-p$ for calculating work input for an ideal gas and s is conjugate to T for calculating thermal energy input.

For a *cyclic reversible* process returning to its initial state characterized by state variables (e.g., u, T, ε), by virtue of (8.79) we have $\oint ds = \oint \left(\frac{\delta q}{T}\right)_{rev} = 0$. Since s is a potential function but q is not, for an *irreversible* process we have $\oint ds = 0$ but $\oint \left(\frac{\delta q}{T}\right)_{irrev} < 0$, as corroborated by experiments. The heat δq entering at temperature T_i provides less entropy input $\delta q/T_i$ than the entropy output $\delta q/T_o$ leaving the same cycle at temperature $T_o < T_i$ (see also [40, Example 6-2]). Since entropy is a potential function, and therefore a state variable, it always satisfies $\oint ds = 0$. Therefore, a negative net entropy supply must be compensated by internal entropy production. The entropy of a system can be raised or lowered by adding or extracting heat (in the form of $\delta q/T$) but it is always raised by internal irreversible processes, such as crack formation, and so on.

Adiabatic systems do not exchange heat with the surroundings ($\delta q = 0$), so the only change in entropy is due to internal irreversibility $\dot{s} \geq 0$, where the equal sign holds for reversible processes only. Note that any system and its surroundings can be made adiabatic by choosing sufficiently large surroundings, e.g., the universe. For an arbitrary system, the total entropy rate is greater than (or equal to) the net entropy input due to heat

$$\dot{s} \geq \frac{r}{T} - \frac{1}{\rho}\nabla \cdot \left(\frac{\mathbf{q}}{T}\right) \tag{8.81}$$

The left-hand side of (8.81) represents the total entropy rate of the system. The right-hand side of (8.81) represents the external entropy supply rate. The difference is the internal entropy production rate

$$\dot{\gamma}_s = \dot{s} - \frac{r}{T} + \frac{1}{\rho}\nabla \cdot \left(\frac{\mathbf{q}}{T}\right) \geq 0 \tag{8.82}$$

Equation 8.82 is called the *local Clausius-Duhem inequality*. Noting that $\nabla \cdot (T^{-1}\mathbf{q}) = T^{-1}\nabla \cdot \mathbf{q} + \mathbf{q}\nabla \, T^{-1}$ results in

$$\dot{\gamma}_s = \dot{s} - \frac{1}{\rho T}\left(\rho r - \nabla \cdot \mathbf{q}\right) + \frac{1}{\rho}\,\mathbf{q}\cdot\nabla\,T^{-1} \geq 0 \qquad (8.83)$$

where the first two terms represent the local entropy production due to local dissipative phenomena, and the last term represents the entropy production due to heat conduction[12] [39]. Assuming it is possible to identify all local dissipative phenomena, their contributions can be written as products of conjugate variables $p_\alpha \dot{s}_\alpha \geq 0$, and (8.83) can be written as

$$\rho T\dot{\gamma}_s = p_\alpha \dot{s}_\alpha + T\mathbf{q}\cdot\nabla T^{-1} \geq 0 \qquad (8.84)$$

where $\alpha = 1\ldots n$, spans the total number of dissipative phenomena considered. Note that the dissipation is a scalar given by the contracted product of a thermodynamic force p_α times the increment of a measurable state variable s_α. The state variable, also called thermodynamic flux, describes univocally the effects of history (e.g., yield, damage) on the material. Note that γ_s is defined as an entropy, not as a dissipation heat, so that it is a potential function, while q is not.

For the particular case of damage due to penny shaped cracks growing self similarly [8], the state variable is the crack area A_c and the thermodynamic force is the energy available to grow the cracks $p_c = G - G_c$, which is equal to the difference between the energy release rate (ERR) G and the critical ERR $G_c = 2\gamma_c$, the latter being equal to twice the surface energy because two surfaces must be created every time a crack appears (see Chapter 10). In this case, the dissipation (heat) is $\rho T\dot{\gamma} = p_c\dot{A}_c$.

From the first law (8.73), considering an adiabatic process $(\rho r - \nabla \cdot \mathbf{q})$ and using the chain rule $\dot{u} = \partial u/\partial\varepsilon : \dot{\varepsilon}$ we have

$$\left[\boldsymbol{\sigma} - \rho\frac{\partial u}{\partial\varepsilon}\right] : \dot{\varepsilon} = 0 \qquad (8.85)$$

Since $\dot{\varepsilon} = 0$ would be a trivial solution, the stress tensor, conjugate to strain, is defined as

$$\boldsymbol{\sigma} = \rho\frac{\partial u}{\partial\varepsilon} \qquad (8.86)$$

The Clausius-Duhem inequality (8.83) for an *isothermal* $\nabla T = 0$ system reduces to

$$\dot{\gamma}_s = \dot{s} - \frac{1}{\rho T}\left(\rho r - \nabla \cdot \mathbf{q}\right) \geq 0 \qquad (8.87)$$

and using the first law we get

$$\rho T\dot{\gamma}_s = \rho T\dot{s} - (\rho\dot{u} - \boldsymbol{\sigma} : \varepsilon) \geq 0 \qquad (8.88)$$

The *Helmholtz free energy (HFE) density* is defined as

$$\psi(T, \varepsilon, s_\alpha) = u - Ts \qquad (8.89)$$

---

[12]Even absent local dissipative phenomena, $\mathbf{q}\cdot\nabla T^{-1} \geq 0$ represents the well-known fact that heat flows opposite to the temperature gradient ∇T.

which is also a potential function. The corresponding extensive function is the Helmholtz free energy[13] $A = \int_\Omega \rho\psi dV$. The rate of change of HFE density is

$$\dot{\psi} = \dot{u} - \dot{T}s - T\dot{s} \tag{8.90}$$

and introducing (8.88), with $\dot{\gamma}_s = 0$ at an equilibrium state, we get

$$\rho\dot{\psi} = -\rho s\dot{T} + \boldsymbol{\sigma} : \dot{\varepsilon} \tag{8.91}$$

from which an alternative definition of stress, conjugate to strain, is found as

$$\boldsymbol{\sigma} = \rho\frac{\partial\psi}{\partial\varepsilon} = \mathbf{C} : \varepsilon \tag{8.92}$$

where the secant elastic stiffness, which is affected by dissipative phenomena, including damage, is defined as

$$\mathbf{C}(s_\alpha) = \rho\frac{\partial^2\psi}{\partial\varepsilon\partial\varepsilon} \tag{8.93}$$

Using the first law (8.73) in the internal entropy production per unit volume, or local Clausius-Duhem inequality (8.83), and expanding $\nabla \cdot (\mathbf{q}T^{-1}) = T^{-1}\nabla \cdot \mathbf{q} + \mathbf{q} \cdot \nabla T^{-1}$, we get

$$\rho T\dot{\gamma}_s = \frac{\mathbf{q}}{T} \cdot \nabla T^{-1} - \rho\left(\dot{\psi} + s\dot{T} - \rho^{-1}\boldsymbol{\sigma} : \dot{\varepsilon}\right) \geq 0 \tag{8.94}$$

Realizing that $\nabla T^{-1} = -\nabla T/T^2$, the Clausius-Duhem inequality becomes

$$T\rho\dot{\gamma}_s = \boldsymbol{\sigma} : \dot{\varepsilon} - \rho\left(\dot{\psi} + s\dot{T}\right) - \frac{\mathbf{q}}{T} \cdot \nabla T \geq 0 \tag{8.95}$$

Since the HFE density is a function of the internal variables ε, T, s_α, we have

$$\dot{\psi} = \left.\frac{\partial\psi}{\partial\varepsilon}\right|_{T,s_\alpha} : \dot{\varepsilon} + \left.\frac{\partial\psi}{\partial T}\right|_{\varepsilon,s_\alpha} \dot{T} + \left.\frac{\partial\psi}{\partial s_\alpha}\right|_{\varepsilon,T} \dot{s}_\alpha \quad ; \quad \alpha = 1\ldots n \tag{8.96}$$

where $\left.\dfrac{\partial}{\partial y}\right|_x$ represents the partial derivative with respect to y at constant x.

Inserting (8.96) into (8.95), using (8.89), (8.92), and $\nabla T^{-1} = -\nabla T/T^2$, the second law can be written as follows

$$\dot{\gamma} = \rho T\dot{\gamma}_s = -\rho\frac{\partial\psi}{\partial s_\alpha}s_\alpha + T\mathbf{q} \cdot \nabla T^{-1} \geq 0 \tag{8.97}$$

where $\dot{\gamma}$ is the heat dissipation rate per unit volume. Comparing (8.97) to (8.84) it becomes clear that $-\rho\partial\psi/\partial s_\alpha = p_\alpha$ are the thermodynamic forces conjugated to s_α, which provides a definition for the thermodynamic forces.

---

[13]The nomenclature of [41] has been used.

The complementary free-energy density, or *Gibbs energy density*, is defined as

$$\chi = \rho^{-1}\boldsymbol{\sigma} : \boldsymbol{\varepsilon} - \psi \qquad (8.98)$$

which is also a potential function. The corresponding extensive function is the Gibbs energy[14] $G = \int_\Omega \rho\chi\, dV$. From (8.98) it follows the definition of strain, conjugate to stress, and the definition of the thermodynamic forces, conjugate to the state variables s_α, as

$$\boldsymbol{\varepsilon} = \rho\frac{\partial\chi}{\partial\boldsymbol{\sigma}} \quad ; \quad p_\alpha = \rho\frac{\partial\chi}{\partial s_\alpha} = -\rho\frac{\partial\psi}{\partial s_\alpha} \qquad (8.99)$$

where s_α includes the damage variables and consequently p_α includes the thermodynamic damage forces (see Example 8.4).

The secant elastic compliance, which is affected by dissipative phenomena, including damage, is defined by

$$S(s_\alpha) = \rho\frac{\partial^2\chi}{\partial\boldsymbol{\sigma}\partial\boldsymbol{\sigma}} \qquad (8.100)$$

Example 8.4 *A damage model able to represent the onset and accumulation of transverse matrix cracks should yield a compliance tensor similar to that obtained experimentally. Experimental evidence shows that the terms S_{22} and S_{66} are the most affected by transversal cracking. Define the form of the Gibbs free energy to yield a compliance matrix that represents the experimentally observed behavior. Using the energy equivalence principle (8.8), obtain the effective stress. Also, compute the thermodynamic forces associated to this model. The model is restricted to represent a laminate in a state of plane stress, under in-plane stress only, and negligible damage along the fiber direction ($D_1 = 0$). Use tensor components of strain ($\varepsilon_1, \varepsilon_2, \varepsilon_6$).*

Solution to Example 8.4 *The following Gibbs free energy is proposed, in expanded form and using Voigt contracted notation, as*

$$\chi = \frac{1}{2\rho}\left[\frac{\sigma_1^2}{E_1} + \frac{\sigma_2^2}{(1-D_2)^2\, E_2} + \frac{\sigma_6^2}{(1-D_6)^2\, G_{12}} - \left(\frac{\nu_{21}}{E_2} + \frac{\nu_{12}}{E_1}\right)\frac{\sigma_1\sigma_2}{1-D_2}\right]$$

where E_1, E_2, ν_{12}, ν_{21} and G_{12} are the in-plane elastic orthotropic properties of a unidirectional lamina where the subscript $()_1$ denotes the fiber direction and $()_2$ denotes the transverse direction. Transverse tension and shear loads create transverse matrix cracks. The associated damage variables D_2 and D_6 can be related to the matrix crack density. The model proposed distinguishes between active (D_{2+}) and passive damage (D_{2-}) variables, corresponding to the opening or closure of transverse matrix cracks, respectively. The determination of the active damage variable is based on the following equation:

$$D_2 = D_{2+}\frac{\langle\sigma_2\rangle}{|\sigma_2|} + D_{2-}\frac{\langle-\sigma_2\rangle}{|\sigma_2|}$$

where $\langle x \rangle$ is defined as $\langle x \rangle := \frac{1}{2}(x + |x|)$.

---

[14]The nomenclature of [41] has been used.

The constitutive model is defined from the derivative of the Gibbs free energy with respect to the stress tensor:

$$\varepsilon = \rho \frac{\partial \chi}{\partial \boldsymbol{\sigma}} = \mathbf{S} : \boldsymbol{\sigma}$$

where the compliance tensor $\mathbf{S}$ is defined as:

$$\mathbf{S} = \rho \frac{\partial^2 \chi}{\partial \boldsymbol{\sigma}^2}$$

Then, the compliance tensor $\mathbf{S}$ can be represented in Voigt notation for a plane stress state as:

$$\mathbf{S} = \begin{bmatrix} \dfrac{1}{E_1} & -\dfrac{\nu_{21}}{E_2\left(1 - D_2\right)} & 0 \\ -\dfrac{\nu_{12}}{E_1\left(1 - D_2\right)} & \dfrac{1}{E_2\left(1 - D_2\right)^2} & 0 \\ 0 & 0 & \dfrac{1}{2G_{12}\left(1 - D_6\right)^2} \end{bmatrix}$$

where the damage variables appear in S_{22} and S_{66}. Using the principle of energy equivalence and (8.67), the compliance matrix can be written as

$$\mathbf{S} = \mathbf{M}^{-1} : \overline{\mathbf{S}} : \mathbf{M}^{-1}$$

where the undamaged compliance is

$$\overline{\mathbf{S}} = \begin{bmatrix} \dfrac{1}{E_1} & -\dfrac{\nu_{21}}{E_2} & 0 \\ -\dfrac{\nu_{12}}{E_1} & \dfrac{1}{E_2} & 0 \\ 0 & 0 & \dfrac{1}{2G_{12}} \end{bmatrix}$$

and where the effective damage tensor $\mathbf{M}$ in Voigt contracted notation is

$$\mathbf{M} = \begin{bmatrix} 1 & 0 & 0 \\ 0 & \left(1 - D_2\right) & 0 \\ 0 & 0 & \left(1 - D_6\right) \end{bmatrix}$$

The stiffness tensor $\mathbf{C}$ is obtained by

$$\mathbf{C} = \mathbf{M} : \overline{\mathbf{C}} : \mathbf{M}$$

where in this example we obtain

$$\mathbf{C} = \begin{bmatrix} \dfrac{E_1}{1 - \nu_{21}\nu_{12}} & \dfrac{\nu_{12}E_2\left(1 - D_2\right)}{1 - \nu_{21}\nu_{12}} & 0 \\ \dfrac{\nu_{21}E_1\left(1 - D_2\right)}{1 - \nu_{21}\nu_{12}} & \dfrac{E_2\left(1 - D_2\right)^2}{1 - \nu_{21}\nu_{12}} & 0 \\ 0 & 0 & 2G_{12}\left(1 - D_6\right)^2 \end{bmatrix}$$

The effective stress tensor $\overline{\boldsymbol{\sigma}}$ is related to the nominal stress tensor $\boldsymbol{\sigma}$ by effective damage tensor $\mathbf{M}$ using $\overline{\boldsymbol{\sigma}} = \mathbf{M}^{-1} : \boldsymbol{\sigma}$, which yields

$$\overline{\boldsymbol{\sigma}}^T = \left\{ \sigma_1, \frac{\sigma_2}{1 - D_2}, \frac{\sigma_6}{1 - D_6} \right\}$$

The thermodynamic forces are obtained by using $Y = \rho \partial \chi / \partial D$. Therefore, the thermodynamic forces for this example can be written in contracted Voigt notation as

$$\mathbf{Y} = \left\{ \begin{array}{c} Y_1 \\ Y_2 \\ Y_6 \end{array} \right\} = \left\{ \begin{array}{c} 0 \\ \dfrac{\sigma_2^{\,2}}{(1 - D_2)^3 E_2} - \dfrac{\sigma_1 \sigma_2 \nu_{12}}{(1 - D_2)^2 E_1} \\ \dfrac{\sigma_6^{\,2}}{(1 - D_6)^3 G_{12}} \end{array} \right\}$$

8.4 Kinetic Law in Three-Dimensional Space

The damage variable $\mathbf{D}$ introduced in Section 8.2 is a state variable that represents the history of what happened to the material. Next, a kinetic equation is needed to predict the evolution of damage in terms of the thermodynamic forces. Kinetic equations can be written directly in terms of internal variables as in (8.21) or as derivatives of potential functions. For three-dimensional problems, it is convenient to derive the kinetic law from a potential function, similar to the flow potential used in plasticity theory.

Two functions are needed. A damage surface $g(\mathbf{Y}(\mathbf{D}), \gamma(\delta)) = 0$ and a convex damage potential $f(\mathbf{Y}(\mathbf{D}), \gamma(\delta)) = 0$ are postulated. The damage surface delimits a region in the space of thermodynamic forces $\mathbf{Y}$ where damage does not occur because the thermodynamic force $\mathbf{Y}$ is inside the surface $g = 0$. The function $\gamma(\delta)$ accomplishes the expansion of g and f needed to model hardening. The damage potential controls the direction of damage evolution (8.102).

If the damage surface and the damage potential are identical ($g = f$), the model is said to be associated and the computational implementation is simplified significantly. For convenience, the damage surface is assumed to be separable in the variables $\mathbf{Y}$ and γ, and written as the sum (see (8.11)-(8.12))

$$g(\mathbf{Y}(\mathbf{D}), \gamma(\delta)) = \hat{g}(\mathbf{Y}(\mathbf{D})) - (\gamma(\delta) + \gamma_0) \tag{8.101}$$

where $\mathbf{Y}$ is the thermodynamic force tensor, $\gamma(\delta)$ is the hardening function, γ_0 is the damage threshold, and δ is the hardening variable.

As a result of damage, $\hat{g}$ may grow but the condition $g \leq 0$ must be satisfied. This is possible by decreasing the value of the hardening function $\gamma(\delta)$, effectively allowing $\hat{g}(\mathbf{Y}(\mathbf{D}))$ to grow. Formally, the hardening function $\gamma(\delta)$ can be derived from the dissipation potential as per (8.99),(8.123), provided the form of the potential can be inferred from knowledge about the hardening process. Alternatively, the form of the function (e.g., polynomial, Prony series, etc.) can be chosen so that the complete model fits adequately the experimental data available. The latter approach is more often followed in the literature, as well as in Section 9.

When $g = 0$, damage occurs, and a kinetic equation is needed to determine the magnitude and components of the damage $\dot{\mathbf{D}}$. This is accomplished by

$$\dot{\mathbf{D}} = \frac{\partial \mathbf{D}}{\partial \mathbf{Y}} = \dot{\lambda} \frac{\partial f}{\partial \mathbf{Y}} \tag{8.102}$$

where $\dot{\lambda}$ yields the magnitude of the damage increment and $\partial f / \partial \mathbf{Y}$ is a direction in $\mathbf{Y}$-space. To find the damage multiplier $\dot{\lambda}$, it is postulated that $\dot{\lambda}$ is also involved in the determination of the rate of change of the hardening variable as follows

$$\dot{\delta} = \dot{\lambda} \frac{\partial f}{\partial \gamma} \qquad (8.103)$$

There are two possible situations regarding g and $\dot{\lambda}$.

1. If $g < 0$, damage is not growing and $\dot{\lambda} = 0$, so $\dot{\mathbf{D}} = 0$.

2. If $g = 0$, damage occurs and $\dot{\lambda} > 0$, so $\dot{\mathbf{D}} > 0$.

These are summarized by the Kuhn-Tucker conditions

$$\dot{\lambda} \geq 0 \quad ; \quad g \leq 0 \quad ; \quad \dot{\lambda} g = 0 \qquad (8.104)$$

The value of $\dot{\lambda}$ can be determined by the consistency condition, which leads to

$$\dot{g} = \frac{\partial g}{\partial \mathbf{Y}} : \dot{\mathbf{Y}} + \frac{\partial g}{\partial \gamma} \dot{\gamma} = 0 ; \quad g = 0 \qquad (8.105)$$

On the other hand, the rates of thermodynamic forces and hardening function can be written as

$$\dot{\mathbf{Y}} = \frac{\partial \mathbf{Y}}{\partial \varepsilon} : \dot{\varepsilon} + \frac{\partial \mathbf{Y}}{\partial \mathbf{D}} : \dot{\mathbf{D}}$$

$$\dot{\gamma} = \frac{\partial \gamma}{\partial \delta} \dot{\delta} \qquad (8.106)$$

or in function of $\dot{\lambda}$, introducing (8.103) and (8.104) into (8.106) as follows

$$\dot{\mathbf{Y}} = \frac{\partial \mathbf{Y}}{\partial \varepsilon} : \dot{\varepsilon} + \dot{\lambda} \frac{\partial \mathbf{Y}}{\partial \mathbf{D}} : \frac{\partial f}{\partial \mathbf{Y}}$$

$$\dot{\gamma} = \frac{\partial \gamma}{\partial \delta} \dot{\lambda} \frac{\partial f}{\partial \gamma} \qquad (8.107)$$

Introducing (8.107) into (8.105) we obtain the following equation

$$\dot{g} = \frac{\partial g}{\partial \mathbf{Y}} : \left[\frac{\partial \mathbf{Y}}{\partial \varepsilon} : \dot{\varepsilon} + \dot{\lambda} \frac{\partial \mathbf{Y}}{\partial \mathbf{D}} : \frac{\partial f}{\partial \mathbf{Y}} \right] + \frac{\partial g}{\partial \gamma} \frac{\partial \gamma}{\partial \delta} \dot{\lambda} \frac{\partial f}{\partial \gamma} = 0 \qquad (8.108)$$

For an associated model, $g = f$, and a damage surface with separable variables $\mathbf{Y}$ and γ, $\partial f / \partial \gamma = \partial g / \partial \gamma = -1$, (8.108) can be written as

$$\dot{g} = \frac{\partial g}{\partial \mathbf{Y}} : \frac{\partial \mathbf{Y}}{\partial \varepsilon} : \dot{\varepsilon} + \left[\frac{\partial g}{\partial \mathbf{Y}} : \frac{\partial \mathbf{Y}}{\partial \mathbf{D}} : \frac{\partial f}{\partial \mathbf{Y}} + \frac{\partial \gamma}{\partial \delta} \right] \dot{\lambda} = 0 \qquad (8.109)$$

Therefore, the damage multiplier $\dot{\lambda}$ can be obtained as

$$\dot{\lambda} = \begin{cases} \mathbf{L}^d : \dot{\varepsilon} & g = 0 \\ 0 & g < 0 \end{cases} \qquad (8.110)$$

where

$$\mathbf{L}^d = -\frac{\dfrac{\partial g}{\partial \mathbf{Y}} : \dfrac{\partial \mathbf{Y}}{\partial \varepsilon}}{\dfrac{\partial g}{\partial \mathbf{Y}} : \dfrac{\partial \mathbf{Y}}{\partial \mathbf{D}} : \dfrac{\partial f}{\partial \mathbf{Y}} + \dfrac{\partial \gamma}{\partial \delta}} \tag{8.111}$$

Equations (8.103), (8.104), and (8.110) yield the pair $\mathbf{D}, \delta$, in rate form as

$$\dot{\mathbf{D}} = \mathbf{L}^d : \frac{\partial f}{\partial Y} : \dot{\varepsilon} \quad ; \quad \dot{\delta} = -\dot{\lambda} \tag{8.112}$$

The tangent constitutive equation can be obtained by differentiation of the constitutive equation $\boldsymbol{\sigma} = \mathbf{C} : \varepsilon$, which yields

$$\dot{\sigma} = \mathbf{C} : \dot{\varepsilon} + \dot{\mathbf{C}} : \varepsilon \tag{8.113}$$

where the last term represents the stiffness reduction. On the other hand, the incremental stress-strain relation for damage evolution in the effective configuration can be expressed using the chain rule as

$$\dot{\sigma} = \frac{\partial \sigma}{\partial \varepsilon} : \dot{\varepsilon} + \frac{\partial \sigma}{\partial \mathbf{D}} : \dot{\mathbf{D}} \tag{8.114}$$

and introducing (8.112) into (8.114) we obtain

$$\dot{\sigma} = \frac{\partial \sigma}{\partial \varepsilon} : \dot{\varepsilon} + \frac{\partial \sigma}{\partial \mathbf{D}} : \mathbf{L}^d : \frac{\partial g}{\partial \mathbf{Y}} \dot{\varepsilon} \quad ; \quad \dot{\mathbf{D}} \geq 0 \tag{8.115}$$

where the second term in (8.115) also describes the stiffness reduction. Therefore,

$$\dot{\mathbf{C}} : \varepsilon = \frac{\partial \sigma}{\partial \mathbf{D}} : \mathbf{L}^d : \frac{\partial g}{\partial \mathbf{Y}} \dot{\varepsilon} \quad ; \quad \dot{\mathbf{D}} \geq 0 \tag{8.116}$$

The tangent constitutive equation can be written as

$$\dot{\sigma} = \mathbf{C}^{ed} : \dot{\varepsilon} \tag{8.117}$$

where the *damaged tangent constitutive tensor*, $\mathbf{C}^{ed}$, can be written as follows

$$\mathbf{C}^{ed} = \begin{cases} \mathbf{C} & \text{if } \dot{\mathbf{D}} \leq 0 \\ \mathbf{C} + \dfrac{\partial \sigma}{\partial \mathbf{D}} : \mathbf{L}^d : \dfrac{\partial g}{\partial \mathbf{Y}} & \text{if } \dot{\mathbf{D}} \geq 0 \end{cases} \tag{8.118}$$

Internal variables $\mathbf{D}, \delta$, and related variables, are found using numerical integration, usually using a return-mapping algorithm as explained in Section 8.4.1.

As explained in Sections 8.1.3 and 8.4, a number of internal material parameters are needed to define the damage surface, damage potential, and hardening functions. These parameters cannot be obtained directly from simple tests, but rather the model is *identified* by adjusting the internal parameters in such a way that model predictions fit well some observed behavior that can be quantified experimentally. Model identification is very specific to the particular model formulation, material, availability of experiments, and feasibility of conducting relevant experiments. Therefore, model identification can be explained only on a case by case basis, as is done in Section 9.3, as well as in Example 8.3.

8.4.1   Return-Mapping Algorithm

A return-mapping algorithm [32, 34, 35] is used to solve for the variables $\dot{\lambda}$, $\dot{\delta}$, $\dot{\mathbf{D}}$, δ and $\mathbf{D}$, in numerically approximated form.

The internal variables are updated by a linearized procedure between two consecutive iterations ($k - 1$ and k). The first-order linearization of (8.109) yields

$$(g)_k - (g)_{k-1} = \left(\frac{\partial g}{\partial \mathbf{Y}} : \frac{\partial \mathbf{Y}}{\partial \mathbf{D}} : \frac{\partial f}{\partial \mathbf{Y}} + \frac{\partial \gamma}{\partial \delta} \right)_{k-1} \Delta \lambda_k = 0 \qquad (8.119)$$

Successful iterations yield $[g]_k = 0$ and

$$\Delta \lambda_k = \frac{-(g)_{k-1}}{\left(\dfrac{\partial g}{\partial \mathbf{Y}} : \dfrac{\partial \mathbf{Y}}{\partial \mathbf{D}} : \dfrac{\partial f}{\partial \mathbf{Y}} + \dfrac{\partial \gamma}{\partial \delta} \right)_{k-1}} \qquad (8.120)$$

The complete algorithm used for a typical integration of constitutive equations is shown next:

1. The strain increment $(\varepsilon)^n$ for step "n" is obtained from the FEM code (discretized equilibrium equation). The incremental and updated strains are easily obtained as

$$(\varepsilon)^n = (\varepsilon)^{n-1} + (\Delta \varepsilon)^n$$

2. The state variables from the previous step "$n-1$" are obtained by starting the return-mapping algorithm, setting the predictor iteration $k = 0$. Therefore,

$$(\mathbf{D})_0^n = (\mathbf{D})^{n-1} \,; \quad (\delta)_0^n = (\delta)^{n-1}$$

3. Thermodynamic forces and damage hardening are computed at this point

$$(\mathbf{Y})_k^n \quad ; \quad (\gamma)_k^n$$

4. The damage threshold is evaluated at this point

$$(g)_k = g \left((\mathbf{Y})_k^n, (\gamma(\delta))_k^n, \gamma_0 \right)$$

Two different conditions define the possible cases:

   (a) If $(g)_k \leq 0$ not damage behavior, then $\Delta \lambda_k = 0$. Therefore, go to (8).

   (b) If $(g)_k > 0$ damage evolution, then $\Delta \lambda_k > 0$. Therefore, go to (5).

5. Damage evolution exists. Starting at iteration $k = k+1$, the damage multiplier is found from $(g)_k = 0$ as

$$\Delta \lambda_k = \frac{-(g)_k}{\left(\dfrac{\partial g}{\partial \mathbf{Y}} \right)_{k-1} : \left(\dfrac{\partial \mathbf{Y}}{\partial \mathbf{D}} \right)_{k-1} : \left(\dfrac{\partial f}{\partial \mathbf{Y}} \right)_{k-1} + \left(\dfrac{\partial \gamma}{\partial \delta} \right)_{k-1}}$$

6. Next, the damage state variables are updated using $\Delta\lambda_k$.

$$\left(D_{ij}\right)_k^n = \left(D_{ij}\right)_{k-1}^n + \Delta\lambda_k \left(\frac{\partial f}{\partial \mathbf{Y}}\right)_{k-1}$$

$$(\delta)_k^n = (\delta)_{k-1}^n + \Delta\lambda_k \left(\frac{\partial f}{\partial \gamma}\right)_{k-1} = (\delta)_{k-1}^n - \Delta\lambda_k$$

7. End of damage linearized process. Go to (3).

8. Set up the stress and damage state variables for the next load increment at each integration point

$$(\mathbf{D})^n = (\mathbf{D})_k^n; \quad (\delta)^n = (\delta)_k^n$$

$$(\boldsymbol{\sigma})^n = (\mathbf{M})^n : \overline{\mathbf{C}} : (\mathbf{M})^n : (\boldsymbol{\varepsilon})^n$$

9. Compute the tangent constitutive tensor

$$\left(\mathbf{C}^{ed}\right)^n = (\mathbf{M})^n : \overline{\mathbf{C}} : (\mathbf{M})^n + \left(\frac{\partial \boldsymbol{\sigma}}{\partial \mathbf{D}}\right)^n : (\mathbf{L}^d)^n : \left(\frac{\partial f}{\partial \mathbf{Y}}\right)^n$$

10. End of the integration algorithm.

Example 8.5 *Following the formulation of Example 8.4, compute all the variables needed to implement the damage model using a return mapping algorithm integration as shown in Section 8.4.1. Implement the damage model in ANSYS using the* USERMAT *capability. Identify model parameters for AS4/8852 carbon/epoxy as shown in Tables 8.1 and 8.2.*

The damage activation function is defined as

$$g := \hat{g} - \hat{\gamma} = \sqrt{\left(1 - \frac{G_{Ic}}{G_{IIc}}\right) \frac{Y_2 E_2}{F_{2t}^2} + \frac{G_{Ic}}{G_{IIc}} \left(\frac{Y_2 E_2}{F_{2t}^2}\right)^2 + \left(\frac{Y_6 G_{12}}{F_6^2}\right)^2} - \hat{\gamma}$$

where G_{Ic} and G_{IIc} are the critical energy release in mode I and in mode II, respectively, F_{2t} and F_6 are the transverse tensile strength and the shear strength, respectively. The damage hardening $\hat{\gamma}$ depends on δ, which is a parameter directly related to the transverse matrix crack density of the lamina, according to

$$\hat{\gamma} = \gamma + \gamma_0 = c_1 \left[\exp\left(\frac{\delta}{c_2}\right) - 1\right] + \gamma_0$$

where γ_0 defines the initial threshold value, c_1 and c_2 are material parameters.

Solution to Example 8.5 *This model represents damage caused by transverse tensile stress and in-plane shear stress. Compression and longitudinal tension have no effect on damage. Therefore, the model is defined in the thermodynamic force space Y_2, Y_6. The shape of the damage surface for AS4/8852 lamina is shown in Figure 8.8.*

To implement the return mapping algorithm shown in Section 8.4.1, expressions for $\partial f/\partial \mathbf{Y}$, $\partial g/\partial \mathbf{Y}$, $\partial f/\partial \gamma$, $\partial g/\partial \gamma$, $\partial \gamma/\partial \delta$, and $\partial \mathbf{Y}/\partial \mathbf{D}$ are needed.

Table 8.1: Elastic and strength properties for AS4/8852 unidirectional lamina

E_1	E_2	G_{12}	ν_{12}	F_{2t}	F_6
171.4 GPa	9.08 GPa	5.29 GPa	0.32	62.29 MPa	92.34 MPa

Table 8.2: Strength, critical energy release, and hardening parameters for AS4/8852 unidirectional lamina

G_{Ic}	G_{IIc}	γ_0	c_1	c_2
170 J/m^2	230 J/m^2	1.0	0.5	-1.8

Assuming $f = g$, the derivative of the potential function and the damage surface w.r.t the thermodynamic forces is given by

$$\frac{\partial g}{\partial \mathbf{Y}} = \frac{\partial f}{\partial \mathbf{Y}} = \left\{ \begin{array}{c} 0 \\ \frac{1}{\hat{g}} \left(\left(1 - \frac{G_{Ic}}{G_{IIc}}\right) \frac{1}{4F_{2t}} \sqrt{\frac{2E_2}{Y_2}} + \frac{G_{Ic}}{G_{IIc}} \frac{E_2}{(F_{2t})^2} \right) \\ \frac{1}{\hat{g}} G_{12} \end{array} \right\}$$

and the derivative of the damage surface w.r.t. the damage hardening function is

$$\frac{\partial g}{\partial \gamma} = \frac{\partial f}{\partial \gamma} = -1$$

Also, the derivative of the hardening function γ w.r.t conjugate variable δ is needed

$$\frac{\partial \gamma}{\partial \delta} = \frac{c_1}{c_2} \exp\left(\frac{\delta}{c_2}\right)$$

Next, the derivative of the thermodynamic forces w.r.t the internal damage variables is written as

$$\frac{\partial \mathbf{Y}}{\partial \mathbf{D}} = \frac{\partial \mathbf{Y}}{\partial \boldsymbol{\sigma}} : \frac{\partial \boldsymbol{\sigma}}{\partial \mathbf{D}}$$

Furthermore, the derivative of the thermodynamic forces with respect to strain is written as

$$\frac{\partial \mathbf{Y}}{\partial \boldsymbol{\varepsilon}} = \frac{\partial \mathbf{Y}}{\partial \boldsymbol{\sigma}} : \frac{\partial \boldsymbol{\sigma}}{\partial \boldsymbol{\varepsilon}} = \frac{\partial \mathbf{Y}}{\partial \boldsymbol{\sigma}} : \mathbf{C}$$

The following are written in contracted notation and multiplied by the Reuter matrix,

$$\frac{\partial \mathbf{Y}}{\partial \boldsymbol{\sigma}} = \left[\begin{array}{ccc} 0 & 0 & 0 \\ \frac{-\sigma_2 \nu_{12}}{(1-D_2)^2 E_1} & \frac{2\sigma_2}{(1-D_2)^3 E_2} - \frac{\sigma_1 \nu_{12}}{(1-D_2)^2 E_1} & 0 \\ 0 & 0 & \frac{4\sigma_6}{(1-D_6)^3 G_{12}} \end{array} \right]$$

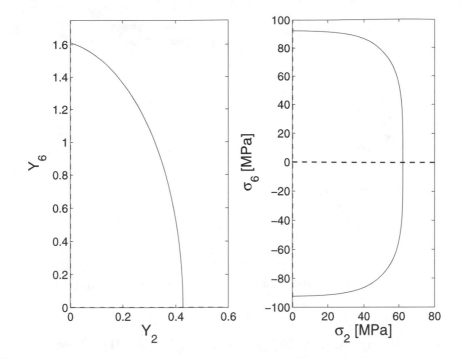

Fig. 8.8: Initial damage surface in thermodynamic and stress spaces.

and

$$\frac{\partial\boldsymbol{\sigma}}{\partial\mathbf{D}} = \begin{bmatrix} 0 & -\dfrac{E_2 v_{12}}{1 - v_{12}v_{21}}\varepsilon_2 & 0 \\[2ex] 0 & -\dfrac{E_1 v_{21}}{1 - v_{12}v_{21}}\varepsilon_1 - \dfrac{2(1 - D_2)E_2 v_{12}}{1 - v_{12}v_{21}}\varepsilon_2 & 0 \\[2ex] 0 & 0 & -8(1 - D_6)\,G_{12}\varepsilon_6 \end{bmatrix}$$

The damage model is implemented in ANSYS using the USERMAT *subroutine* usermatps.f, *available in [19], which can be used in conjunction with plain stress element (*PLANE182 *or* PLANE183*) and laminate shells (*SHELL181*). The model response under one-dimensional transverse stress and under only in-plane shear stress is shown in Figure 8.9.*

8.5   Damage and Plasticity

For polymer matrix composites reinforced by strong and stiff fibers, it will be shown in Section 9 that the damage and its conjugate thermodynamic force can be described by second-order tensors $\mathbf{D}$ and $\mathbf{Y}$. Furthermore, the hardening processes that take place during plasticity and damage imply additional *dissipation*, so that

$$\rho\pi = T\rho\pi_s = \boldsymbol{\sigma} : \dot{\boldsymbol{\varepsilon}}^p + R\dot{p} + \mathbf{Y} : \dot{\mathbf{D}} + \gamma\dot{\delta} \tag{8.121}$$

where (R, p) is the thermodynamic force-flux pair associated to plastic hardening, and (γ, δ) is the thermodynamic force-flux pair associated to damage hardening, and $\rho\pi$ is the dissipation heat due to irreversible phenomena.

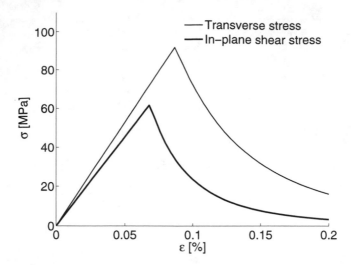

Fig. 8.9: Response under one-dimensional transverse stress and under in-plane shear stress.

For the particular case of (8.121), from (8.92) and (8.99), the following definitions for the thermodynamic forces are obtained

$$\boldsymbol{\sigma} = \rho\frac{\partial\psi}{\partial\varepsilon} = -\rho\frac{\partial\psi}{\partial\varepsilon^p} \quad \varepsilon = \rho\frac{\partial\chi}{\partial\boldsymbol{\sigma}} \quad \mathbf{Y} = -\rho\frac{\partial\psi}{\partial\mathbf{D}} = \rho\frac{\partial\chi}{\partial\mathbf{D}} \tag{8.122}$$

as well as definitions for the hardening equations

$$\gamma = \rho\frac{\partial\chi}{\partial\delta} = -\rho\frac{\partial\psi}{\partial\delta} = \rho\frac{\partial\pi}{\partial\delta} \qquad R = \rho\frac{\partial\chi}{\partial p} = -\rho\frac{\partial\psi}{\partial p} = \rho\frac{\partial\pi}{\partial p} \tag{8.123}$$

The additive decomposition [8]

$$\varepsilon = \varepsilon^e + \varepsilon^p \tag{8.124}$$

can be rewritten taking into account that the elastic component of strain can be calculated from stress and compliance, so that

$$\varepsilon = \mathbf{S} : \boldsymbol{\sigma} + \varepsilon^p \tag{8.125}$$

Therefore, the strain-stress law in incremental and rate form are

$$\delta\varepsilon = \mathbf{S} : \delta\boldsymbol{\sigma} + \delta\mathbf{S} : \boldsymbol{\sigma} + \delta\varepsilon^p$$
$$\dot{\varepsilon} = \mathbf{S} : \dot{\boldsymbol{\sigma}} + \dot{\mathbf{S}} : \boldsymbol{\sigma} + \dot{\varepsilon}^p \tag{8.126}$$

showing that an increment of strain has three contributions, elastic, damage, and plastic. The elastic strain occurs as a direct result of an increment in stress, the damage strain is caused by the increment in compliance as the material damages, and the plastic strain occurs at constant compliance. The elastic unloading stiffness does not change due to plasticity but it reduces due to damage. Following this argument, it is customary [42] to assume that the free energy and complementary free energy can be separated as follows

$$\psi(\varepsilon, \varepsilon^p, p, \mathbf{D}, \delta) = \psi^e(\varepsilon^e, \mathbf{D}, \delta) + \psi^p(\varepsilon^p, p)$$
$$\chi(\sigma, \varepsilon^p, p, \mathbf{D}, \delta) = \chi^e(\sigma, \mathbf{D}, \delta) + \chi^p(\varepsilon^p, p) \tag{8.127}$$

Suggested Problems

Problem 8.1 *Using the formulation and properties of Example 8.2, obtain a graphical representation of the evolution strain vs. nominal stress (ε vs. σ) and the evolution of strain vs. effective stress (ε vs. $\overline{\sigma}$) for a point on the top surface of the beam and for another point on the bottom surface of the beam. Comment on the graphs obtained.*

Problem 8.2 *Implement in ANSYS a one-dimensional CDM active in the x_1-direction only. Use 2D plane stress constitutive equations, and implement them into subroutine* usermatps.f *as a USERMAT. Leave the x_2-direction, Poisson's, and shear terms as linear elastic with no damage. Verify the program by recomputing Example 8.2 and the plots obtained in Problem 8.1. Note that to obtain the same values, the Poisson's ratio should be set to zero.*

Problem 8.3 *The Gibbs free energy is defined in expanded form and using Voigt contracted notation, as:*

$$\chi = \frac{1}{2\rho} \left[\frac{\sigma_1^2}{(1-D_1)^2 E_1} + \frac{\sigma_2^2}{(1-D_2)^2 E_2} + \frac{\sigma_6^2}{(1-D_1)(1-D_2) G_{12}} - \left(\frac{\nu_{21}}{E_2} + \frac{\nu_{12}}{E_1} \right) \frac{\sigma_1 \sigma_2}{(1-D_1)(1-D_2)} \right]$$

where E_1, E_2, ν_{12}, ν_{21} and G_{12} are the in-plane elastic orthotropic properties of a unidirectional lamina where the subindex $()_1$ denotes the fiber direction and $()_2$ denotes the transverse direction. (a) Obtain the secant constitutive equations, $\mathbf{C}$ and $\mathbf{S}$, using the given Gibbs free energy. (b) Obtain the thermodynamic forces Y_1 and Y_2 associated to D_1 and D_2. (c)If $\mathbf{M}$ is represented using Voigt contracted notation and multiplied by a Reuter matrix as

$$\mathbf{M} = \begin{bmatrix} (1-D_1) & 0 & 0 \\ 0 & (1-D_2) & 0 \\ 0 & 0 & \sqrt{1-D_1}\sqrt{1-D_2} \end{bmatrix}$$

check if this definition of $\mathbf{M}$ can be used as the damage effect tensor in a damage model using the principle of energy equivalence. Justify and comment on your conclusion.

Problem 8.4 *The damage activation function, for the model shown in Problem 8.3, is defined as*

$$g := \hat{g} - \hat{\gamma} = \sqrt{Y_1^2 H_1 + Y_2^2 H_2} - (\gamma + \gamma_0)$$

where H_1 and H_2 are model parameters that depend on elastic and strength material properties, and Y_1 and Y_2 are the thermodynamic forces associated to the damage variables D_1 and D_2, respectively. The damage hardening depends on δ according to

$$\hat{\gamma} = \gamma + \gamma_0 = c_1 \left[\exp\left(\frac{\delta}{c_2} \right) - 1 \right] + \gamma_0$$

Table 8.3: Elastic properties for composite material lamina

E_1	E_2	G_{12}	ν_{12}
171.4 GPa	9.08 GPa	5.29 GPa	0.32

Table 8.4: Identification model parameters for composite material lamina

H_1	H_2	γ_0	c_1	c_2
0.024	8.36	1.0	1.5	-2.8

where γ_0 defines the initial threshold value, c_1 and c_2 are material parameters. All necessary material parameters are shown in Tables 8.4 and 8.4.

Using a flow chart diagram, describe the algorithm, with all necessary steps, to implement this model as a constitutive subroutine in a finite element code. Then, compute the analytic expressions that are necessary to implement the model. Also program the algorithm using USERMAT *capability in ANSYS. Use the subroutine* usermatps.f. *Finally, plot a single curve of apparent stress σ_2 vs. apparent strain ε_2 for a RVE loaded only with ε_2 using the ANSYS program. Append the command list to your report as an Appendix.*

References

[1] J. Lemaitre, and A. Plumtree, Application of Damage Concepts to Predict Creep-Fatigue Failures, American Society of Mechanical Engineers (Paper) (78-PVP-26) (1978) 10–26.

[2] I. N. Rabotnov, in Rabotnov: Selected Works - Problems of the Mechanics of a Deformable Solid Body, Moscow Izdatel Nauka.

[3] L. M. Kachanov, On the Creep Fracture Time, Izv. Akad. Nauk USSR 8 (1958) 26–31.

[4] N. R. Hansen, and H. L. Schreyer, A Thermodynamically Consistent Framework for Theories of Elastoplasticity Coupled with Damage, International Journal of Solids and Structures 31 (3) (1994) 359–389.

[5] J. Janson, and J. Hult, Fracture Mechanics and Damage Mechanics - A Combined Approach, Journal Mec. Theor. Appl. 1 (1977) S18S28.

[6] E. J. Barbero, and K. W. Kelly, Predicting High Temperature Ultimate Strength of Continuous Fiber Metal Matrix Composites, Journal of Composite Materials 27 (12) (1993) 1214–1235.

[7] K. W. Kelly, and E. Barbero, Effect of Fiber Damage on the Longitudinal Creep of a CFMMC, International Journal of Solids and Structures 30 (24) (1993) 3417–3429.

[8] J. R. Rice, Constitutive Equations in Plasticity, MIT Press, Cambridge, MA, 1975, Ch. Continuum Mechanics and Thermodynamics of Plasticity in Relation to Microscale Deformation Mechanisms, pp. 23–79.

[9] J. C. Simo, and T. J. R. Hughes, Computational Inelasticity, Springer, Berlin, 1998.

[10] D. Krajcinovic, J. Trafimow, and D. Sumarac, Simple Constitutive Model for a Cortical Bone, Journal of Biomechanics 20 (8) (1987) 779–784.

[11] D. Krajcinovic, and D. Fanella, Micromechanical Damage Model for Concrete, Engineering Fracture Mechanics 25 (5-6) (1985) 585–596.

[12] D. Krajcinovic, Damage Mechanics, Mechanics of Materials 8 (2-3) (1989) 117–197.

[13] A. M. Neville, Properties of Concrete, 2nd Edition, Wiley, New York, 1973.

[14] ACI, American Concrete Institute.

[15] B. W. Rosen, The Tensile Failure of Fibrous Composites, AIAA Journal 2 (11) (1964) 1985–1911.

[16] W. Weibull, A Statistical Distribution Function of Wide Applicability, Journal of Applied Mechanics 18 (1951) 293–296.

[17] Wikipedia, The Gamma Function (2006).
http://en.wikipedia.org/wiki/Gamma_function

[18] E. J. Barbero, Introduction to Composite Materials Design, Taylor & Francis, Philadelphia, 1998.

[19] E. J. Barbero, Web resource (2006) http://www.mae.wvu.edu/barbero/feacm/

[20] B. W. Rosen, Fiber Composite Materials, American Society for Metals, Metals Park, OH, 1965, Ch. 3.

[21] A. S. D. Wang, A Non-Linear Microbuckling Model Predicting the Compressive Strength of Unidirectional Composites, in: ASME Winter Annual Meeting, Vol. WA/Aero-1, ASME, 1978.

[22] J. S. Tomblin, E. J. Barbero, and L. A. Godoy, Imperfection Sensitivity of Fiber Micro-Buckling in Elastic-Nonlinear Polymer-Matrix Composites, International Journal of Solids and Structures 34 (13) (1997) 1667–1679.

[23] E. Barbero, and J. Tomblin, A Damage Mechanics Model for Compression Strength of Composites, International Journal of Solids and Structures 33 (29) (1996) 4379–4393.

[24] E. J. Barbero, Prediction of Compression Strength of Unidirectional Polymer Matrix Composites, Journal of Composite Materials 32 (5) (1998) 483–502.

[25] D. C. Lagoudas, and A. M. Saleh, Compressive Failure Due to Kinking of Fibrous Composites, Journal of Composite Materials 27 (1) (1993) 83–106.

[26] P. Steif, A Model for Kinking in Fiber Composites. I. Fiber Breakage via Micro-Buckling, International Journal of Solids and Structures 26 (5-6) (1990) 549–61.

[27] S. W. Yurgartis, and S. S. Sternstein, Experiments to Reveal the Role of Matrix Properties and Composite Microstructure in Longitudinal Compression Strength, in: ASTM Special Technical Publication, no. 1185, ASTM, Philadelphia, PA, 1994, 193–204.

[28] C. Sun, and A. W. Jun, Effect of Matrix Nonlinear Behavior on the Compressive Strength of Fiber Composites, in: American Society of Mechanical Engineers, Applied Mechanics Division, AMD, Vol. 162, ASME, New York, 1993, 91–101.

[29] D. Adams, and E. Lewis, Current Status of Composite Material Shear Test Methods, SAMPE Journal 31 (1) (1995) 32–41.

[30] A. Maewal, Postbuckling Behavior of a Periodically Laminated Medium in Compression, International Journal of Solids and Structures 17 (3) (1981) 335–344.

[31] L. M. Kachanov, Rupture Time Under Creep Conditions Problems of Continuum Mechanics, SIAM (1958) 202–218.

[32] E. J. Barbero, and P. Lonetti, Damage Model for Composites Defined in Terms of Available Data, Mechanics of Composite Materials and Structures 8 (4) (2001) 299–315.

[33] E. J. Barbero, and P. Lonetti, An Inelastic Damage Model for Fiber Reinforced Laminates, Journal of Composite Materials 36 (8) (2002) 941–962.

[34] P. Lonetti, R. Zinno, F. Greco, and E. J. Barbero, Interlaminar Damage Model for Polymer Matrix Composites, Journal of Composite Materials 37 (16) (2003) 1485–1504.

[35] P. Lonetti, E. J. Barbero, R. Zinno, and F. Greco, Erratum: Interlaminar Damage Model for Polymer Matrix Composites (Journal of Composite Materials 37: 16 (1485-1504)), Journal of Composite Materials 38 (9) (2004) 799–800.

[36] E. J. Barbero, F. Greco, and P. Lonetti, Continuum Damage-Healing Mechanics with Application to Self-Healing Composites, International Journal of Damage Mechanics 14 (1) (2005) 51–81.

[37] S. Murakami, Mechanical Modeling of Material Damage, Journal of Applied Mechanics 55 (1988) 281–286.

[38] J. M. Smith, and H. C. Van Ness, Introduction to Chemical Engineering Thermodynamics, 3rd Edition, McGraw-Hill, New York, 1975.

[39] L. E. Malvern, Introduction to the Mechanics of a Continuous Medium, Prentice-Hall, Upper Saddle River, NJ, 1969.

[40] Y. A. Cengel, and M. A. Boles, Thermodynamics: An Engineering Approach, 3rd Edition, McGraw-Hill, New York, 1998.

[41] I. Mills, T. Cvitas, K. Homann, N. Kallay, and K. Kuchitsu, Quantities, Units and Symbols in Physical Chemistry, Oxford (1993)
http://www.iupac.org/publications/books/author/mills.html

[42] J. Lubliner, Plasticity Theory, Collier Macmillan, New York, 1990.

Chapter 9

A Damage Model for Fiber Reinforced Composites

J. A. Mayugo[1] and E. J. Barbero[2]

The objective of this chapter is to illustrate the concepts introduced in Chapter 8, by developing a complete theoretical model and its finite element implementation. The model includes in-plane and interlaminar damage evolution as well as coupled damage and plasticity. The identification of internal parameters from experimental data is developed in detail for a particular case. Model predictions are compared with experimental results available in the literature.

9.1 Theoretical Formulation

9.1.1 Damage and Effective Spaces

A second-order damage tensor, $\mathbf{D}$, is used to represent damage (Section 8.2). Damage and integrity tensors are used to describe a mapping between the effective, $\overline{\mathbb{C}}$, and damaged (actual), $\mathbb{C}$, configurations by a linear operator, f, as $f : \mathbb{C} \to \overline{\mathbb{C}}$ [1] as explained in Section 8.2.

9.1.2 Thermodynamic Formulation

The remaining constitutive equations are derived by thermodynamic principles and complementary laws. The Helmholtz free energy density (8.89) is assumed to be separable into elastic energy terms and additional terms related to the evolution of the internal parameters. In other words, the Helmholtz free energy density is

[1]Universitat de Girona, Spain
[2]West Virginia University, USA.

postulated to be the sum of two terms, the strain energy density $\varphi\left(\varepsilon_{ij}, \varepsilon_{ij}^p, D_{ij}\right)$ and the dissipation term $\pi\left(\delta, p\right)$

$$\psi = \psi\left(\varepsilon_{ij}, \varepsilon_{ij}^p, p, D_{ij}, \delta\right) = \varphi\left(\varepsilon_{ij}, \varepsilon_{ij}^p, D_{ij}\right) + \pi\left(\delta, p\right) \tag{9.1}$$

where ψ is the Helmholtz free energy density, ε_{ij}^p and D_{ij} are the plastic strains and the second-order damage tensor, respectively and p and δ are the hardening variables. The strain energy density (8.75) is defined as

$$\varphi = \frac{1}{2}\left(\varepsilon_{ij} - \varepsilon_{ij}^p\right) C_{ijkl}\left(\varepsilon_{kl} - \varepsilon_{kl}^p\right) = \frac{1}{2}\varepsilon_{ij}^e C_{ijkl}\varepsilon_{kl}^e = \frac{1}{2}\varepsilon_{ij}^e \sigma_{ij} \tag{9.2}$$

and can be written in both the damaged and effective configurations as

$$\varphi = \frac{1}{2}\sigma_{ij}\varepsilon_{ij}^e = \frac{1}{2}M_{ijkl}\bar{\sigma}_{kl}M_{ijrs}^{-1}\bar{\varepsilon}_{rs}^e = \frac{1}{2}I_{klrs}\bar{\sigma}_{kl}\bar{\varepsilon}_{rs}^e = \frac{1}{2}\bar{\sigma}_{kl}\bar{\varepsilon}_{kl}^e = \bar{\varphi} \tag{9.3}$$

where it is clear that energy is an invariant in all configurations. The dissipation term can be separated into two terms, which express the evolution of the damage and plastic-strain surfaces

$$\pi\left(\delta, p\right) = \psi_d(\delta) + \psi_p(p) \tag{9.4}$$

Here, we postulate that the damage energy dissipation density $\psi_d(\delta)$ and the plastic-strain energy dissipation density $\psi_p(p)$ can be represented by Prony series

$$\psi_d(\delta) = \sum_{i=1}^{m} \alpha_i^d \left[-\beta_i^d \exp\left(\frac{\delta}{\beta_i^d}\right) + \delta\right] \tag{9.5}$$

$$\psi_p(p) = \sum_{i=1}^{n} \alpha_i^p \left[-\beta_i^p \exp\left(\frac{p}{\beta_i^p}\right) + p\right] \tag{9.6}$$

where α_i^d, β_i^d, α_i^p and β_i^p are material parameters. Since the function $\pi\left(\delta, p\right)$ is assumed to be convex, its second derivative must be positive. By satisfying the Clausius-Duhem inequality, thus assuring non-negative dissipation, the following stress and thermodynamic forces are defined

$$\sigma_{ij} = \frac{\partial\psi}{\partial\varepsilon_{ij}^e} = C_{ijkl}\left(\varepsilon_{kl} - \varepsilon_{kl}^p\right) = C_{ijkl}\varepsilon_{kl}^e \tag{9.7}$$

$$Y_{ij} = -\frac{\partial\psi}{\partial D_{ij}} = -\frac{1}{2}\left(\varepsilon_{kl} - \varepsilon_{kl}^p\right)\frac{\partial C_{klpq}}{\partial D_{ij}}\left(\varepsilon_{pq} - \varepsilon_{pq}^p\right) = -\frac{1}{2}\varepsilon_{kl}^e \frac{\partial C_{klpq}}{\partial D_{ij}}\varepsilon_{pq}^e \tag{9.8}$$

as well as hardening functions

$$\gamma(\delta) = -\frac{\partial\psi}{\partial\delta} = \sum_{i=1}^{m} \alpha_i^d \left[\exp\left(\frac{\delta}{\beta_i^d}\right) - 1\right] \tag{9.9}$$

$$R(p) = -\frac{\partial \psi}{\partial p} = \sum_{i=1}^{n} \alpha_i^p \left[\exp\left(\frac{p}{\beta_i^p}\right) - 1 \right] \tag{9.10}$$

where σ, $\mathbf{Y}$, γ and R are the thermodynamic forces associated to the internal state variables ε, $\mathbf{D}$, p, and δ. It is worth noting that the second-order tensor $\mathbf{Y}$ includes the $Y_{33} = Y_3$ component that represents the energy release rate for the interlaminar shear damage and through-the-thickness normal damage.

Explicit expressions for the thermodynamic forces (9.8) can be written (see Appendix C) in terms of effective strain as

$$Y_{11} = \frac{1}{\Omega_1^2} \left(\overline{C}_{11} \bar{\varepsilon}_1^2 + \overline{C}_{12} \bar{\varepsilon}_2 \bar{\varepsilon}_1 + \overline{C}_{13} \bar{\varepsilon}_3 \bar{\varepsilon}_1 + 2\overline{C}_{55} \bar{\varepsilon}_5^2 + 2\overline{C}_{66} \bar{\varepsilon}_6^2 \right)$$

$$Y_{22} = \frac{1}{\Omega_2^2} \left(\overline{C}_{22} \bar{\varepsilon}_2^2 + \overline{C}_{12} \bar{\varepsilon}_2 \bar{\varepsilon}_1 + \overline{C}_{23} \bar{\varepsilon}_3 \bar{\varepsilon}_2 + 2\overline{C}_{44} \bar{\varepsilon}_4^2 + 2\overline{C}_{66} \bar{\varepsilon}_6^2 \right) \tag{9.11}$$

$$Y_{33} = \frac{1}{\Omega_3^2} \left(\overline{C}_{33} \bar{\varepsilon}_3^2 + \overline{C}_{13} \bar{\varepsilon}_3 \bar{\varepsilon}_1 + \overline{C}_{23} \bar{\varepsilon}_3 \bar{\varepsilon}_2 + 2\overline{C}_{44} \bar{\varepsilon}_4^2 + 2\overline{C}_{55} \bar{\varepsilon}_5^2 \right)$$

The evolution of the internal variables can be defined by (8.102)-(8.103) leading to the following flow rules, which describe the development of the inelastic effects

$$\dot{D}_{ij} = \dot{\lambda}^d \frac{\partial g^d}{\partial Y_{ij}} \quad ; \quad \dot{\delta} = \dot{\lambda}^d \frac{\partial g^d}{\partial \gamma}$$

$$\dot{\bar{\varepsilon}}_{ij}^p = \dot{\lambda}^p \frac{\partial g^p}{\partial \bar{\sigma}_{ij}} \quad ; \quad \dot{p} = \dot{\lambda}^p \frac{\partial g^p}{\partial R} \tag{9.12}$$

where $\dot{\lambda}^p$ and $\dot{\lambda}^d$ are the plastic-strain and damage multipliers, respectively. The plastic-strain (yield) surface g^p and damage surface g^d are defined in the next section.

9.1.3 Damage and Plastic Strain

An anisotropic damage criterion expressed in tensor form, introducing two fourth-order tensors, $\mathbf{B}$ and $\mathbf{J}$, is discussed in this section. It defines a multiaxial limit surface in the thermodynamic force space, $\mathbf{Y}$, that bounds the damage domain. The damage evolution is defined by a damage potential associated to the damage surface and by an isotropic hardening function. The proposed damage surface g^d is given by

$$g^d = \left(\hat{Y}_{ij}^N J_{ijhk} \hat{Y}_{hk}^N \right)^{1/2} + \left(Y_{ij}^S B_{ijhk} Y_{hk}^S \right)^{1/2} - (\gamma(\delta) + \gamma_0) \tag{9.13}$$

where γ_0 is the initial damage threshold value and $\gamma(\delta)$ defines the hardening proposed in (9.9).

The Y^N and Y^S are the thermodynamic forces due to normal strains and the thermodynamic forces due to shear strains, respectively. The Y^N can be defined as

$$Y_{ij}^N = -\frac{1}{2} \varepsilon_{kl}^e \delta_{kl} \delta_{mn} \frac{\partial C_{mnpq}}{\partial D_{ij}} \delta_{pq} \delta_{rs} \varepsilon_{rs}^e \tag{9.14}$$

The simple addition of $\mathbf{Y}^N$ and $\mathbf{Y}^S$ is the thermodynamic force tensor, $\mathbf{Y}$, defined in (9.8). Therefore, $\mathbf{Y}^S$ can be written as

$$Y_{ij}^S = Y_{ij} - Y_{ij}^N \tag{9.15}$$

The tensor $\hat{\mathbf{Y}}^{\mathbf{N}}$ is the mapped thermodynamic force associated to normal strains and is defined as

$$\hat{Y}_{ij}^N = \hat{A}_{ijkl}Y_{kl}^N \tag{9.16}$$

The tensor $\hat{\mathbf{A}}$ is a diagonal transformation tensor proposed here to represent different behavior in tension and compression while preserving convexity of the damage threshold surface in stress-strain and thermodynamic force spaces. In principal directions, it can be written as

$$\hat{A}_{ijkl} = \delta_{im}\delta_{jm}\delta_{km}\delta_{lm} + (A_m - 1)\,\delta_{im}\delta_{jm}\delta_{km}\delta_{lm}r_m \tag{9.17}$$

where

$$r_m = \frac{1}{2}\left(1 - \frac{\varepsilon_m}{|\varepsilon_m|}\right) = \begin{cases} 0 & if \quad \sigma_m > 0 \\ 1 & if \quad \sigma_m < 0 \end{cases} \tag{9.18}$$

with $m = 1, 2, 3$. The values of ε_m are the values of the normal strain in the principal material directions. The three values A_m represent the relation between damage thresholds in uniaxial compression and uniaxial extension in the principal material directions. The diagonal fourth-order positive defined tensors, $\hat{\mathbf{A}}$, $\mathbf{B}$, and $\mathbf{J}$ are determined by available experimental data for unidirectional composite materials, as shown in Section 9.3.

On the other hand, the plastic-strain evolution is modeled by classical plasticity formulation [2]. An associate flow rule is assumed in the effective stress space, coupling plasticity and damage effects. The plastic-strain (yield) surface g^p is a function of the thermodynamic forces in the effective configuration $(\bar{\sigma}, R)$. Therefore, the plastic-strain surface, which accounts for thickness terms, is

$$g^p = \sqrt{f_{ij}\bar{\sigma}_i\bar{\sigma}_j + f_i\bar{\sigma}_i} - (R\,(p) + R_0) \tag{9.19}$$

where $(i = 1, 2...6)$, R_0 is the initial plastic-strain threshold and R is defined by (9.10). The coefficients f_i and f_{ij} are obtained from the strength properties of the composite as in (9.58-9.59) in Section 9.3.

9.1.4 Evolution Functions

By using the additive decomposition hypothesis (8.124), which splits the total strain rate into elastic and plastic contributions, the rates of stress, thermodynamic forces, and both hardening functions can be written as

$$\dot{\bar{\sigma}}_{ij} = \frac{\partial\bar{\sigma}_{ij}}{\partial\varepsilon_{kl}}\dot{\varepsilon}_{kl} + \frac{\partial\bar{\sigma}_{ij}}{\partial\varepsilon_{kl}^p}\dot{\varepsilon}_{kl}^p$$

$$\dot{Y}_{ij} = \frac{\partial Y_{ij}}{\partial\varepsilon_{kl}}\dot{\varepsilon}_{kl} + \frac{\partial Y_{ij}}{\partial\varepsilon_{kl}^p}\dot{\varepsilon}_{kl}^p + \frac{\partial Y_{ij}}{\partial D_{kl}}\dot{D}_{kl} \tag{9.20}$$

$$\dot{\gamma} = \frac{\partial\gamma}{\partial\delta}\dot{\delta} \quad ; \qquad \dot{R} = \frac{\partial R}{\partial p}\dot{p}$$

The incremental stress-strain relations for evolution of damage and plastic-strain can be obtained by introducing (9.12) into (9.20) as follows

$$\dot{\bar{\sigma}}_{ij} = \frac{\partial \bar{\sigma}_{ij}}{\partial \varepsilon_{kl}} \dot{\varepsilon}_{kl} + \frac{\partial \bar{\sigma}_{ij}}{\partial \varepsilon_{kl}^p} M_{klmn}^{-1} \dot{\lambda}^p \frac{\partial g^p}{\partial \bar{\sigma}_{mn}}$$

$$\dot{Y}_{ij} = \frac{\partial Y_{ij}}{\partial \varepsilon_{kl}} \dot{\varepsilon}_{kl} + \frac{\partial Y_{ij}}{\partial \varepsilon_{kl}^p} M_{klmn}^{-1} \dot{\lambda}^p \frac{\partial g^p}{\partial \bar{\sigma}_{mn}} + \frac{\partial Y_{ij}}{\partial D_{kl}} \dot{\lambda}^d \frac{\partial g^d}{\partial Y_{kl}} \tag{9.21}$$

$$\dot{\gamma} = \frac{\partial \gamma}{\partial \delta} \dot{\lambda}^d \frac{\partial g^d}{\partial \gamma} \quad ; \quad \dot{R} = \frac{\partial R}{\partial p} \dot{\lambda}^p \frac{\partial g^p}{\partial R}$$

Both consistency conditions, one for damage ($\dot{g}^d = g^d = 0$), another for plastic-strain ($\dot{g}^p = g^p = 0$), must be used in order to find the damage and plastic-strain consistency multipliers $\dot{\lambda}^d$ and $\dot{\lambda}^p$ that define the evolution of internal variables. The consistency conditions lead to

$$\dot{g}^d = \frac{\partial g^d}{\partial Y_{ij}} \dot{Y}_{ij} + \frac{\partial g^d}{\partial \gamma} \dot{\gamma} = 0 \; ; \; g^d = 0 \tag{9.22}$$

$$\dot{g}^p = \frac{\partial g^p}{\partial \bar{\sigma}_{ij}} \dot{\bar{\sigma}}_{ij} + \frac{\partial g^p}{\partial R} \dot{R} = 0 \; ; \; g^p = 0 \tag{9.23}$$

Introducing (9.21) into (9.22) and (9.23) we obtain the following linear system of equations

$$\dot{g}^d = \frac{\partial g^d}{\partial Y_{ij}} \left[\frac{\partial Y_{ij}}{\partial \varepsilon_{kl}} \dot{\varepsilon}_{kl} + \frac{\partial Y_{ij}}{\partial \varepsilon_{kl}^p} M_{klmn}^{-1} \dot{\lambda}^p \frac{\partial g^p}{\partial \bar{\sigma}_{mn}} + \frac{\partial Y_{ij}}{\partial D_{kl}} \dot{\lambda}^d \frac{\partial g^d}{\partial Y_{kl}} \right] + \frac{\partial g^d}{\partial \gamma} \frac{\partial \gamma}{\partial \delta} \dot{\lambda}^d \frac{\partial g^d}{\partial \gamma} = 0$$

$$\dot{g}^p = \frac{\partial g^p}{\partial \bar{\sigma}_{ij}} \left[\frac{\partial \bar{\sigma}_{ij}}{\partial \varepsilon_{kl}} \dot{\varepsilon}_{kl} + \frac{\partial \bar{\sigma}_{ij}}{\partial \varepsilon_{kl}^p} M_{klmn}^{-1} \dot{\lambda}^p \frac{\partial g^p}{\partial \bar{\sigma}_{mn}} \right] + \frac{\partial g^p}{\partial R} \frac{\partial R}{\partial p} \dot{\lambda}^p \frac{\partial g^p}{\partial R} = 0 \tag{9.24}$$

Rewriting (9.24) we get

$$-\frac{\partial g^d}{\partial Y_{ij}} \frac{\partial Y_{ij}}{\partial \varepsilon_{kl}} \dot{\varepsilon}_{kl} = \left[\frac{\partial g^d}{\partial Y_{ij}} \frac{\partial Y_{ij}}{\partial D_{kl}} \frac{\partial g^d}{\partial Y_{kl}} + \frac{\partial g^d}{\partial \gamma} \frac{\partial \gamma}{\partial \delta} \frac{\partial g^d}{\partial \gamma} \right] \dot{\lambda}^d +$$

$$+ \left[\frac{\partial g^d}{\partial Y_{ij}} \frac{\partial Y_{ij}}{\partial \varepsilon_{kl}^p} M_{klmn}^{-1} \frac{\partial g^p}{\partial \bar{\sigma}_{mn}} \right] \dot{\lambda}^p \tag{9.25}$$

$$-\frac{\partial g^p}{\partial \bar{\sigma}_{ij}} \frac{\partial \bar{\sigma}_{ij}}{\partial \varepsilon_{kl}} \dot{\varepsilon}_{kl} = \left[\frac{\partial g^p}{\partial \bar{\sigma}_{ij}} \frac{\partial \bar{\sigma}_{ij}}{\partial \varepsilon_{kl}^p} M_{klmn}^{-1} \frac{\partial g^p}{\partial \bar{\sigma}_{mn}} + \frac{\partial g^p}{\partial R} \frac{\partial R}{\partial p} \frac{\partial g^p}{\partial R} \right] \dot{\lambda}^p$$

which provides a system of two equations in the two unknown multipliers $\dot{\lambda}^d$ and $\dot{\lambda}^p$ as

$$-b_{ij}^1 \dot{\varepsilon}_{ij} = a^{11} \dot{\lambda}^d + a^{12} \dot{\lambda}^p$$
$$-b_{ij}^2 \dot{\varepsilon}_{ij} = a^{21} \dot{\lambda}^d + a^{22} \dot{\lambda}^p \tag{9.26}$$

where the second-order tensors $\mathbf{b}^1$ and $\mathbf{b}^2$ are

$$b_{mn}^1 = \left[\frac{\partial g^d}{\partial Y_{ij}} \frac{\partial Y_{ij}}{\partial \varepsilon_{mn}} \right]$$

$$b_{mn}^2 = \left[\frac{\partial g^p}{\partial \bar{\sigma}_{ij}} M_{ijkl}^{-1} \frac{\partial \sigma_{kl}}{\partial \varepsilon_{mn}} \right] \tag{9.27}$$

and the scalars a^{11}, a^{22}, a^{12} and a^{21} are given by

$$a^{11} = \left[\frac{\partial g^d}{\partial Y_{ij}} \frac{\partial Y_{ij}}{\partial D_{kl}} \frac{\partial g^d}{\partial Y_{kl}} + \frac{\partial g^d}{\partial \gamma} \frac{\partial \gamma}{\partial \delta} \frac{\partial g^d}{\partial \gamma} \right]$$

$$a^{22} = \left[\frac{\partial g^p}{\partial \bar{\sigma}_{ij}} M^{-1}_{ijkl} \frac{\partial \sigma_{kl}}{\partial \varepsilon^p_{mn}} M^{-1}_{mnpq} \frac{\partial g^p}{\partial \bar{\sigma}_{pq}} + \frac{\partial g^p}{\partial R} \frac{\partial R}{\partial p} \frac{\partial g^p}{\partial R} \right] \qquad (9.28)$$

$$a^{12} = \left[\frac{\partial g^d}{\partial Y_{ij}} \frac{\partial Y_{ij}}{\partial \varepsilon^p_{kl}} M^{-1}_{klmn} \frac{\partial g^p}{\partial \bar{\sigma}_{mn}} \right]$$

$$a^{21} = 0$$

The damage and plastic-strain multipliers $\dot{\lambda}^d$ and $\dot{\lambda}^p$ are obtained by solving the linear system of equations (9.26) as

$$\dot{\lambda}^d = \begin{cases} L^d_{ij} \dot{\varepsilon}_{ij}, & g^d = 0 \\ 0, & g^d < 0 \end{cases}$$
$$\dot{\lambda}^p = \begin{cases} L^p_{ij} \dot{\varepsilon}_{ij}, & g^p = 0 \\ 0, & g^p < 0 \end{cases} \qquad (9.29)$$

where the second-order tensors $\mathbf{L}^d$ and $\mathbf{L}^p$ are

$$L^d_{ij} = \frac{-a^{22} b^1_{ij} + a^{12} b^2_{ij}}{a^{11} a^{22} - a^{12} a^{21}}$$

$$L^p_{ij} = \frac{-a^{11} b^2_{ij} + a^{21} b^1_{ij}}{a^{11} a^{22} - a^{12} a^{21}} \qquad (9.30)$$

Using (9.29) we can rewrite (9.12) as

$$\dot{D}_{kl} = L^d_{rs} \frac{\partial g^d}{\partial Y_{kl}} \dot{\varepsilon}_{rs} \qquad (9.31)$$

$$\dot{\varepsilon}^p_{kl} = L^p_{rs} M^{-1}_{klmn} \frac{\partial g^p}{\partial \bar{\sigma}_{mn}} \dot{\varepsilon}_{rs} \qquad (9.32)$$

From (9.7), the incremental stress-strain relations for damage and plastic-strain evolution in the effective configuration can be expressed as

$$\dot{\sigma}_{ij} = \frac{\partial \sigma_{ij}}{\partial \varepsilon_{kl}} \dot{\varepsilon}_{kl} + \frac{\partial \sigma_{ij}}{\partial \varepsilon^p_{kl}} \dot{\varepsilon}^p_{kl} + \frac{\partial \sigma_{ij}}{\partial D_{kl}} \dot{D}_{kl} \qquad (9.33)$$

Introducing (9.31) and (9.32) into (9.33) we obtain

$$\dot{\sigma}_{ij} = \frac{\partial \sigma_{ij}}{\partial \varepsilon_{rs}} \dot{\varepsilon}_{rs} + \frac{\partial \sigma_{ij}}{\partial \varepsilon^p_{kl}} L^p_{rs} M^{-1}_{klmn} \frac{\partial g^p}{\partial \bar{\sigma}_{mn}} \dot{\varepsilon}_{rs} +$$

$$+ \frac{\partial \sigma_{ij}}{\partial D_{kl}} L^d_{rs} \frac{\partial g^d}{\partial Y_{kl}} \dot{\varepsilon}_{rs} \quad ; \quad \dot{D}_{kl}, \dot{\varepsilon}^p_{kl} \geq 0 \qquad (9.34)$$

or

$$\dot{\sigma}_{ij} = C^{epd}_{ijrs} \dot{\varepsilon}_{rs} \tag{9.35}$$

where the *elasto-plastic damaged tangent constitutive tensor*, C^{epd}_{ijrs}, can be written as

$$C^{epd}_{ijrs} = \frac{\partial\sigma_{ij}}{\partial\varepsilon_{rs}} + \frac{\partial\sigma_{ij}}{\partial\varepsilon^p_{kl}} M^{-1}_{klmn} L^p_{rs} \frac{\partial g^p}{\partial\bar{\sigma}_{mn}} + \frac{\partial\sigma_{ij}}{\partial D_{kl}} L^d_{rs} \frac{\partial g^d}{\partial Y_{kl}} \quad ; \quad \dot{D}_{kl}, \dot{\varepsilon}^p_{kl} \geq 0 \tag{9.36}$$

Hence, C^{epd}_{ijrs} depends on the evolution of the strain tensor $\dot{\varepsilon}_{ij}$ and the current state represented by thermodynamic forces $(Y_{ij}, \sigma_{ij}, \gamma, R)$ or by the conjugate kinematic variables $(D_{ij}, \varepsilon^p_{ij}, \delta, p)$. While the first two terms in (9.36) are the classical terms for plasticity, the third term describes the stiffness reduction due to an increment of damage. Moreover, the tangent constitutive equation can be obtained from differentiation of (8.66) as

$$\dot{\sigma}_{ij} = C_{ijkl} \dot{\varepsilon}^e_{kl} + \dot{C}_{ijkl} \varepsilon^e_{kl} \tag{9.37}$$

Using the additive decomposition hypothesis (8.124) we get

$$\dot{\sigma}_{ij} = C_{ijkl} \left(\dot{\varepsilon}_{kl} - \dot{\varepsilon}^p_{kl} \right) + \dot{C}_{ijkl} \varepsilon^e_{kl} \tag{9.38}$$

and using (9.31) we have

$$\dot{\sigma}_{ij} = C_{ijkl} \left(I_{klrs} - L^p_{rs} M^{-1}_{klmn} \frac{\partial g^p}{\partial\bar{\sigma}_{mn}} \right) \dot{\varepsilon}_{rs} + \dot{C}_{ijkl} \varepsilon^e_{kl} \tag{9.39}$$

The last term of (9.34) and the last term of (9.39), which describe the stiffness reduction, must be equal to each other. Therefore

$$\dot{C}_{ijrs} \varepsilon^e_{rs} = \frac{\partial\sigma_{ij}}{\partial D_{kl}} L^d_{rs} \frac{\partial g^d}{\partial Y_{kl}} \dot{\varepsilon}_{rs} \tag{9.40}$$

Therefore, the tangent constitutive equation can be expressed by substituting (9.40) into (9.39) as

$$C^{epd}_{ijrs} = C_{ijkl} \left(I_{klrs} - L^p_{rs} M^{-1}_{klmn} \frac{\partial g^p}{\partial\bar{\sigma}_{mn}} \right) \dot{\varepsilon}_{rs} + \frac{\partial\sigma_{ij}}{\partial D_{kl}} L^d_{rs} \frac{\partial g^d}{\partial Y_{kl}} \tag{9.41}$$

This expression takes different forms according to the loading/unloading conditions derived from the Kuhn-Tucker relations as follows

$$C^{epd}_{ijrs} = \begin{cases} C_{ijrs} & \text{if } \dot{D} \leq 0 \text{ and } \dot{\varepsilon}^p \leq 0 \\[2mm] C_{ijrs} + \dfrac{\partial\sigma_{ij}}{\partial D_{kl}} L^d_{rs} \dfrac{\partial g^d}{\partial Y_{kl}} & \text{if } \dot{D} \geq 0 \text{ and } \dot{\varepsilon}^p \leq 0 \\[4mm] C_{ijkl} \left(I_{klrs} - L^p_{rs} M^{-1}_{klmn} \dfrac{\partial g^p}{\partial\bar{\sigma}_{mn}} \right) & \text{if } \dot{D} \leq 0 \text{ and } \dot{\varepsilon}^p \geq 0 \\[4mm] C_{ijkl} \left(I_{klrs} - L^p_{rs} M^{-1}_{klmn} \dfrac{\partial g^p}{\partial\bar{\sigma}_{mn}} \right) + \\[2mm] + \dfrac{\partial\sigma_{ij}}{\partial D_{kl}} L^d_{rs} \dfrac{\partial g^d}{\partial Y_{kl}} & \text{if } \dot{D} \geq 0 \text{ and } \dot{\varepsilon}^p \geq 0 \end{cases} \tag{9.42}$$

The nonlinear problem is solved by a return-mapping algorithm [3, 4, 5] described in the next section.

9.2   Numerical Implementation

A displacement-based finite element formulation is used. The geometry is discretized by the finite element method. The material nonlinearity is tracked at each Gauss integration point. The return-mapping algorithm of Section 8.4.1 is here augmented to integrate the rate equations taking into account plasticity as well as damage.

Damage and plastic-deformation, both state variables, are updated by a linearized procedure between two consecutive iterations ($k - 1$ and k). With a first-order linearization of (9.24), and using (9.28), we can write

$$
\begin{aligned}
(G^d)_k - (G^d)_{k-1} &= (a^{11})_{k-1}\,\Delta\lambda_k^d + (a^{12})_{k-1}\,\Delta\lambda_k^p \\
(G^p)_k - (G^p)_{k-1} &= (a^{21})_{k-1}\,\Delta\lambda_k^d + (a^{22})_{k-1}\,\Delta\lambda_k^p
\end{aligned}
\tag{9.43}
$$

Successful convergence at iteration k yields $(G^d)_k = 0$ and $(G^p)_k = 0$. Using (9.28.d) the following linear system is obtained

$$
\begin{aligned}
-(G^d)_{k-1} &= (a^{11})_{k-1}\,\Delta\lambda_k^d + (a^{12})_{k-1}\,\Delta\lambda_k^p \\
-(G^p)_{k-1} &= (a^{22})_{k-1}\,\Delta\lambda_k^p
\end{aligned}
\tag{9.44}
$$

from which we can solve first the plastic-strain multiplier $\Delta\lambda_k^p$ using the second equation of the linear system. The plastic-strain state variables are updated using $\Delta\lambda_k^p$ and the iteration is continued until $\Delta\lambda_k^p \approx 0$. Next solve for $\Delta\lambda_k^d$ using the first equation of the linear system which is uncoupled from plastic-strain because $\Delta\lambda_k^p = 0$.

The complete algorithm used for the integration of the constitutive equation, is shown next.

1. The displacement increment $(\Delta u_i)^n$ for step "n" is obtained from the discretized equilibrium equation. The incremental and updated strains are easily obtained as

$$
\begin{aligned}
(\Delta\varepsilon_{ij})^n &= \tfrac{1}{2}\left((\Delta u_{i,j})^n + (\Delta u_{j,i})^n\right) \\
(\varepsilon_{ij})^n &= (\varepsilon_{ij})^{n-1} + (\Delta\varepsilon_{ij})^n
\end{aligned}
$$

2. The state variables from the previous step "$n-1$" are obtained by starting the return-mapping algorithm with damage and plastic-strains, setting the predictor iteration $k = 0$. Therefore

$$
\begin{aligned}
\left(D_{ij}\right)_0^n &= \left(D_{ij}\right)^{n-1} ; \quad (\delta)_0^n = (\delta)^{n-1} \\
\left(\varepsilon_{ij}^p\right)_0^n &= \left(\varepsilon_{ij}^p\right)^{n-1} ; \quad (p)_0^n = (p)^{n-1}
\end{aligned}
$$

3. The effective stress tensor and the plastic-strain hardening are computed at this point

$$
\left(\overline{\sigma}_{ij}\right)_k^n ; \quad (R)_k^n
$$

4. The plastic-strain threshold is evaluated at this point

$$(G^p)_k = g^p \left((\overline{\sigma}_{ij})_k^n , (R(p))_k^n , R_0 \right)$$

Two different conditions define the possible cases:

(a) If $(G^p)_k \leq 0$ not plastic-strain evolution, then $\Delta \lambda_k^p = 0$. Therefore, go to (8).

(b) If $(G^p)_k > 0$ plastic-strain evolution, then $\Delta \lambda_k^p > 0$. Therefore, go to (5).

5. Plastic-strain evolution exists. Starting at iteration $k = k + 1$. The plastic-strain multiplier is found from $(G^p)_k = 0$ as

$$\Delta \lambda_k^p = -\frac{(G^p)_{k-1}}{(a^{22})_{k-1}}$$

6. Next, the plastic-strain state variables are updated using $\Delta \lambda_k^p$, as

$$\left(\varepsilon_{ij}^p \right)_k^n = \left(\varepsilon_{ij}^p \right)_{k-1}^n + \Delta \lambda_k^p \left(\frac{\partial g^p}{\partial \overline{\sigma}_{ij}} \right)_{k-1}$$

$$(p)_k^n = (p)_{k-1}^n + \Delta \lambda_k^p \left(\frac{\partial g^p}{\partial R} \right)_{k-1} = (p)_{k-1}^n - \Delta \lambda_k^p$$

7. End of plastic-strain linearized process. Go to (3).

8. The thermodynamic forces and damage hardening are computed at this point

$$\left(Y_{ij} \right)_k^n ; \quad \left(\hat{Y}_{ij}^N \right)_k^n ; \quad \left(Y_{ij}^S \right)_k^n ; \quad (\gamma)_k^n$$

9. The damage threshold is evaluated at this point

$$(G^d)_k = g^d \left(\left(\hat{Y}_{ij}^N \right)_k^n , (Y_{ij}^S)_k^n , (\gamma(\delta))_k^n , \gamma_0 \right)$$

Two different conditions define the possible cases:

(a) If $(G^d)_k \leq 0$ no damage behavior, then $\Delta \lambda_k^d = 0$. Therefore, go to (13).

(b) If $(G^d)_k > 0$ damage evolution, then $\Delta \lambda_k^d > 0$. Therefore, go to (10).

10. Damage evolution exists. Starting at iteration $k = k+1$, the damage multiplier is found from $\left(G^d\right)_k = 0$ as

$$\Delta\lambda_k^d = -\frac{\left(G^d\right)_{k-1}}{\left(a^{11}\right)_{k-1}}$$

11. Next, the damage state variables are updated using $\Delta\lambda_k^d$.

$$\left(D_{ij}\right)_k^n = \left(D_{ij}\right)_{k-1}^n + \Delta\lambda_k^d \left(\frac{\partial f^d}{\partial Y_{ij}}\right)_{k-1}$$

$$(\delta)_k^n = (\delta)_{k-1}^n + \Delta\lambda_k^d \left(\frac{\partial f^d}{\partial \gamma}\right)_{k-1} = (\delta)_{k-1}^n - \Delta\lambda_k^d$$

12. End of damage linearized process. Go to (8).

13. Set up the stress, damage, and plastic-strain state variables for the next load increment at each integration point

$$\left(D_{ij}\right)^n = \left(D_{ij}\right)_k^n ; \quad (\delta)^n = (\delta)_k^n$$
$$\left(\varepsilon_{ij}^p\right)^n = \left(\varepsilon_{ij}^p\right)_k^n ; \quad (p)^n = (p)_k^n$$
$$\left(\sigma_{ij}\right)^n = \left(M_{ijkl}\right)^n \overline{C}_{klrs} \left(M_{rstu}\right)^n \left((\varepsilon_{tu})^n - (\varepsilon_{tu}^p)^n\right)$$

14. Compute the tangent constitutive tensor

$$\left(C_{ijrs}^{epd}\right)^n = \left[\left(M_{ijkl}\right)^n \overline{C}_{klmn} \left(M_{mnpq}\right)^n\right]$$
$$\left[I_{pqrs} - \left(M_{pqtu}^{-1}\right)^n (L_{rs}^p)^n \left(\frac{\partial g^p}{\partial \overline{\sigma}_{tu}}\right)^n\right]$$
$$+ \left(\frac{\partial \sigma_{ij}}{\partial D_{pq}}\right)^n (L_{rs}^d)^n \left(\frac{\partial f^d}{\partial Y_{pq}}\right)^n$$

15. End of the integration algorithm.

9.3   Model Identification

The model uses a number of internal parameters that are explicitly related to experimental material properties. The damage domain is defined by diagonal fourth-order tensors $\hat{\mathbf{A}}$, $\mathbf{B}$, and $\mathbf{J}$. The idea is to compare the Tsai-Wu criterion with the damage domain in the effective stress space, obtaining a linear system with solutions that characterize uniquely the coefficients in the $\hat{\mathbf{A}}$ tensor (A_1, A_2, and A_3), the $\mathbf{B}$ tensor (B_1, B_2, and B_3), and the $\mathbf{J}$ tensor (J_1, J_2, and J_3). To obtain a convex damage surface, all coefficients must be non-negative. In this way these coefficients can be obtained in terms of the elastic properties and strength values under uniaxial loading of the unidirectional lamina, which for the most part are available in the literature. The proposed identification procedure along with the damage surface described by (9.13) does not require experimental data for matrix-dominated shear loading, which are very difficult to obtain. Note that the Voigt contracted notation and the tensor shear strain notation are used, $\epsilon_\alpha = 1/2\gamma_\alpha$ which $\alpha = 4, 5, 6$.

9.3.1 Damage Surface Shear Coefficients

The coefficients B_1, B_2, and B_3 represent the effect of damage caused by shear loading on the shape of the damage surface g^d in (9.13). They are directly related to available material properties obtained by shear tests of unidirectional lamina, i.e., the undamaged shear moduli (G_{12} and G_{13}), the damaged shear moduli at imminent failure ($G^*{}_{12}$ and $G^*{}_{13}$, see Figure 9.1),and fiber-dominated shear strengths (F_6 and F_5). Note that at failure we can assume that $(\gamma^* + \gamma_0) = 1$ in (9.13). Considering two different shear loading cases, when the failure of the material is imminent, in-plane shear loading ($\sigma_6 \neq 0$, $\sigma_1 = \sigma_2 = \sigma_3 = \sigma_5 = \sigma_4 = 0$), and longitudinal-thickness loading ($\sigma_5 \neq 0$, $\sigma_1 = \sigma_2 = \sigma_3 = \sigma_6 = \sigma_4 = 0$), and substituting in (9.13), the damage surface g^d reduces to

$$\sqrt{\frac{B_1}{\Omega_{1s}^4} + \frac{B_2}{\Omega_{2s}^4}\frac{2\overline{S}_{66}}{\Omega_{1s}^2\Omega_{2s}^2}} F_6^2 = (\gamma^* + \gamma_0) = 1 \tag{9.45}$$

$$\sqrt{\frac{B_1}{\Omega_{1s}^4} + \frac{B_3}{\Omega_{3s}^4}\frac{2\overline{S}_{55}}{\Omega_{1s}^2\Omega_{3s}^2}} F_5^2 = (\gamma^* + \gamma_0) = 1 \tag{9.46}$$

where Ω_{1s}, Ω_{2s}, Ω_{3s}, are the critical values of the integrity tensor at shear failure in the longitudinal, transverse, and thickness directions, respectively, and $\overline{S}_{55}$, $\overline{S}_{66}$ are coefficients in the undamaged compliance tensor, which can be written in terms of undamaged shear moduli in usual way [6] as $\overline{S}_{66} = 1/\overline{G}_{12}$ and $\overline{S}_{55} = 1/\overline{G}_{13}$.

Experimental evidence reveals a highly nonlinear behavior for a fiber-reinforced lamina subject to pure shear. B_2 and B_3 represent the effect on the damage surface of matrix damage caused by shear loading. B_1 represents the effect of fiber damage under shear loading. Experimental data suggest that a lamina subject to pure shear experiences mostly matrix damage and has negligible effect on fiber damage [7]. Therefore, it is reasonable to assume that under shear loading $B_1 = 0$ and fiber integrity $\Omega_1 = 1$. Therefore,

$$\Omega_1^2\Omega_2^2 = \Omega_2^2 = \frac{C_{66}}{\overline{C}_{66}} = \frac{G_{12}}{\overline{G}_{12}} \quad ; \quad \Omega_1^2\Omega_3^2 = \Omega_3^2 = \frac{C_{55}}{\overline{C}_{55}} = \frac{G_{13}}{\overline{G}_{13}} \tag{9.47}$$

Just prior to failure, the damaged shear stiffness G_i^* can be obtained from the unloading shear stress-strain path (Figure 9.1), so that (9.47) becomes

$$\Omega_{2s}^2 = \frac{G_{12}^*}{\overline{G}_{12}} \quad ; \quad \Omega_{3s}^2 = \frac{G_{13}^*}{\overline{G}_{13}} \tag{9.48}$$

where Ω_{2s}, Ω_{3s} are the ratios between the damaged G_i^* and the undamaged $\overline{G}_i$ shear modulus for in-plane shear and longitudinal-transverse shear.

Solving (9.45) and (9.46) results in

$$B_2 = \left(\frac{G_{12}^*}{G_{12}}\right)^4 \frac{1}{4\overline{S}_{66}^2} \frac{1}{F_6^4} \quad ; \quad B_2 > 0 \tag{9.49}$$

$$B_3 = \left(\frac{G_{13}^*}{G_{13}}\right)^4 \frac{1}{4\overline{S}_{55}^2} \frac{1}{F_5^4} \quad ; \quad B_3 > 0 \tag{9.50}$$

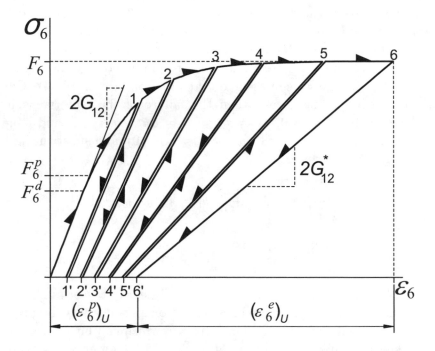

Fig. 9.1: Representation of in-plane stress-strain shear response with several load and unload cycles. Six pairs of tension-unrecoverable strain (i - i') are shown.

9.3.2 Damage Surface Normal Coefficients

The coefficients A_1, A_2, A_3, J_1, J_2 and J_3 represent the effect on the damage surface of damage caused by axial loading. They are directly related to material properties obtained by tensile and compression tests of a unidirectional lamina in the principal material directions. For transverse isotropic materials, the parameters in the thickness direction (denoted by subscript 3) take the same values of those in the transverse direction (denoted by subscript 2). The elastic properties (Young's moduli E_1, E_2 and E_3, and Poisson coefficients ν_{12}, ν_{13} and ν_{23}) and normal strengths (F_{1t}, F_{1c}, F_{2t}, F_{2c}, F_{3t} and F_{3c}) are obtained from uniaxial tests. Also it is necessary to associate critical integrity values and hardening values to these failure states.

Critical values of the integrity tensor $\mathbf{\Omega}$ can be easily estimated by formulas based on experimental results or estimated by analytical procedures [8, 5].

For the undamaged material, g^d defines an initial reduced elastic domain with $\gamma = \gamma_0$. At failure, the elastic domain becomes larger and $\gamma = \gamma(\delta_c) + \gamma_0$, where $\gamma(\delta_c)$ are the critical hardening associated to each failure mechanism. Following the procedure leading to (9.68) but now for cases of uniaxial normal loading results in the following

$$\delta_{1t} = -\frac{1}{\sqrt{J_1}}\left(1 - \Omega_{1t}^2\right) \qquad \delta_{1c} = -\frac{1}{\sqrt{J_1 A_1}}\left(1 - \Omega_{1c}^2\right)$$

$$\delta_{2t} = -\frac{1}{\sqrt{J_2}}\left(1 - \Omega_{2t}^2\right) \qquad \delta_{2c} = -\frac{1}{\sqrt{J_2 A_2}}\left(1 - \Omega_{2c}^2\right) \qquad (9.51)$$

$$\delta_{3t} = -\frac{1}{\sqrt{J_3}}\left(1 - \Omega_{3t}^2\right) \qquad \delta_{3c} = -\frac{1}{\sqrt{J_3 A_3}}\left(1 - \Omega_{3c}^2\right)$$

Writing the g^d function for longitudinal uniaxial tension and compression of the unidirectional lamina $(\sigma_1 \neq 0, \sigma_2 = \sigma_3 = \sigma_6 = \sigma_5 = \sigma_4 = 0)$, the damage surface g^d becomes

$$\sqrt{J_1}\frac{\overline{S}_{11}}{\Omega_{1t}^6}F_{1t}^2 - (\gamma(\delta_{1t}) + \gamma_0) = 0 \qquad (9.52)$$

$$\sqrt{J_1 A_1}\frac{\overline{S}_{11}}{\Omega_{1c}^6}F_{1c}^2 - (\gamma(\delta_{1c}) + \gamma_0) = 0 \qquad (9.53)$$

where F_{1t} and F_{1c} are the longitudinal tensile strength and longitudinal compressive strength of a single composite lamina, $\overline{S}_{11}$ is a coefficient in the undamaged compliance matrix, the quantities Ω_{1t} and Ω_{1c} are the critical values of the integrity tensor at failure for a state of longitudinal stress, while the values J_1 and A_1 are the unknown coefficients.

For a unidirectional lamina under transverse tensile stress $(\sigma_2 \neq 0, \sigma_1 = \sigma_3 = \sigma_6 = \sigma_5 = \sigma_4 = 0)$ the function g^d reduces to

$$\sqrt{J_2}\frac{\overline{S}_{22}}{\Omega_{2t}^6}F_{2t}^2 - (\gamma(\delta_{2t}) + \gamma_0) = 0 \qquad (9.54)$$

$$\sqrt{J_2 A_2}\frac{\overline{S}_{22}}{\Omega_{2c}^6}F_{2c}^2 - (\gamma(\delta_{2c}) + \gamma_0) = 0 \qquad (9.55)$$

and in the thickness direction $(\sigma_3 \neq 0, \sigma_1 = \sigma_2 = \sigma_6 = \sigma_5 = \sigma_4 = 0)$ to

$$\sqrt{J_3}\frac{\overline{S}_{33}}{\Omega_{3t}^6}F_{3t}^2 - (\gamma(\delta_{3c}) + \gamma_0) = 0 \qquad (9.56)$$

$$\sqrt{J_3 A_3}\frac{\overline{S}_{33}}{\Omega_{3c}^6}F_{3c}^2 - (\gamma(\delta_{3c}) + \gamma_0) = 0 \qquad (9.57)$$

where F_{2t}, F_{2c}, F_{3t} and F_{3c} are the strength values and $\overline{S}_{22}$, $\overline{S}_{33}$ are coefficients in the undamaged compliance tensor. The quantities Ω_{2t}, Ω_{2c}, Ω_{3t} and Ω_{3c} are the critical values of the integrity tensor at failure for a state of transverse stress.

Substituting (9.51) in (9.52)-(9.57), six independent nonlinear equations are obtained from which the six components of $\mathbf{A}$ and $\mathbf{J}$ can be easily found.

9.3.3 Plastic-Strain Surface

The plastic-strain surface is defined by (9.19). A three-dimensional Tsai-Wu criterion shape is chosen due to its ability to represent different behavior among the different load paths in stress space. The coefficients f_i are defined as

$$f_1 = \frac{1}{\overline{F}_{1t}} - \frac{1}{\overline{F}_{1c}}; \qquad f_2 = \frac{1}{\overline{F}_{2t}} - \frac{1}{\overline{F}_{2c}}; \qquad f_3 = \frac{1}{\overline{F}_{3t}} - \frac{1}{\overline{F}_{3c}};$$

$$f_{11} = \frac{1}{\overline{F}_{1t}\overline{F}_{1c}}; \qquad f_{22} = \frac{1}{\overline{F}_{2t}\overline{F}_{2c}}; \qquad f_{33} = \frac{1}{\overline{F}_{3t}\overline{F}_{3c}}; \qquad (9.58)$$

$$f_{44} = \frac{1}{\overline{F}_4^2}; \qquad f_{55} = \frac{1}{\overline{F}_5^2}; \qquad f_{66} = \frac{1}{\overline{F}_6^2};$$

and f_{ij} take the following form

$$f_{23} \cong -\tfrac{1}{2}(f_{22}f_{33})^{1/2}; \qquad f_{13} \cong -\tfrac{1}{2}(f_{11}f_{33})^{1/2}; \qquad f_{12} \cong -\tfrac{1}{2}(f_{11}f_{22})^{1/2} \qquad (9.59)$$

The parameters $\overline{F}_{it}$, $\overline{F}_{ic}$, and $\overline{F}_i$ are the effective strength values. That is, the strength values in effective configuration. They are defined as:

$$\overline{F}_{1t} = \frac{F_{1t}}{\Omega_{1t}}; \qquad \overline{F}_{2t} = \frac{F_{2t}}{\Omega_{2t}}; \qquad \overline{F}_{3t} = \frac{F_{3t}}{\Omega_{3t}};$$

$$\overline{F}_{1c} = \frac{F_{1c}}{\Omega_{1c}}; \qquad \overline{F}_{2c} = \frac{F_{2c}}{\Omega_{2c}}; \qquad \overline{F}_{3c} = \frac{F_{3c}}{\Omega_{3c}}; \qquad (9.60)$$

$$\overline{F}_4 = F_4 \frac{\overline{G}_{12}}{G_{12}^*}\frac{\overline{G}_{13}}{G_{13}^*}; \qquad \overline{F}_5 = F_5\frac{\overline{G}_{13}}{G_{13}^*}; \qquad \overline{F}_6 = F_6\frac{\overline{G}_{12}}{G_{12}^*}$$

where the parameters F_{it} and F_{ic} (with $i = 1, 2, 3$), and F_i with (with $i = 4, 5, 6$) are the strength values in tension, compression, in-plane shear, and out-of-plane shear for a composite lamina. G_i^* and $\overline{G}_i$ (with $i = 12, 13$) are the damaged shear modulus and the undamaged shear modulus, respectively. These values are tabulated in the literature, or they can be easily obtained following standardized test methods [9].

9.3.4 Hardening Functions

The plastic-strain (yield) evolution and the damage evolution are defined by two hardening functions (9.9) and (9.10). The in-plane shear stress-strain curve of a single composite lamina is used to identify these hardening evolution functions. Typical in-plane stress-strain shear behavior is shown in Figure 9.1. In order to identify the damage hardening parameters α_i^d, β_i^d and γ_0, and the plastic-strain hardening parameters α_i^p, β_i^p and R_0, it is also necessary to define two initial thresholds in the in-plane shear stress-strain response (Figure 9.1). The highest shear stress without significant plastic-strain is the initial plastic-strain threshold F_6^p. The highest shear stress without significant damage is the initial damage threshold F_6^d. Experimental evidence suggests that plastic-strains do not appear before damage [10]. Therefore, it is assumed that

$$F_6^d < F_6^p \tag{9.61}$$

but the identification process could be easily changed to reflect the opposite situation (i.e., $F_6^d > F_6^p$).

In a state of in-plane shear load, the damage surface function g^d proposed in (9.13) takes the form

$$g^d = \sqrt{B_{22}} Y_{33}^S - (\gamma(\delta) + \gamma_0) = 0 \tag{9.62}$$

and substituting (9.9) we get

$$g^d = \sqrt{B_{22}} \frac{2\overline{S}_{66}}{\Omega_{2s}^4} \sigma_6^2 - \left(\sum_{i=1}^{n} \alpha_i^d \left[\exp\left(\delta/\beta_i^d \right) - 1 \right] \right) - \gamma_0 = 0 \tag{9.63}$$

Similarly, the yield function (9.19) of a unidirectional lamina loaded with pure in-plane shear becomes

$$g^p = \frac{\bar{\sigma}_6}{\overline{F}_6} - (R(p) + R_0) = 0 \tag{9.64}$$

and substituting (9.10) we get

$$g^p = \frac{\bar{\sigma}_6}{\overline{F}_6} - \left(\sum_{i=1}^{n} \alpha_i^p \left[\exp\left(p/\beta_i^p \right) - 1 \right] \right) - R_0 = 0 \tag{9.65}$$

Next, the parameters of both hardening functions are determined explicitly. Using the flow rule (9.12), it is possible to obtain the relationship between the evolution of the damage state variables $\dot{D}_{ij}$ and the hardening damage variables $\dot{\delta}$, as well as the relationship between the evolution of the plastic-strain $\dot{\bar{\varepsilon}}_{ij}^p$ and the hardening variable $\dot{p}$. Using the identification procedure presented here, using in-plane shear load only, the derivatives of (9.63) and (9.65) are constants and take the form

$$\frac{\partial g^d}{\partial Y_{ij}} = \sqrt{B_{22}} \quad ; \quad \frac{\partial g^d}{\partial \gamma} = -1 \quad ; \quad \frac{\partial g^p}{\partial \sigma_{ij}} = \frac{1}{\overline{F}_6} \quad ; \quad \frac{\partial g^p}{\partial R} = -1 \tag{9.66}$$

Therefore, the ratios between the evolution laws, (9.12), become

$$\frac{\dot{D}_{ij}}{\dot{\delta}} = \frac{\dot{\lambda}^d \dfrac{\partial g^d}{\partial Y_{ij}}}{\dot{\lambda}^d \dfrac{\partial g^d}{\partial \gamma}} = -\sqrt{B_{22}} \quad ; \quad \frac{\dot{\bar{\varepsilon}}_{ij}^p}{\dot{p}} = \frac{\dot{\lambda}^p \dfrac{\partial g^p}{\partial \sigma_{ij}}}{\dot{\lambda}^p \dfrac{\partial g^p}{\partial R}} = -\frac{1}{\overline{F}_6} \tag{9.67}$$

The integration of (9.67) with trivial initial conditions allows us to find the hardening variables δ and p explicitly as functions of damage and plastic-strain.

Therefore, for in-plane shear load only, the relationships between the state variables and the hardening state variables are

$$\delta = -\frac{1}{\sqrt{B_{22}}}d_2 = -\frac{1}{\sqrt{B_{22}}}\left(1 - \Omega_2^2\right)$$
$$p = -\bar{F}_6 \, \bar{\varepsilon}_6^p \tag{9.68}$$

Substituting (9.68) in (9.63) and (9.65) we get

$$g^d = \sqrt{B_{22}}\frac{2\bar{S}_{66}}{\Omega_{2s}^4}\sigma_6^2 - \left(\sum_{i=1}^{n}\alpha_i^d\left[\exp\left(-\frac{\left(1 - \Omega_2^2\right)}{\sqrt{B_{22}}\beta_i^d}\right) - 1\right]\right) - \gamma_0 = 0 \tag{9.69}$$

$$g^p = \frac{\sigma_6}{\bar{F}_6} - \left(\sum_{i=1}^{n}\alpha_i^p\left[\exp\left(-\frac{\bar{F}_6\bar{\varepsilon}_6^p}{\beta_i^p}\right) - 1\right]\right) - R_0 = 0 \tag{9.70}$$

When the material is under in-plane shear stress lower than the threshold value F_6^d, the material is undamaged, hence $\delta = 0$, $\Omega_2 = 1$, and the hardening function become $\gamma(\delta) = 0$. Therefore, from (9.69) the parameter γ_0 can be determined easily when the shear load is equal to F_6^d as

$$\gamma_0 = \sqrt{B_{22}}\frac{2\bar{S}_{66}}{\Omega_{2s}^4}(F_6^d)^2 \tag{9.71}$$

In the same way, when the virgin material is loaded with a pure shear stress below the plastic-strain threshold F_6^p, the plastic-strain hardening variable p is equal to zero and the plastic-strains are also zero. Consequently the hardening function $R(p)$ is equal to zero. The initial threshold value for plastic-strain R_0 can be determined with (9.70) as

$$R_0 = \frac{\bar{F}_6^p}{\bar{F}_6} \tag{9.72}$$

and $\overline{F}_6$ is given in (9.60) and $\overline{F}_6^p$ is the effective shear plastic-strain threshold

$$\bar{F}_6^p = \frac{1}{\Omega_2^p}F_6^p \tag{9.73}$$

where the value of integrity Ω_2^p at the shear plastic-strain threshold can be obtained from the in-plane shear strain-stress response (Figure 9.1).

Substituting (9.71) into (9.69) and solving for the positive value of σ_6 yields the relation between the in-plane shear stress and the integrity Ω_2 as

$$\sigma_6 = \left[\left(\sum_{i=1}^{n}\alpha_i^d\left[\exp\left(-\frac{\left(1 - \Omega_2^2\right)}{\sqrt{B_{22}}\beta_i^d}\right) - 1\right]\right)\frac{\Omega_2^4}{2\bar{S}_{66}\sqrt{B_{22}}} + (F_6^d)^2\right]^{1/2} \tag{9.74}$$

In the same way, it is possible to obtain the relation between the effective in-plane shear stress and the effective in-plane plastic-strain from (9.70) and (9.72) as

$$\bar{\sigma}_6 = \bar{F}_6\left(\sum_{i=1}^{n}\alpha_i^p\left[\exp\left(-\frac{\bar{F}_6\bar{\varepsilon}_6^p}{\beta_i^p}\right) - 1\right]\right) + \bar{F}_6^p \tag{9.75}$$

The parameters α_i^d, β_i^d, α_i^p and β_i^p can be obtained by adjusting (9.74) and (9.75) to the experimental data in the form σ_6 vs. Ω_2 and $\bar{\sigma}_6$ vs. $\bar{\varepsilon}_6^p$, respectively, which can be easily derived from experimental data in the form σ_6 vs. ε_6. Cycling loading-unloading of a typical unidirectional or $[0/90]_s$ laminate in shear is shown in Figure 9.1. With each load-unload cycle, one pair $\sigma_6 - \varepsilon_6^p$ and one pair $\sigma_6 - \Omega_2$ are obtained (pairs i - i' in Figure 9.1).

For each loading to σ_6, a value of plastic-strain ε_6^p is obtained upon unloading. These are reproduced in Figure 9.2.(a). The elastic recovery during unloading are used to create Figure 9.2.(b). Three zones can be distinguished: (i) linear elastic, (ii) damage without plastic-strain, and (iii) damage with plastic-strain. The integrity variable Ω_2 on each zone is shown in Figure 9.2.(c) and represented by

$$\Omega_2 = \begin{cases} 1 & \sigma_6 < F_6^d \\[2mm] \sqrt{1 - \dfrac{F_6^d}{\varepsilon_6 G_{12}}} & F_6^d < \sigma_6 < F_6^p \\[3mm] \sqrt{1 - \dfrac{F_6^d}{(\varepsilon_6 - \varepsilon_6^p)G_{12}}} & F_6^d < \sigma_6 < F_6 \end{cases} \qquad (9.76)$$

where G_{12} is the undamaged shear modulus.

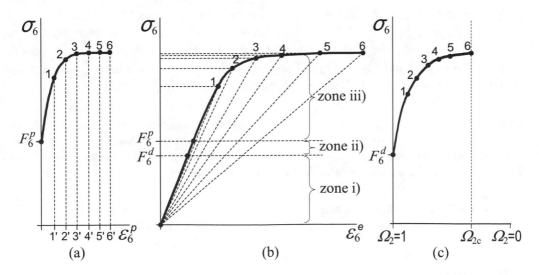

Fig. 9.2: Experimental data mapped onto integrity, plastic-strain, and elastic-strain spaces: (a) stress vs. plastic strain, (b) stress vs. elastic strain, and (c) stress vs integrity.

The relation between shear stress σ_6 and the integrity variable Ω_2 calculated with (9.76) is used to adjust the parameters α_i^d and β_i^d so that (9.74) is satisfied.

Since Ω_2 can be calculated with (9.76), the effective stress and strain can be calculated as

$$\bar{\sigma}_6 = \frac{\sigma_6}{\Omega_2} \qquad\qquad \bar{\varepsilon}_6^p = \Omega_2 \varepsilon_6^p \qquad (9.77)$$

Table 9.1: Elastic properties for a T300/914 unidirectional lamina

E_1	$E_2 = E_3$	$G_{12} = G_{13}$	G_{23}	ν_{12}
142.0 GPa	10.3 GPa	6.42 GPa	3.71 GPa	0.21

Table 9.2: Strength and critical integrity properties for a T300/914 unidirectional lamina

F_{1t}	F_{1c}	$F_{2t} = F_{3t}$	$F_{2c} = F_{3c}$	$F_6 = F_5$	F_4
1830 MPa	1096 MPa	57.0 MPa	57.0 MPa	89.1 MPa	78.0 MPa

D_{1t}^{cr}	D_{1c}^{cr}	$D_{2t}^{cr} = D_{3t}^{cr}$	$D_{2c}^{cr} = D_{3c}^{cr}$
0.1161	0.1109	0.5	0.5

and used to adjust the parameters α_i^p and β_i^p so that (9.75) is satisfied.

In summary, the identification process for a transversely isotropic material requires: (i) five undamaged elastic properties E_1, $E_2 = E_3$, $\nu_{12} = \nu_{13}$, $G_{12} = G_{13}$, and G_{23} (or ν_{23} using $G_{23} = E_2/2(1 + \nu_{23})$), (ii) six strength values (F_{1t}, F_{1c}, $F_{2t} = F_{3t}$, $F_{2c} = F_{3c}$, F_4 and $F_5 = F_6$), (iii) the damaged in-plane shear moduli at imminent failure ($G^*_{12} = G^*_{13}$), and iv) estimates for the critical values of the integrity tensor, Ω_{1t}, Ω_{1c}, $\Omega_{2t} = \Omega_{3t}$ and $\Omega_{2c} = \Omega_{3c}$. The critical integrity values are obtained from the critical damage values (8.32), (8.42), (8.60), and so on.

Example 9.1 *Experimental data for T300/914 Carbon/Epoxy [11] are used to identify the model. The elastic properties of T300/914 lamina are given in Table 9.1. The strength and critical integrity values are shown in Table 9.2. Compare the experimental response for T300/914 Carbon/Epoxy available in [11] with the model response.*

Solution to Example 9.1 *The data obtained from shear response of the material, available in [11], is shown in the Figure 9.3. Using the experimental data the damage and plastic-strain thresholds are obtained, we get $F_6^d = F_6^p = 17.8$ MPa. Using the last unloading path we get $G^*_{12} = 3.41$ MPa and $(\varepsilon_6^p)^U = 2.01$ %.*

The internal parameters shown in Tables 9.3 and 9.4 define the shape of the damage and plastic-strain surfaces and the hardening functions. They are obtained, in terms of the values in Tables 9.1 and 9.2 and the shear response shown in Figure 9.3, using the formulation presented in the previous section and the procedure described next.

First, the components of $\mathbf{B}$ are computed with (9.49), (9.50) and $B_1 = 0$. The effective strengths are computed using (9.60). With the damage and plastic thresholds F_6^d, F_6^p, obtained from Figure 9.3, the hardening thresholds γ_0, R_0 are computed with (9.71) and (9.72).

The load and unload path pairs in Figure 9.3 allow us to determine the evolution of integrity Ω_2, total strain ε_6, elastic strain ε_6^e, and plastic strain ε_6^p as a function of the applied shear stress σ_6.

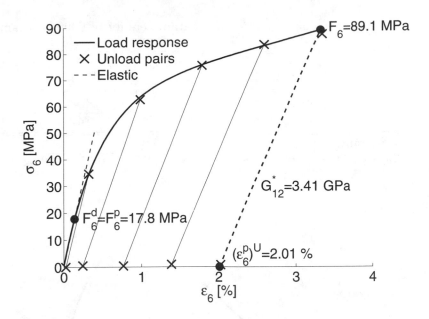

Fig. 9.3: Shear response parameters of a T300/914 Carbon/Epoxy to identify the damage/plastic-strain model.

Table 9.3: Model internal damage parameters for a T300/914 unidirectional lamina

Damage	J_1	$J_2 = J_3$	B_1	$B_2 = B_3$	A_1	$A_2 = A_3$
surface	$6.905 \ 10^{-3}$	0.2040	0	0.2085	1.191	1

Damage	γ_0	α_1^d	β_1^d
hardening	$11.29 \ 10^{-3}$	$36.47 \ 10^{-3}$	-0.3195

Only one term of the Prony series in (9.9) is needed to obtain a good fit of (9.74), which becomes

$$\sigma_6 = \left(\alpha_1^d \left[\exp\left(-\frac{(1 - \Omega_2^2)}{\sqrt{B_2}\beta_1^d} \right) - 1 \right] \frac{\Omega_2^4}{2\overline{S}_{66}\sqrt{B_2}} + (F_6^d)^2 \right)^{1/2} \qquad (9.78)$$

The internal material parameters α_1^d and β_1^d are adjusted using experimental values of σ_6 vs. Ω_2 from sequential load-unload paths as shown in Figure 9.2.(c). Model predictions are compared to experimental data in Figure 9.4. Experimentally, it is not possible to test beyond the critical damage, but the numerical model is able to calculate the response after this point.

Then, (9.51)-(9.57) allow us to obtain the components of $\hat{\mathbf{A}}$ and $\mathbf{J}$. Finally, the plastic-strain hardening parameters are obtained. The experimental data shown in Figure 9.3 are first mapped onto effective space using (9.77) and shown in Figure 9.5. Two different approximations of the Prony series are used. The first is obtained using only one exponential

Table 9.4: Model internal plastic-strain parameters for a T300/914 lamina

Yield surface (MPa)	$\mathbf{F_{1t}}$ 1956	$\mathbf{F_{1c}}$ 1950	$\mathbf{F_{2t}} = \mathbf{F_{3t}}$ 80.6	$\mathbf{F_{2c}} = \mathbf{F_{3c}}$ 80.6	$\mathbf{F_4}$ 146.8	$\mathbf{F_5} = \mathbf{F_6}$ 122.2
Plastic-strain hardening (1 term)	$\mathbf{R_0}$ 0.1458	α_1^P -0.7981	β_1^P 0.2742			
Plastic-strain hardening (2 terms)	$\mathbf{R_0}$ 0.1458	α_1^P -0.3724	β_1^P 0.0750	α_2^P -0.5554	β_2^P 0.9206	

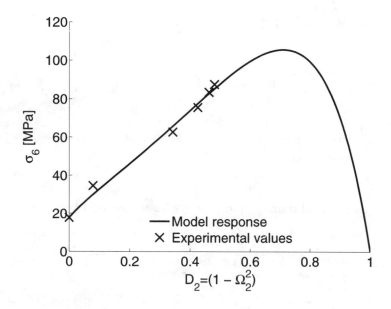

Fig. 9.4: Adjusting in-plane damage hardening with in-plane shear stress-strain data for T300/914 Carbon/Epoxy.

term in (9.10), so that (9.75) becomes

$$\overline{\sigma}_6 = \overline{F}_6 \alpha_{1a}^p \left[\exp\left(-\frac{\overline{F}_6 \overline{\varepsilon}_6^p}{\beta_{1a}^p} \right) - 1 \right] + \overline{F}_6^p \tag{9.79}$$

which allows us to obtain the parameters α_{1a}^p and β_{1a}^p. Using two exponential terms in the Prony series in (9.10), (9.75) becomes

$$\overline{\sigma}_6 = \overline{F}_6 \alpha_{1b}^p \left[\exp\left(-\frac{\overline{F}_6 \overline{\varepsilon}_6^p}{\beta_{1b}^p} \right) - 1 \right] + \overline{F}_6 \alpha_{2b}^p \left[\exp\left(-\frac{\overline{F}_6 \overline{\varepsilon}_6^p}{\beta_{2b}^p} \right) - 1 \right] + \overline{F}_6^p \tag{9.80}$$

which allows us to obtain the parameters α_{1b}^p, β_{1b}^p, α_{2b}^p and β_{2b}^p. A nonlinear optimization method has been used to calculate the hardening parameters so that (9.79) and (9.80) adjust the experimental data in effective space (Figure 9.5). The model responses using one and two exponential terms in the Prony series are compared with the experimental data in Figure 9.5. This completes the identification of the model.

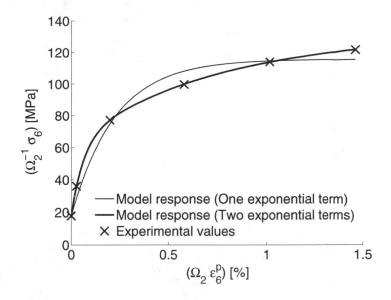

Fig. 9.5: Adjusting in-plane plastic-strain hardening with in-plane shear stress-strain data for T300/914 Carbon/Epoxy.

Once all the internal parameters in the model are completely identified, the model can be used to predict the response of different laminates under different load conditions, provided the fiber volume fraction and the nature of the constituent materials are not changed. The constitutive equations have been implemented in a finite element code. The response under in-plane shear load computed using the finite element code is compared to experimental data for unidirectional composite in Figure 9.6. For additional validation, the response a $[\pm 45]_{2S}$ laminate under tension along the x-direction is compared to experimental data [11] in Figure 9.7.

9.4   Laminate Damage

First, provision is made to model the stress redistribution in the laminate that occurs as a result of the reduction in stiffness due to damage. When a laminate is under bending, the strain distribution is linear through the thickness. Therefore, the material damages uniformly through the thickness of the laminate, and the material properties vary continuously through the thickness of each ply.

The stress-strain relations in the material coordinates for a single lamina are

$$
\begin{bmatrix} \sigma_1 \\ \sigma_2 \\ \sigma_6 \end{bmatrix}^{z_k} = \begin{bmatrix} Q_{11}^d & Q_{12}^d & 0 \\ Q_{12}^d & Q_{22}^d & 0 \\ 0 & 0 & Q_{66}^d \end{bmatrix}^{z_k} \begin{bmatrix} \varepsilon_1 - \varepsilon_1^p \\ \varepsilon_2 - \varepsilon_2^p \\ \gamma_6 - \gamma_6^p \end{bmatrix}^{z_k} \tag{9.81}
$$

where the superscripts d, p, refer to damage and plasticity effects, respectively. In the framework of the classical plate theory (CPT, see Section 3.1.2) the kinematic variables are the mid-plane strain and curvatures. The elastic-damaged terms of the reduced stiffness matrix are defined by a linear function of the material property

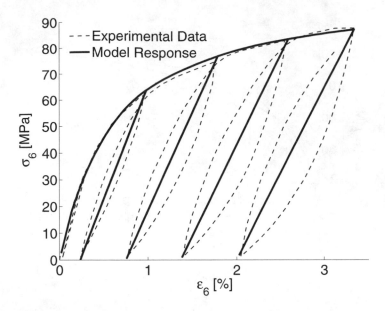

Fig. 9.6: Model and experimental response [11] of a unidirectional T300/914 Carbon/Epoxy laminate under pure shear.

from the top and bottom surface of the k-th lamina as

$$
\overline{Q_{ij}^d}(z) = \overline{Q_{ij}^d}(z_k^b) + \frac{\overline{Q_{ij}^d}(z_k^t) - \overline{Q_{ij}^d}(z_k^b)}{\left(z_k^t - z_k^b\right)}\left(z - z_k^b\right) \tag{9.82}
$$

where an over-bar indicates quantities in the global coordinate system of the laminate, and superscripts t, b refer to top and bottom, respectively. Such quantities are obtained by standard coordinate transformation of (9.81) as shown in [6, 5.49]. The tension σ_i over the lamina is a linear function of the top and bottom values of the stress [6], given by

$$
\sigma_i(z) = \sigma_i(z_k^b) + \frac{\sigma_i(z_k^t) - \sigma_i(z_k^b)}{\left(z_k^t - z_k^b\right)}\left(z - z_k^b\right) \tag{9.83}
$$

To assemble the total stiffness of the material we use the definition of force and moment resultants (3.8). Therefore, the laminate constitutive equations become

$$
\begin{bmatrix} N_x \\ N_y \\ N_{xy} \\ M_x \\ M_y \\ M_{xy} \end{bmatrix} = \begin{bmatrix} A_{11} & A_{12} & A_{16} & B_{11} & B_{12} & B_{16} \\ & A_{22} & A_{26} & B_{12} & B_{22} & B_{26} \\ & & A_{66} & B_{16} & B_{26} & B_{66} \\ & & & D_{11} & D_{12} & D_{16} \\ & & & & D_{22} & D_{26} \\ & & & & & D_{66} \end{bmatrix} \begin{bmatrix} \varepsilon_x^0 - \varepsilon_x^{0p} \\ \varepsilon_y^0 - \varepsilon_y^{0p} \\ \gamma_{xy}^0 - \gamma_{xy}^{0p} \\ \kappa_x^0 - \kappa_x^{0p} \\ \kappa_y^0 - \kappa_y^{0p} \\ \kappa_{xy}^0 - \kappa_{xy}^{0p} \end{bmatrix} \tag{9.84}
$$

where the coefficients are computed in terms of the damaged values of the reduced stiffness coefficients in global coordinates $\overline{Q_{i,j}^d}$ using the standard equations

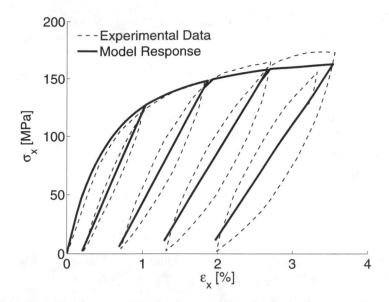

Fig. 9.7: Model and experimental response [11] of a $[\pm 45]_{2S}$ T300/914 Carbon/Epoxy laminate under tension along the x-direction.

[6, (6.16)]. Noting that the damage and plastic effects are piecewise linear functions through the thickness of the laminate, the following explicit equations are obtained for the coefficients of the laminate stiffness matrices

$$A_{ij} = \frac{1}{t_k}\left(\frac{\overline{Q^d_{ij}}\left(z^t_k\right) + \overline{Q^d_{ij}}\left(z^b_k\right)}{2}\right)\left(z^t_k - z^b_k\right)^2$$

$$B_{ij} = \frac{1}{6}\frac{1}{t_k}\left\{\begin{array}{c}\left(z^t_k\right)^3\left[2\overline{Q^d_{ij}}\left(z^t_k\right) + \overline{Q^d_{ij}}\left(z^b_k\right)\right] + \left(z^b_k\right)^3\left[\overline{Q^d_{ij}}\left(z^t_k\right) + 2\overline{Q^d_{ij}}\left(z^b_k\right)\right] \\ -3\cdot z^t_k z^b_k\left[\overline{Q^d_{ij}}\left(z^t_k\right) z^t_k + \overline{Q^d_{ij}}\left(z^b_k\right) z^b_k\right]\end{array}\right.$$

$$D_{ij} = \frac{1}{12}\frac{1}{t_k}\left\{\begin{array}{c}\left(z^t_k\right)^4\left[3\overline{Q^d_{ij}}\left(z^t_k\right) + \overline{Q^d_{ij}}\left(z^b_k\right)\right] + \left(z^b_k\right)^4\left[\overline{Q^d_{ij}}\left(z^t_k\right) + 3\overline{Q^d_{ij}}\left(z^b_k\right)\right] \\ -4z^b_k Z^t_k\left[\overline{Q^d_{ij}}\left(z^t_k\right)\left(z^t_k\right)^3 + \overline{Q^d_{ij}}\left(z^b_k\right)\left(z^b_k\right)^3\right]\end{array}\right.$$

(9.85)

References

[1] B. Luccioni, and S. Oller, A Directional Damage Model, Computer Methods in Applied Mechanics and Engineering 192 (9-10) (2003) 1119–1145.

[2] J. Lubliner, On the Thermodynamic Foundations of Non-Linear Solid Mechanics, International Journal of Non-Linear Mechanics 7 (3) (1972) 237–54.

[3] P. Lonetti, R. Zinno, F. Greco, and E. J. Barbero, Interlaminar Damage Model for Polymer Matrix Composites, Journal of Composite Materials 37 (16) (2003) 1485–1504.

[4] P. Lonetti, E. J. Barbero, R. Zinno, and F. Greco, Erratum: Interlaminar Damage Model for Polymer Matrix Composites (Journal of Composite Materials 37: 16 (1485-1504)), Journal of Composite Materials 38 (9) (2004) 799–800.

[5] E. J. Barbero, and L. De Vivo, Constitutive Model for Elastic Damage in Fiber-Reinforced PMC Laminae, International Journal of Damage Mechanics 10 (1) (2001) 73–93.

[6] E. J. Barbero, Introduction to Composite Materials Design, Taylor & Francis, Philadelphia, 1999.

[7] E. J. Barbero, and E. A. Wen, Compressive Strength of Production Parts Without Compression Testing, ASTM Special Technical Publication (1383) (2000) 470–489.

[8] E. J. Barbero, and P. Lonetti, An Inelastic Damage Model for Fiber Reinforced Laminates, Journal of Composite Materials 36 (8) (2002) 941–962.

[9] M. M. Schwartz, Composite Materials: Properties, Non-destructive Testing, and Repair, Vol. 1, Prentice Hall, Upper Saddle River, NJ, 1997.

[10] C. T. Herakovich, Mechanics of Fibrous Composites, Wiley, New York, 1998.

[11] P. Ladeveze, and E. Le Dantec, Damage Modelling of the Elementary Ply for Laminated Composites, Composite Science and Technology 43 (1992) 257–267.

[12] E. J. Barbero, Web resource (2006) http://www.mae.wvu.edu/barbero/feacm/

Chapter 10

Delaminations

D. Bruno,[1] F. Greco,[1] and E. J. Barbero[2]

Delamination is a frequent mode of failure affecting the structural performance of composite laminates. The interface between layers offers a low-resistance path for crack growth because the bonding between two adjacent layers depends only on matrix properties. Delamination may originate from manufacturing imperfections, cracks produced by fatigue or low velocity impact, stress concentration near geometrical/material discontinuity such as joints and free edges, or due to high interlaminar stresses.

In laminates loaded in compression, the delaminated layers may buckle and propagate due to interaction between delamination growth and buckling. The presence of delaminations (Figure 10.1) may reduce drastically the buckling load and the compressive strength of the composite laminates [1]. Delaminations may also be driven by buckling in laminates under transverse loading [2]. The analysis of delamination buckling requires the combination of geometrically nonlinear structural analysis with fracture mechanics.

According to its shape, delaminations are classified into through-the-width or strip [2]-[9], circular [9]-[15], elliptic [16], rectangular [17] or arbitrary [18, 19]. Depending on its location through the laminate thickness, delaminations are classified into thin film, symmetric split [1, 4, 5], and general [6, 9, 12, 13, 15]. In addition, analysis of combined buckling and growth for composite laminates containing multiple delaminations under in-plane compressive loading has been carried out [20, 21]. Experimental results on delamination buckling are presented in [22, 23].

Other delamination configurations which have been investigated in the literature are the beam-type delamination specimens subjected to bending, axial, and shear loading [22]-[28] which form the basis for experimental methods used to measure interlaminar fracture strength under pure mode I, mode II, and mixed mode conditions in composites, adhesive joints, and other laminated materials (Figure 10.2).

---

[1]Università della Calabria, Italy.
[2]West Virginia University, USA.

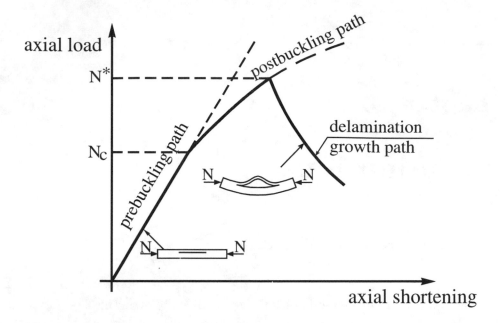

Fig. 10.1: Delamination buckling in a compressed laminate.

In plates with piezoelectric sensors or actuators, an imperfect bonding between the piezoelectric lamina and the base plate may grow under mechanical and/or electrical loading. As a consequence, the adaptive properties of the smart system can be significantly reduced since de-bonding results in significant changes to the static or dynamic response [29, 30]. Finally, delamination growth may be caused by dynamic effects, such as vibration and impact. For instance, the dynamics effects resulting from the inertia of the laminate on the growth process resulting from the buckling of the delamination has been investigated for a circular delamination and time-dependent loadings [31].

Delaminations in layered plates and beams have been analyzed by using both cohesive damage models and fracture mechanics. A cohesive damage model implements interfacial constitutive laws defined in terms of damage variables and a damage evolution law. Cohesive damage elements are usually inserted between solid elements [32]-[34] or beam/shell elements [34].

In the context of the fracture mechanics approach, which is the methodology followed in this chapter, the propagation of an existing delamination is analyzed by comparing the amount of energy release rate (ERR) with the fracture toughness of the interface. When mixed mode conditions are involved, the decomposition of the total ERR into mode I, mode II, and mode III components becomes necessary due to the mixed-mode dependency of interface toughness [35]. A number of fracture mechanics-based models have been proposed in the literature to study delamination, including three-dimensional models [36, 37] and simplified beam-like models [1, 3, 28, 38, 39].

Three-dimensional elasticity can be used to represent accurately the local 3D

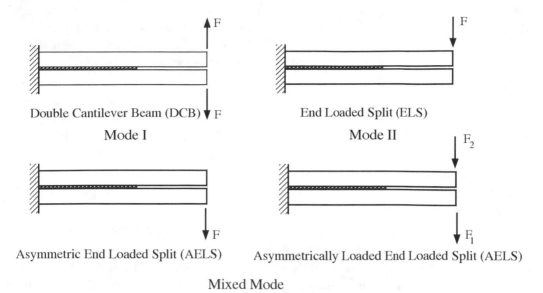

Fig. 10.2: Beam-type delamination specimens.

stress field in the neighborhood of the delamination front. Due to the high gradient of stress and strain states in the neighborhood of the delamination front, solid finite elements and a refined mesh at the delamination front are needed [40]-[45]. Computation of the ERR mode components requires many layers of solid elements through the laminate thickness and a very fine mesh along the in-plane directions in the neighborhood of the delamination front, especially for complex laminates due to anisotropic properties, lay-up sequence, geometry, and loading. Therefore, the computational cost of the analysis is high. Moreover, additional complications arise due to the oscillatory behavior of stress and displacement fields near the delamination front, which are predicted by the three-dimensional elasticity theory [41].

Beam-like models based on the Euler-Bernoulli beam theory provide a simplified analysis tool for the global behavior of delaminated plates. Various extensions of these models are available. For example, the adhesion between the layers of a double cantilever beam is modeled by means of an elastic foundation in [46], while a refined plate theory is used to model the layers in [47].

Most classical delamination studies based on beam or plate theories assume that the laminated plate is essentially composed of two plate elements in the delaminated region and a single plate element in the undelaminated region [9, 10, 12, 19, 21]. This is equivalent to using a two-layer plate model, namely two plate elements bonded in the undelaminated region, using the Kirchhoff plate kinematics (see Section 3.1.2), which prescribes equal rotations for the plates in the undelaminated zone [4, 27, 48].

Classic delamination models based on plate theories give a good estimate of total ERR. Unfortunately, they may provide inaccurate results when utilized to separate ERR into mode components except in particular cases [28]. In fact, a notable underestimation of the actual ERR mode components may arise since usually shear effects are neglected and the local crack tip strain state is not accurately described.

Shear deformability has a notable effect on the ERR evaluation in laminates due to their low shear to bending stiffness ratio [25, 48]. As a consequence, the classical plate-based delamination models have been improved in several ways.

The global-local analysis concept illustrated in [49]-[51] aims to reduce the cost of computation of full 3D methods and to obtain a more accurate mode partition with respect to beam-plate models. In this context, numerical solutions to integral equations [26, 49, 50] or finite element models [51, 52] are used in the literature. The total ERR is determined from an overall analysis in which the structure is modeled by an assembly of plates and the mode partition is performed by means of numerical solution of the local continuum model in a small element containing the delamination front.

The global-local approach has shown its effectiveness for both 2D [53] and 3D [54, 55] delamination problems and it is known as the crack tip element (CTE) method [52]. Since the classical plate theory is generally used, contributions from shear deformations of delaminated and undelaminated members to ERR are usually neglected. Although the original solution in [49, 50] is improved in [55, 56], the CTE method cannot be used for general delaminations and one must resort to full three-dimensional FEA, which may involve high computational effort. In addition, delaminations have been modeled by using sublaminates governed by transverse shear deformable laminate theory, thus obtaining a reasonable approximation to the mode separation solution [57].

An intermediate approach between models based on classical plate theory and on continuum analysis has been proposed both in combination with numerical solutions or analytical procedures [24, 25, 27, 48, 58, 59, 60, 61] by coupling interface elements and plate theories. This technique achieves two basic objectives. First, it retains the simplicity of classic delamination models by operating in terms of plate variables. Second, it leads to accurate and direct mode partition.

The interface technique is also able to take into account nonlinear effects due to bridging mechanisms or to damage and to easily incorporate unilateral contact conditions [59, 62, 63]. In [48], the undelaminated region of the laminate is modeled as two first-order shear deformable plates, instead of a single plate element as in classical models. A linear elastic interface model is introduced to reconstruct inter-laminar stresses and thus to compute ERR mode components. In this way, shear deformations are incorporated in the solution and a refined ERR computation is obtained.

In the context of a plate-based model, the accuracy of mode partition is influenced by the laminate kinematics assumptions (see Section 3.1). Therefore, a more reliable mode partition can be obtained by using improved plate kinematical models. Consequently the two-layer plate kinematical model is refined in [64] by developing a coupled interface-multilayer approach, showing that mode partition may be performed accurately by introducing an appropriate number of plate models in each sublaminate. Moreover, the complications related to oscillatory behavior of ERRs are avoided using interface and plate variables, since the interface model may be interpreted as a very thin resin layer embedding the delamination. Finally, some refined plate theories are proposed to include the possibility of de-bonding between laminate layers. For example, the layer-wise theory was introduced in [65] incorporating delamination in composite laminates.

Fracture mechanics allows us to predict the growth of a preexisting crack or defect. In a homogeneous and isotropic body subjected to a generic loading condition, a crack tends to grow by kinking in a direction such that a pure mode I condition at its tip is maintained. On the contrary, delaminations in laminated composites are constrained to propagate in its own plane because the toughness of the interface is relatively low in comparison to that of the adjoining material. Therefore a delamination crack propagates with its advancing tip in mixed mode condition and, consequently, requires a fracture criterion including all three mode components.

The theory of crack growth may be developed by using one of two approaches. First, the Griffith energetic (or global) approach introduces the concept of ERR G as the energy available for fracture on one hand, and the critical surface energy G_c as the energy necessary for fracture on the other hand. Alternatively, the Irwin (local) approach is based on the stress intensity factor concept, which represents the energy stress field in the neighborhood of the crack tip. These two approaches are equivalent and, therefore, the energy criterion may be rewritten in terms of stress intensity factors.

A number of path independent integrals have been proposed to calculate the ERR, such as the J-integral [66]. Calculating these integrals along special paths in terms of plate variables, one can write the ERR as a function of stress resultants acting upon sections adjoining the crack border.

10.1   Two-Dimensional Delamination

The problem of a through-the-width delamination crack is considered in this section. Its generalization to the case of a plane crack of arbitrary geometry is given in Section 10.2.

10.1.1   Energy Release Rate (ERR)

Consider a linear elastic solid occupying a region B in its reference configuration (Figure 10.3). The solid, referred to a fixed system of Cartesian coordinate xyz has a uniform thickness h in the z-direction with its volume represented by $\Omega = A \times (-h/2, h/2)$, with A denoting the trace of the solid in the x-y plane (Figure 10.3).

Suppose that the body is in a two-dimensional deformation state (plane strain or plane stress) with a crack with faces parallel to the x-z plane, with length ℓ along the x-axis. Under these assumptions, it is sufficient to consider only the variables x and y. The body illustrated in Figure 10.3 is in a small-deformation regime under the action of surface forces $\boldsymbol{t}_p(\lambda)$ prescribed on a portion of the boundary, whose trace in the x-y plane is ∂A^t, and of displacements $\boldsymbol{u}_p(\lambda)$ imposed on the complementary part. The objective in this section is to analyze the quasi-static

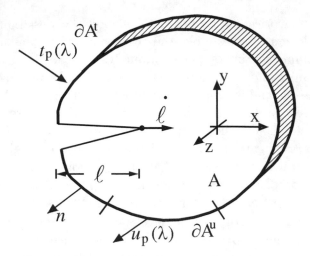

Fig. 10.3: Two-dimensional cracked body.

propagation of a crack described by a monotonically varying time-like parameter τ. The load parameter λ is assumed constant during crack propagation. For the sake of simplicity, equations in the following calculations are written for a unit thickness (namely, $h = 1$) and body forces are omitted.

Energy is dissipated by the system in order to create the two new surfaces of an advancing crack. Experiential evidence indicates that this process is irreversible since the crack cannot be healed without some external action or effort [67]. The second law of thermodynamics in the form of Clausius-Duhem inequality (8.95) for isothermal process ($T = \text{const.}$) becomes $\sigma : \dot{\varepsilon} - \rho\dot{\psi} \geq 0$.

For the sake of discussion, let's assume a crack with area A_c advancing in a self-similar fashion and let's take its area to represent the state of fracture of the system. In this case, the Helmholtz free energy $\psi = \psi(\varepsilon, A_c)$ is a function of the strain ε and the crack area. Then, using (8.97) for an isothermal process ($\nabla T = 0$), yields an expression for the dissipation rate $\dot{\Gamma}$ leaving the body to create new crack area,

$$\dot{\gamma} = p_\alpha \dot{s}_\alpha = f_c \dot{A}_c \geq 0 \quad ; \quad \dot{\Gamma} = \int_\Omega \dot{\gamma}\, d\Omega \qquad (10.1)$$

where $p_\alpha = f_c$ is the thermodynamic force driving the advance of the crack at a rate $\dot{s}_\alpha = \dot{A}_c$, and $\dot{\gamma}$ is the heat dissipation rate per unit volume. In this chapter a dot over a variable denotes total differentiation with respect to the time-like parameter τ.

Now, the first law of thermodynamics[3] (8.69) allows us to perform an energy balance. Since $\dot{\Gamma}$ accounts for all the dissipation taking place during an elastic fracture, the internal energy rate $\dot{U}$ in (8.69) reduces to the strain energy rate, i.e., the volume integral of (8.75). The heat entering the body $\dot{Q}$ in (8.69) reduces to

---

[3]Note that the classical thermodynamics sign convention is respected, with positive heat being that added to the system, positive work being done by the system on its surroundings, and positive internal energy rate signifying an increase in the internal energy of the system.

minus the dissipation rate $\dot{\Gamma}$ leaving the body to create new surfaces. Still, the rate of work done by the body on its surroundings W, is equal to minus the rate of work done by the surroundings on the body,

$$\dot{W} = - \int_{\partial A^t} t_p(\lambda) \cdot \dot{u} \, dS \tag{10.2}$$

The dot between the vectors denotes the scalar product, dS is the element of arc, and ∂A^t denotes the boundary of the body where the surface tractions are applied. Therefore, the first law of thermodynamics (8.69) can be written as

$$\dot{\Gamma}(\mathbf{u}, \ell, \lambda) = -\left(\dot{U} + \dot{W}\right) = -\dot{\Pi}_e \tag{10.3}$$

The term in parenthesis is the potential energy Π_e of an elastic (and thus reversible) system. It consists of the strain energy rate minus the work done by the environment on the system. Unlike (10.1), equation (10.3) provides a convenient operational formula to compute the dissipation rate, which is a function of the displacement vector field $\mathbf{u}$, the crack length ℓ and the load parameter λ.

Note that in (10.3) the total potential energy must be calculated along the equilibrium path for the displacements at fixed crack length. Therefore, under the assumption of positiveness of the strain energy, the displacements must satisfy the minimum of the total potential energy and the potential energy can be expressed in a reduced form

$$\dot{\Pi}_r\left(\ell, \lambda\right) = \dot{\Pi}_e\left(\mathbf{u}\left(\ell, \lambda\right), \ell, \lambda\right) = \min \dot{\Pi}_e\left(\mathbf{u}^*, \ell, \lambda\right) = -G(\ell)\dot{\ell} \tag{10.4}$$

where explicit dependence on the displacements has been eliminated by equilibrium requirements. In (10.4) $\boldsymbol{u}^*$ denotes an admissible displacement field which satisfies the geometric boundary conditions on ∂A^u, and the ERR associated with the crack length ℓ, namely the energy available for unit area crack growth. Using the chain rule $\partial/\partial t = \partial/\partial \ell \, \dot{\ell}$ on (10.4) leads to an operational definition for the ERR as

$$G\left(\ell\right) = -\frac{\partial}{\partial \ell}\Pi_r(\ell, \lambda) \tag{10.5}$$

10.1.2 Modes of Fracture

The individual modes of fracture are illustrated in Figure 10.4, where the mode III condition is also shown arising from out-of-plane shear forces, which may be present in the case of 3D delamination problems.

ERR is a global measure of the energy available at the crack tip, but it does not represent the way in which the crack could advance. To this end, the opening (or mode I) G_I, and the sliding (or mode II) G_{II} components of the ERR (denoted by G in this chapter) can be defined as the work done by the normal and shear interface tractions through the corresponding interface relative displacements, as the crack advances. This decomposition is exact only when the crack is embedded in a homogeneous material, but it is only approximate for an interface crack between dissimilar materials [26]. Assuming the crack to be embedded in a thin layer between the bi-material system avoids ambiguity in the ERR mode components definitions [41]. Mode I and II ERR components are related to the singular normal and shear stress, respectively. These act close to the crack tip, with amplitudes expressed by means of stress intensity factors K_I and K_{II}. The singular traction distribution on the line ahead of the crack takes the following form when expressed in terms of stress intensity factors

$$\sigma_{yy} = K_I (2\pi r)^{-\frac{1}{2}}, \quad \sigma_{yx} = K_{II} (2\pi r)^{-\frac{1}{2}}, \quad \sigma_{yz} = K_{III} (2\pi r)^{-1/2} \qquad (10.6)$$

As a consequence, the relationship between energy release mode components and stress intensity factors assumes the following form

$$G_I = \tilde{E}^{-1} K_I^2, \quad G_{II} = \tilde{E}^{-1} K_{II}^2, \quad G_{III} = \frac{K_{III}^2}{2\mu} \qquad (10.7)$$

where μ is the shear modulus and

$$\tilde{E} = \left\{ \begin{array}{ll} \dfrac{E}{1 - \nu^2} & \text{(plane strain)}, \quad \varepsilon_z = 0) \\ E & \text{(plane stress)}, \quad \sigma_z = 0 \end{array} \right\} \qquad (10.8)$$

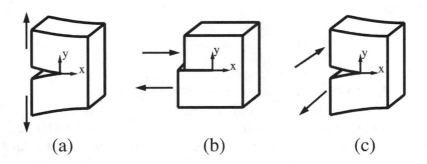

(a)   (b)   (c)

Fig. 10.4: Fracture modes of delamination growth: (a) opening mode, I, (b) sliding mode, II, (c) tearing mode, III.

10.1.3   Crack Propagation

From the energy balance introduced in the previous section, the following fracture criterion is obtained

$$\begin{array}{ll} G < G_c, \quad \dot{\ell} = 0, & \text{(no propagation)} \\ G = G_c, \quad \dot{\ell} \geq 0, & \text{(possible propagation)} \end{array} \qquad (10.9)$$

where G_c is the critical ERR, that is the resistance that must be overcome for a unit delamination growth, equal to Γ. Experimental evidence shows that the critical ERR often depends on the individual modes of fracture and it is a function of the relative amount of mode II to mode I energy release rate. The mixed-mode dependence is especially evident for interface cracks in composite systems. Unlike cracks in homogeneous media, which follow a path maintaining a pure mode I at their tip, delamination at the interface involves coupled fracture modes. Various phenomenological delamination criteria have been proposed in the literature. An example is the power-law criterion

$$\left(\frac{G_I}{G_I^c}\right)^a + \left(\frac{G_{II}}{G_{II}^c}\right)^\beta \geq 1 \tag{10.10}$$

where G_I^c, G_{II}^c, and α, β are the interlaminar fracture toughness and the mixed mode fracture parameters, respectively, which must be determined experimentally, for instance by fitting (10.10) to test results (for details see [26]). Determination of the individual components of ERR is a complex but indispensable task due to the mixed-mode dependence of the delamination criterion.

10.2 Delamination in Composite Plates

Consider a laminate that contains a single in-plane delamination crack of area Ω_D with a smooth front $\partial \Omega_D$, as shown in Figure 10.5.[4] The laminate thickness is denoted by h_0. The x-y plane is taken to be the midplane of the laminate, and the z-axis is taken positive downward from the midplane.

10.2.1 Sublaminate Modeling

The delamination plane separates the delaminated structure into two *sublaminates* of thickness h_1, h_2, both of which are assumed small compared to the in-plane dimensions. Each sublaminate is represented by an assemblage of first order shear deformable (FSDT) plate elements (see Table 3.3) bonded by zero-thickness interfaces in the transverse direction as shown in Figure 10.6. The upper sublaminate is subdivided into n_u plates and the lower one into n_l plates. In the examples, the notation $(n_u - n_l)$ indicates the level of refinement of the discretization through the thickness.

The first plate element is at the bottom and the thickness of the ith plate element is denoted by t_i. Each plate element may, in turn, represent one or several physical fiber-reinforced plies with their material axes arbitrarily oriented. Adhesion between the plates inside each sublaminate is enforced by using constraint equations (CE) implemented through Lagrangian multipliers. Accordingly, the displacements in the ith plate element, in terms of a global reference system located at the laminate

---

[4]Note that in this chapter, the z-coordinate is oriented downward, but otherwise the coordinate system in Figure (10.5) is identical to the one defined in Figure (3.3).

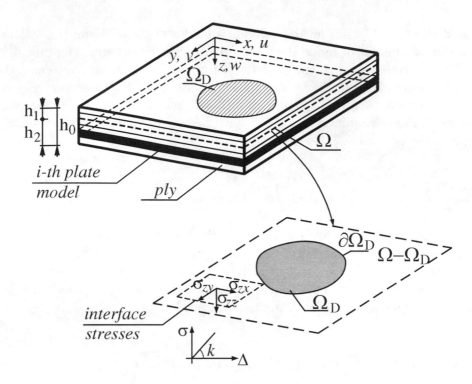

Fig. 10.5: Delaminated composite plate.

midsurface, are expressed by

$$
\begin{aligned}
u_i\,(x,y,z) &= u_i^0\,(x,y) + (z - z_i) \cdot \phi_{xi}\,(x,y) \\
v_i\,(x,y,z) &= v_i^0\,(x,y) + (z - z_i) \cdot \phi_{yi}\,(x,y) \\
w_i(x,y,z) &= w_i^0\,(x,y)
\end{aligned}
\tag{10.11}
$$

where u_i, v_i refer to the in-plane displacements, and w_i to the transverse displacements through the thickness of the ith plate element, u_i^0, v_i^0, w_i^0, are the displacements at the mid-surface of the ith plate element, respectively, and ϕ_{xi}, ϕ_{yi} denote rotations of transverse normals about y and x, respectively. In addition, z_i denotes the coordinate along the z-direction of the ith mid-plane. This is equivalent to modeling each sublaminate with generalized laminated plate theory (GLPT, [71]-[74]). However, the presentation in this section is geared towards using standard first order shear deformation theory (FSDT) finite elements available in commercial FEA packages, and to join these elements at the interfaces inside each sublaminate using CE or rigid links characterized by two nodes and three degrees of freedom at each node. FSDT is described in Section 3.1.1.

At the reference surfaces, the membrane strain vector ϵ_i, the curvature κ_i, and

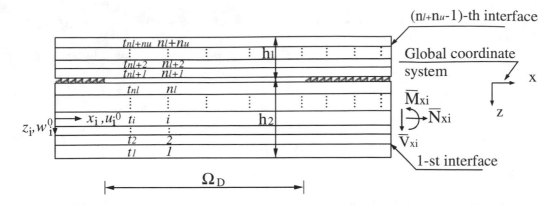

Fig. 10.6: Laminate subdivision in plate elements.

transverse shear strains γ_i, respectively, are defined as

$$
\left\{ \begin{array}{c} \epsilon_{xxi} \\ \epsilon_{yyi} \\ \gamma_{xyi} \end{array} \right\} = \left\{ \begin{array}{c} \dfrac{\partial u_i^0}{\partial x} \\[2mm] \dfrac{\partial v_i^0}{\partial y} \\[2mm] \dfrac{\partial u_i^0}{\partial y} + \dfrac{\partial v_i^0}{\partial x} \end{array} \right\} , \left\{ \begin{array}{c} \kappa_{xxi} \\ \kappa_{yyi} \\ \kappa_{xyi} \end{array} \right\} = \left\{ \begin{array}{c} \dfrac{\partial \phi_{xi}}{\partial x} \\[2mm] \dfrac{\partial \phi_{yi}}{\partial y} \\[2mm] \dfrac{\partial \phi_{xi}}{\partial y} + \dfrac{\partial \phi_{yi}}{\partial x} \end{array} \right\} , \left\{ \begin{array}{c} \gamma_{yzi} \\ \gamma_{xzi} \end{array} \right\} = \left\{ \begin{array}{c} \phi_{yi} + \dfrac{\partial w_i^0}{\partial y} \\[2mm] \phi_{xi} + \dfrac{\partial w_i^0}{\partial x} \end{array} \right\}
$$

(10.12)

The constitutive relations between stress resultants (3.7) and corresponding strains (3.6) are

$$
\left\{ \begin{array}{c} \{N_i\} \\ \{M_i\} \end{array} \right\} = \left[\begin{array}{cc} [A_i] & [B_i] \\ [B_i] & [D_i] \end{array} \right] \left\{ \begin{array}{c} \{\epsilon_i\} \\ \{\kappa_i\} \end{array} \right\}
$$

(10.13)

$$
\{V_i\} = [H_i]\{\gamma_i\}
$$

where $\{N_i\} = \{N_{xxi}, N_{yyi}, N_{xyi}\}^T$ are the membrane force resultants, $\{M_i\} = \{M_{xxi}, M_{yyi}, M_{xyi}\}^T$ are the moment resultants and $\{V_i\} = \{V_{yzi}, V_{xzi}\}^T$ the transverse shear force resultants. In addition, $[A_i]$, $[D_i]$, $[B_i]$ denote the classical extensional stiffness, bending stiffness, and bending-extension coupling stiffness, respectively, whereas $[H_i]$ are the transverse shear stiffness (3.9)

$$
([A_i],\,[B_i],\,[D_i]) = \int_{-t_{i/2}}^{t_{i/2}} \left[\overline{Q}_i\,(z_i)\right]\, \left(1, z_i, z_i^2\right) dz_i
$$

(10.14)

$$
[H_i]_{jk} = \int_{-t_{i/2}}^{t_{i/2}} K_j K_k \left[\overline{Q}_i^*\right]_{jk} (z_i)\, dz_i
$$

(10.15)

where $\overline{Q}_i\,(z_i)$ is the transformed reduced stiffness matrix and $\overline{Q}_i^*$ is the transformed interlaminar shear stiffness matrix [75, Section 5.4.3]. Moreover, K_i are the shear correction coefficients which are usually set to $\sqrt{5/6}$ or else calculated as $[H_i]_{jk}$ as in (3.9).

Inside each sublaminate the displacement continuity requirements between any two adjacent plate elements, i and $i+1$ with $(i \neq n_l)$ lead to

$$\begin{aligned}
\triangle u_i^0 &= u_i^0 - \frac{t_i}{2}\phi_{xi} - u_{i+1}^0 - \frac{t_{i+1}}{2}\phi_{xi+1} = 0 \\
\triangle v_i^0 &= v_i^0 - \frac{t_i}{2}\phi_{yi} - v_{i+1}^0 - \frac{t_{i+1}}{2}\phi_{yi+1} = 0 \\
\triangle w_i^0 &= w_i^0 - w_{i+1}^0 = 0
\end{aligned} \tag{10.16}$$

These are enforced by Lagrange multipliers, which in this case represent interlaminar stresses.

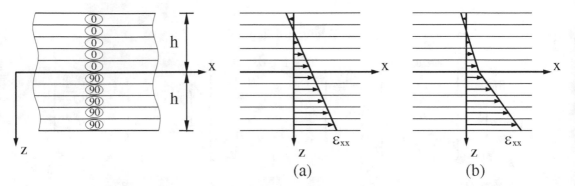

Fig. 10.7: Laminate stacking sequence and solution to models (a) and (b) of Example 10.1.

Example 10.1 *A $[0_5/90_5]_T$ laminate with dimensions $2a = 2b = 250$ mm and $2h = 25$ mm is subjected to in-plane load $N_{xx} = 1.0$ N/mm at the mid-surface. The ply thickness is $t_k = 2.5$ mm and the material properties are given in Table 10.1. Plot the through-the-thickness deformation ε_{xx} at the element centroid using two alternate modeling approximations: (a) one ANSYS element SHELL181 to represent the whole thickness, and (b) one ANSYS element SHELL181 to encompass each of the groups of plies with the same angle, and joining them with CE.*

Table 10.1: Material properties of the double cantilever beam (DCB) specimen

Exx = 126.0 GPa	$\nu_{xy} = 0.263$	Gxy = 1.070 GPa
Eyy = 9.50 GPa	$\nu_{xz} = 0.263$	Gxz = 1.070 GPa
Ezz = 9.50 GPa	$\nu_{yz} = 0.270$	Gxy = 0.8063 GPa

Solution to Example 10.1 *According to model (a), one ANSYS element SHELL181 is used to represent the whole thickness, all strain components vary linearly through the laminate thickness as shown in Figure 10.7. The model (b), using one ANSYS element SHELL181*

for each group of plies with the same angle, leads to a continuous variation for all strain components with a different linear variation in different elements as shown in Figure 10.7. In model (b) the two shell elements are joined by imposing displacement continuity conditions (10.18) at the interface by means of CE. The laminate stacking sequence and the solution to models (a) and (b) are shown in Figure 10.7.

Model (a) *is obtained by using the ANSYS input command sequence shown below, also available in [76]. Mesh dimensions are in mm.*

```
/NOPR              ! SUPPRESSES EXPANDED INTERPRETED INPUT DATA LISTING
/PMETH,OFF,0       ! ACTIVATES THE p-METHOD SOLUTION OPTIONS IN THE GUI

KEYW,PR_SET,1      ! SETS THE KEYWORD USED BY THE GUI FOR
                   ! STRUCTURAL CONTEXT FILTERING
KEYW,PR_STRUC,1
KEYW,PR_THERM,0
KEYW,PR_FLUID,0
KEYW,PR_ELMAG,0
KEYW,MAGNOD,0
KEYW,MAGEDG,0
KEYW,MAGHFE,0
KEYW,MAGELC,0
KEYW,PR_MULTI,0
KEYW,PR_CFD,0


/PREP7             ! ENTERS THE MODEL CREATION PREPROCESSOR

ANTYPE,STATIC      ! SPECIFIES  THE ANALYSIS TYPE AND RESTART STATUS

LOCAL,11,0,0,0,0,0,0,0, ! DEFINES A LOCAL COORD. SYSTEM ALIGNED WITH
                        ! THE GLOBAL ONE BY LOCATION AND ORIENTATION

ET,1,SHELL181           ! DEFINES SHELL181 AS A LOCAL ELEMENT TYPE.
                        ! THIS WILL BE USED TO MESH THE PLATE MODEL
KEYOPT,1,1,0            ! SETS SHELL181 KEY OPTIONS
KEYOPT,1,3,2
KEYOPT,1,8,0
KEYOPT,1,9,0
KEYOPT,1,10,0


MPTEMP,,,,,,,,      ! DEFINES A TEMPERATURE TABLE FOR MATERIAL PROPERTIES
MPTEMP,1,0
MPDATA,EX,1,,126E+3     ! DEFINES MATERIAL PROPERTY DATA TO BE
                        ! ASSOCIATED WITH THE TEMPERATURE TABLE
MPDATA,EY,1,,9.5E+3
MPDATA,EZ,1,,9.5E+3
MPDATA,PRXY,1,,0.263
MPDATA,PRYZ,1,,0.27
MPDATA,PRXZ,1,,0.263
MPDATA,GXY,1,,1.07E+3
MPDATA,GYZ,1,,0.8063E+3
```

```
MPDATA,GXZ,1,,1.07E+3

SECTYPE,1,SHELL,,          ! DEFINES SHELL SECTION PROPERTY DATA TO BE
                           ! ASSOCIATED WITH THE PLATE MODEL
SECDATA,12.5,1,90,3
SECDATA,12.5,1,0,3
SECOFFSET,MID
SECCONTROL,,,,,,,

K,1,1,1,1,                 ! DEFINES ALL THE NECESSARY KEYPOINTS
                           ! FOR THE MODEL'S GEOMETRY
K,2,251,1,1,
K,3,1,251,1,
K,4,251,251,1,

LSTR,1,3                   ! GENERATES ALL THE NECESSARY LINES
                           ! FOR THE MODEL'S GEOMETRY
LSTR,1,2
LSTR,2,4
LSTR,3,4
LSEL,S,LINE,,1,4,1,0       ! SELECTS A SUBSET OF LINES
LESIZE,ALL,,,2,,,,,1       ! SPECIFIES THE DIVISIONS AND
                           ! SPACING RATIO ON UNMESHED LINES
LSEL,ALL,,,,,,,

AL,1,2,3,4  ! GENERATES AN AREA BOUNDED BY PREVIOUSLY DEFINED LINES

TYPE,1       ! SETS SHELL181 AS THE ELEMENT TYPE ATTRIBUTE POINTER
MAT,1        ! SETS THE ELEMENT MATERIAL ATTRIBUTE POINTER
REAL,        ! SETS THE ELEMENT REAL CONSTANTS SET ATTRIBUTE POINTER
ESYS,11      ! SETS THE ELEMENT COORDINATE SYSTEM ATTRIBUTE POINTER
SECNUM,1     ! SETS THE ELEMENT SECTION ATTRIBUTE POINTER
AMAP,1,1,2,4,3! GENERATES 2D-MAPPED MESH BASED ON SPEC. AREA CORNERS

NSEL,S,LOC,X,250.09,251.01,,0        ! SELECTS A SUBSET OF NODES
NSEL,R,LOC,Z,0.09,1.01,,0
D,ALL,UZ,0,,,,,,,,,        ! DEFINES DOF CONSTRAINTS AT PREVIOUSLY
                ! DEFINED NODES CONSTRAINING THE Z TRANSLATION
NSEL,ALL,,,,,,
NSEL,S,LOC,X,0.99,1.01,,0
NSEL,R,LOC,Z,0.99,1.01,,0
D,ALL,UZ,0,,,,UX,UY,,,,   ! DEFINES DOF CONSTRAINTS AT PREVIOUSLY
        ! DEFINED NODES CONSTRAINING THE Z, X AND Y TRANSLATIONS
NSEL,ALL,,,,,,            ! REACTIVATES SUPPRESSED NODES

LSEL,S,LINE,,1,3,2,0
SFL,ALL,PRES,-1,          ! APPLY Nx=1 AT THE TWO EDGES X=0,2a
LSEL,ALL,,,,,,,

FINISH                    ! EXITS NORMALLY FROM A PROCESSOR
```

```
/SOL                    ! ENTERS THE SOLUTION PROCESSOR
/STATUS,SOLU
SOLVE                   ! STARTS A SOLUTION

FINISH                   ! EXITS NORMALLY FROM A PROCESSOR
/POST1              ! ENTERS THE DATABASE RESULTS POSTPROCESSOR
GPLOT
PLDISP,2
/VIEW,1,1,1,1
/ANG,1
/REP,FAST


AVPRIN,0,,  ! FILLS A TABLE ELEMENT SUMMABLE MISCELLANEOUS DATA
            ! CONTAINING MEMBRANE AXIAL STRAIN IN THE X-DIRECTION
ETABLE,EPS_11,SMISC,9
AVPRIN,0,,  ! FILLS A TABLE ELEMENT SUMMABLE MISCELLANEOUS DATA
            ! CONTAINING CURVATURE IN THE X-DIRECTION
ETABLE,CHI_11,SMISC,12


SET,LIST,999
SET,,, ,,, ,1
PRETAB,EPS_11,CHI_11
```

Model (b) *is obtained by using the ANSYS input command sequence shown below, also available in [76]. Mesh dimensions are in mm.*

```
/NOPR          ! SUPPRESSES EXPANDED INTERPRETED INPUT DATA LISTING
/PMETH,OFF,0! ACTIVATES THE p-METHOD SOLUTION OPTIONS IN THE GUI

KEYW,PR_SET,1   ! SETS THE KEYWORD USED BY GUI FOR STRUCTURAL
                ! CONTEXT FILTERING
KEYW,PR_STRUC,1
KEYW,PR_THERM,0
KEYW,PR_FLUID,0
KEYW,PR_ELMAG,0
KEYW,MAGNOD,0
KEYW,MAGEDG,0
KEYW,MAGHFE,0
KEYW,MAGELC,0
KEYW,PR_MULTI,0
KEYW,PR_CFD,0
/PREP7          ! ENTERS THE MODEL CREATION PREPROCESSOR

ANTYPE,STATIC   ! SPECIFIES  THE ANALYSIS TYPE AND RESTART STATUS

LOCAL,11,0,0,0,0,0,0,0, ! DEFINES A LOCAL COORDINATE SYSTEM ALIGNED
                ! WITH THE GLOBAL ONE BY LOCATION AND ORIENTATION

ET,1,SHELL181   ! DEFINES SHELL181 AS LOCAL ELEMENT TYPE.
                ! THIS WILL BE USED TO MESH THE PLATE MODEL
KEYOPT,1,1,0    ! SETS SHELL181 KEY OPTIONS
```

```
KEYOPT,1,3,2
KEYOPT,1,8,0
KEYOPT,1,9,0
KEYOPT,1,10,0

R,1,12.5,12.5,12.5,12.5,0,0,    ! DEFINES SHELL181 REAL CONSTANTS
RMORE,,,,,,,
R,2,12.5,12.5,12.5,12.5,90,0,   ! DEFINES SHELL181 REAL CONSTANTS
RMORE,,,,,,,

MPTEMP,,,,,,,,   ! DEFINES A TEMPERATURE TABLE FOR MATERIAL PROPERTIES
MPTEMP,1,0
MPDATA,EX,1,,126E+3      ! DEFINES MATERIAL PROPERTY DATA TO BE
                        ! ASSOCIATED WITH THE TEMPERATURE TABLE
MPDATA,EY,1,,9.5E+3
MPDATA,EZ,1,,9.5E+3
MPDATA,PRXY,1,,0.263
MPDATA,PRYZ,1,,0.27
MPDATA,PRXZ,1,,0.263
MPDATA,GXY,1,,1.07E+3
MPDATA,GYZ,1,,0.8063E+3
MPDATA,GXZ,1,,1.07E+3

K,1,1,1,1,   ! DEFINES ALL NECESSARY KEYPOINTS FOR MODEL'S GEOMETRY
K,2,251,1,1,
K,3,1,251,1,
K,4,251,251,1,

LSTR,1,3     ! GENERATES ALL NECESSARY LINES FOR MODEL'S GEOMETRY
LSTR,1,2
LSTR,2,4
LSTR,3,4
LSEL,S,LINE,,1,4,1,0    ! SELECTS A SUBSET OF LINES
LESIZE,ALL,,,2,,,,,1    ! SPECIFIES DIVISIONS AND SPACING RATIO
                        ! ON UNMESHED LINES
LSEL,ALL,,,,,,,

AL,1,2,3,4  ! GENERATES AN AREA BOUNDED BY PREVIOUSLY DEFINED LINES

TYPE,1      ! SETS SHELL181 AS THE ELEMENT TYPE ATTRIBUTE POINTER
MAT,1       ! SETS THE ELEMENT MATERIAL ATTRIBUTE POINTER
REAL,1      ! SETS THE ELEMENT REAL CONSTANTS SET ATTRIBUTE POINTER
ESYS,11     ! SETS THE ELEMENT COORDINATE SYSTEM ATTRIBUTE POINTER
SECNUM,     ! SETS THE ELEMENT SECTION ATTRIBUTE POINTER
AMAP,1,1,2,4,3  ! GENERATES A 2D-MAPPED MESH BASED ON
                ! SPECIFIED AREA CORNERS

ASEL,S,AREA,,1,1,1,0
AGEN,2,ALL,,,0,0,12.5,,1
ASEL,ALL,,,,,,,,
```

```
TYPE,1        ! SETS SHELL181 AS THE ELEMENT TYPE ATTRIBUTE POINTER
MAT,1         ! SETS THE ELEMENT MATERIAL ATTRIBUTE POINTER
REAL,2        ! SETS THE ELEMENT REAL CONSTANTS SET ATTRIBUTE POINTER
ESYS,11       ! SETS THE ELEMENT COORDINATE SYSTEM ATTRIBUTE POINTER
SECNUM,       ! SETS THE ELEMENT SECTION ATTRIBUTE POINTER
AMAP,2,5,8,7,6  ! GENERATES A 2D-MAPPED MESH BASED ON
                ! SPECIFIED AREA CORNERS

NUMCMP,NODE

CE,1,0,1,UX,-1,10,UX,1,1,ROTY,-6.25 ! DEFINES ALL CONSTRAINT EQNS.
CE,1,,10,ROTY,-6.25
CE,2,0,1,UY,-1,10,UY,1,1,ROTX,6.25
CE,2,,10,ROTX,6.25
CE,3,0,1,UZ,1,10,UZ,-1,,,

CE,4,0,8,UX,-1,17,UX,1,8,ROTY,-6.25
CE,4,,17,ROTY,-6.25
CE,5,0,8,UY,-1,17,UY,1,8,ROTX,6.25
CE,5,,17,ROTX,6.25
CE,6,0,8,UZ,1,17,UZ,-1,,,

CE,7,0,6,UX,-1,15,UX,1,6,ROTY,-6.25
CE,7,,15,ROTY,-6.25
CE,8,0,6,UY,-1,15,UY,1,6,ROTX,6.25
CE,8,,15,ROTX,6.25
CE,9,0,6,UZ,1,15,UZ,-1,,,

CE,10,0,3,UX,-1,12,UX,1,3,ROTY,-6.25
CE,10,,12,ROTY,-6.25
CE,11,0,3,UY,-1,12,UY,1,3,ROTX,6.25
CE,11,,12,ROTX,6.25
CE,12,0,3,UZ,1,12,UZ,-1,,,

CE,13,0,9,UX,-1,18,UX,1,9,ROTY,-6.25
CE,13,,18,ROTY,-6.25
CE,14,0,9,UY,-1,18,UY,1,9,ROTX,6.25
CE,14,,18,ROTX,6.25
CE,15,0,9,UZ,1,18,UZ,-1,,,

CE,16,0,7,UX,-1,16,UX,1,7,ROTY,-6.25
CE,16,,16,ROTY,-6.25
CE,17,0,7,UY,-1,16,UY,1,7,ROTX,6.25
CE,17,,16,ROTX,6.25
CE,18,0,7,UZ,1,16,UZ,-1,,,

CE,19,0,2,UX,-1,11,UX,1,2,ROTY,-6.25
CE,19,,11,ROTY,-6.25
CE,20,0,2,UY,-1,11,UY,1,2,ROTX,6.25
```

```
CE,20,,11,ROTX,6.25
CE,21,0,2,UZ,1,11,UZ,-1,,,

CE,22,0,5,UX,-1,14,UX,1,5,ROTY,-6.25
CE,22,,14,ROTY,-6.25
CE,23,0,5,UY,-1,14,UY,1,5,ROTX,6.25
CE,23,,14,ROTX,6.25
CE,24,0,5,UZ,1,14,UZ,-1,,,

CE,25,0,4,UX,-1,13,UX,1,4,ROTY,-6.25
CE,25,,13,ROTY,-6.25
CE,26,0,4,UY,-1,13,UY,1,4,ROTX,6.25
CE,26,,13,ROTX,6.25
CE,27,0,4,UZ,1,13,UZ,-1,,,

NSEL,S,LOC,X,250.09,251.01,,0
NSEL,R,LOC,Z,0.09,1.01,,0
D,ALL,UZ,0,,,,,,,,,        ! DEFINES DOF CONSTRAINTS AT PREVIOUSLY
                   ! DEFINED NODES CONSTRAINING THE Z TRANSLATION
NSEL,ALL,,,,,,
NSEL,S,LOC,X,0.99,1.01,,0
NSEL,R,LOC,Z,0.99,1.01,,0
D,ALL,UZ,0,,,,UX,,,,,    ! DEFINES DOF CONSTRAINTS AT PREVIOUSLY
    ! DEFINED NODES CONSTRAINING THE Z, X AND Y TRANSLATIONS
NSEL,ALL,,,,,,

LSEL,S,LINE,,1,7,2,0
SFL,ALL,PRES,-0.5,        ! APPLY Nx=0.5 AT THE TWO EDGES X=0,2a
                         ! FOR THE TWO SUBLAMINATES
LSEL,ALL,,,,,,,

FINISH
/SOL
/STATUS,SOLU
SOLVE

FINISH
/POST1
GPLOT
PLDISP,2
/VIEW,1,1,1,1
/ANG,1
/REP,FAST

AVPRIN,0,,
ETABLE,EPS_11,SMISC,9
AVPRIN,0,,
ETABLE,EPS_22,SMISC,10
AVPRIN,0,,
ETABLE,CHI_11,SMISC,12
```

```
AVPRIN,0,,
ETABLE,CHI_22,SMISC,13

SET,LIST,999
SET,,, ,,, ,1
PRETAB,EPS_11,EPS_22,CHI_11,CHI_22
```

10.2.2 Delamination Modeling

Perfect adhesion is assumed in the undelaminated region $\Omega - \Omega_D$, whereas sublaminates are free to deflect along the delaminated region Ω_D but not to penetrate each other. To enforce adhesion, a linear interface model [25, 48], is introduced along Ω-Ω_D. The constitutive equation of the interface involves two stiffness parameters, k_z, k_{xy}, imposing displacement continuity in the thickness and in-plane directions, respectively, by treating them as penalty parameters. The relationship between the components of the traction vector $\{\sigma\}$ acting at the lower surface of the upper sublaminate, σ_{zx}, σ_{zy} and σ_{zz}, in the out-of-plane (z) and in-plane (x and y) directions, respectively, and the corresponding components of relative interface displacement vector $\{\Delta\}$, Δu, Δv and Δw, are expressed as

$$\{\sigma\} = [K]\{\Delta\} \tag{10.17}$$

or in matrix form as

$$\begin{Bmatrix} \sigma_{zx} \\ \sigma_{zy} \\ \sigma_{zz} \end{Bmatrix} = \begin{bmatrix} k_{xy} & 0 & 0 \\ 0 & k_{xy} & 0 \\ 0 & 0 & k_z \end{bmatrix} \begin{Bmatrix} \Delta u \\ \Delta v \\ \Delta w \end{Bmatrix} \tag{10.18}$$

Relative opening and sliding displacements are evaluated as the difference between displacements at the interface between the lower and upper sublaminate. Interface elements are implemented in ANSYS as type COMBIN14.

Example 10.2 *Add a mid-plane delamination that spans the region $a \leq x \leq 2a$; $0 \leq y \leq 2b$, at the midplane of the laminate in Example 10.1. Model the delaminated and the undelaminated regions.*

Solution to Example 10.2 *Interface elements, necessary to connect the two sublaminates in the undelaminated region $\Omega - \Omega_D$, are implemented by using a combination of CE and spring elements (ANSYS COMBIN14). The linear elastic interface has been modeled only along the delamination front, whereas the undelaminated portion of the interface has been modeled only by means of CE. For each mid-plane node of the upper plate model, three coincident nodes located on the lower surface of the plate are created, and constrained to deform as embedded in a plate segment normal to the mid-plane by means of CE. Similarly, for each mid-plane node of the lower plate model, three coincident nodes located on the upper surface are created. Three COMBIN14 elements connected to the three pairs of coincident nodes placed at the delamination plane are then introduced, each one acting in a different translational direction.*

*   **The FE model for this example** is obtained by using the ANSYS input command sequence shown below, also available in [76].*

```
/NOPR          ! SUPPRESSES THE EXPANDED INTERPRETED INPUT DATA LISTING
/PMETH,OFF,0! ACTIVATES THE p-METHOD SOLUTION OPTIONS IN THE GUI

KEYW,PR_SET,1     ! SETS THE KEYWORD USED BY THE GUI FOR
                  ! STRUCTURAL CONTEXT FILTERING
KEYW,PR_STRUC,1
KEYW,PR_THERM,0
KEYW,PR_FLUID,0
KEYW,PR_ELMAG,0
KEYW,MAGNOD,0
KEYW,MAGEDG,0
KEYW,MAGHFE,0
KEYW,MAGELC,0
KEYW,PR_MULTI,0
KEYW,PR_CFD,0


/PREP7                      ! ENTERS THE MODEL CREATION PREPROCESSOR

ANTYPE,STATIC    ! SPECIFIES THE ANALYSIS TYPE AND RESTART STATUS

LOCAL,11,0,0,0,0,0,0,0, ! DEFINES A LOCAL COORDINATE SYSTEM ALIGNED
                  ! WITH THE GLOBAL ONE BY LOCATION AND ORIENTATION


ET,1,SHELL181    ! DEFINES SHELL181 AS A LOCAL FINITE ELEMENT TYPE.
                  ! THIS WILL BE USED TO MESH THE PLATE MODEL
KEYOPT,1,1,0     ! SETS SHELL181 KEY OPTIONS
KEYOPT,1,3,2
KEYOPT,1,8,0
KEYOPT,1,9,0
KEYOPT,1,10,0
R,1,12.5,12.5,12.5,12.5,0,0,      ! DEFINES SHELL181 REAL CONSTANTS
RMORE,,,,,,,
R,2,12.5,12.5,12.5,12.5,90,0,     ! DEFINES SHELL181 REAL CONSTANTS
RMORE,,,,,,,


ET,2,COMBIN14    ! DEFINES COMBIN14 AS A LOCAL FINITE ELEMENT TYPE.
                  ! THIS WILL BE USED TO MODEL INTERFACE
KEYOPT,2,1,0     ! SETS COMBIN14 KEY OPTIONS
KEYOPT,2,2,1     ! ACTIVATE THE UX DOF FOR THE SPRING ELEMENT
KEYOPT,2,3,0


ET,3,COMBIN14    ! DEFINES COMBIN14 AS A LOCAL FINITE ELEMENT TYPE
KEYOPT,3,1,0     ! SETS COMBIN14 KEY OPTIONS
KEYOPT,3,2,2     ! ACTIVATE THE UY DOF FOR THE SPRING ELEMENT
KEYOPT,3,3,0


ET,4,COMBIN14    ! DEFINES COMBIN14 AS A LOCAL FINITE ELEMENT TYPE
KEYOPT,4,1,0     ! SETS COMBIN14 KEY OPTIONS
KEYOPT,4,2,3     ! ACTIVATE THE UZ DOF FOR THE SPRING ELEMENT
KEYOPT,4,3,0
```

```
R,3,1E+8,0,0,    ! DEFINES COMBIN14 REAL CONSTANTS,
! THE FIRST CONSTANT REPRESENTS THE STIFFNESS PARAMETER VALUE K
! FOR A NODE IN THE INTERIOR PART OF THE DELAMINATION FRONT
R,4,50000000,0,0,    ! DEFINES COMBIN14 REAL CONSTANTS, THE FIRST
! CONSTANT REPRESENTS THE STIFFNESS  PARAMETER
! VALUE K FOR A NODE AD THE EDGE OF THE DELAMINATION FRONT

MPTEMP,,,,,,,,! DEFINES A TEMPERATURE TABLE FOR MATERIAL PROPERTIES
MPTEMP,1,0
MPDATA,EX,1,,126E+3 ! DEFINES MATERIAL PROPERTY DATA TO BE
                    ! ASSOCIATED WITH THE TEMPERATURE TABLE
MPDATA,EY,1,,9.5E+3
MPDATA,EZ,1,,9.5E+3
MPDATA,PRXY,1,,0.263
MPDATA,PRYZ,1,,0.27
MPDATA,PRXZ,1,,0.263
MPDATA,GXY,1,,1.07E+3
MPDATA,GYZ,1,,0.8063E+3
MPDATA,GXZ,1,,1.07E+3

K,1,1,1,1,  ! DEFINES ALL THE NECESSARY KEYPOINTS
            ! FOR THE MODEL'S GEOMETRY
K,2,251,1,1,
K,3,1,251,1,
K,4,251,251,1,

LSTR,1,3              ! GENERATES ALL THE NECESSARY LINES
                     ! FOR THE MODEL'S GEOMETRY
LSTR,1,2
LSTR,2,4
LSTR,3,4
LSEL,S,LINE,,1,4,1,0    ! SELECTS A SUBSET OF LINES
LESIZE,ALL,,,2,,,,,1    ! SPECIFIES THE DIVISIONS AND SPACING RATIO
                       ! ON UNMESHED LINES
LSEL,ALL,,,,,,,

AL,1,2,3,4  ! GENERATES AN AREA BOUNDED BY PREVIOUSLY DEFINED LINES

TYPE,1       ! SETS SHELL181 AS THE ELEMENT TYPE ATTRIBUTE POINTER
MAT,1        ! SETS THE ELEMENT MATERIAL ATTRIBUTE POINTER
REAL,1       ! SETS THE ELEMENT REAL CONSTANTS SET ATTRIBUTE POINTER
ESYS,11      ! SETS THE ELEMENT COORDINATE SYSTEM ATTRIBUTE POINTER
SECNUM,      ! SETS THE ELEMENT SECTION ATTRIBUTE POINTER
AMAP,1,1,2,4,3  ! GENERATES A 2D-MAPPED MESH BASED ON SPECIFIED
                ! AREA CORNERS

ASEL,S,AREA,,1,1,1,0
AGEN,2,ALL,,,0,0,12.5,,1
ASEL,ALL,,,,,,,
```

```
TYPE,1  ! SETS SHELL181 AS THE ELEMENT TYPE ATTRIBUTE POINTER
MAT,1   ! SETS THE ELEMENT MATERIAL ATTRIBUTE POINTER
REAL,2  ! SETS THE ELEMENT REAL CONSTANTS SET ATTRIBUTE POINTER
ESYS,11 ! SETS THE ELEMENT COORDINATE SYSTEM ATTRIBUTE POINTER
SECNUM, ! SETS THE ELEMENT SECTION ATTRIBUTE POINTER
AMAP,2,5,8,7,6

NSEL,S,LOC,X,125.0,127.0,,0 ! SELECTS THE NECESSARY SUBSET
! OF NODES OF THE LOWER PLATE MODEL TO CREATE THE DELAMINATION FRONT
NSEL,R,LOC,Z,0.0,2.0,,0
! GENERATES A COPY OF ADDITIONAL NODES PLACED ON THE UPPER SURFACE
! OF THE LOWER PLATE MODEL FROM A PATTERN OF PREVIOUSLY DEFINED NODES
NGEN,2,18,ALL,,,0,0,6.25,1
NSEL,ALL,,,,,,,           ! REACTIVATES SUPPRESSED NODES
NUMCMP,NODE               ! COMPRESS THE NUMBERING OF NODES

NSEL,S,LOC,X,124.0,127.0,,0 ! SELECTS THE SUBSET OF COPIED NODES
NSEL,R,LOC,Z,7.24,7.26,,0
NGEN,6,18,ALL,,,0,0,0,1     ! GENERATES 5 COPIES OF ADDITIONAL NODES
! FROM A PATTERN OF PREVIOUSLY DEFINED NODES
NSEL,ALL,,,,,,            ! REACTIVATES SUPPRESSED NODES
NUMCMP,NODE               ! COMPRESS THE NUMBERING OF NODES

CE,1,0,1,UX,-1,10,UX,1,1,ROTY,-6.25 ! DEFINES ALL CONSTRAINT
                                    ! EQUATIONS IN THE MODEL
CE,1,,10,ROTY,-6.25
CE,2,0,1,UY,-1,10,UY,1,1,ROTX,6.25
CE,2,,10,ROTX,6.25
CE,3,0,1,UZ,1,10,UZ,-1,,,

CE,4,0,2,UX,-1,11,UX,1,2,ROTY,-6.25
CE,4,,11,ROTY,-6.25
CE,5,0,2,UY,-1,11,UY,1,2,ROTX,6.25
CE,5,,11,ROTX,6.25
CE,6,0,2,UZ,1,11,UZ,-1,,,

CE,7,0,3,UX,-1,12,UX,1,3,ROTY,-6.25
CE,7,,12,ROTY,-6.25
CE,8,0,3,UY,-1,12,UY,1,3,ROTX,6.25
CE,8,,12,ROTX,6.25
CE,9,0,3,UZ,1,12,UZ,-1,,,

CE,10,0,8,UX,-1,20,UX,1,8,ROTY,-6.25
CE,11,0,8,UY,-1,26,UY,1,8,ROTx,6.25
CE,12,0,8,UZ,-1,32,UZ,1,,,

! FOR A FIXED X-Y POSITION ALONG THE DELAMINATION FRONT,
! DEFINE THE CONNECTIONS BETWEEN NODES BY MEANS OF
! SPRING ELEMENTS AND CONSTRAINT EQUATIONS
ESYS, 11    ! SETS THE ELEMENT COORDINATE SYSTEM ATTRIBUTE POINTER
```

```
REAL,4         ! SETS THE ELEMENT REAL CONSTANTS SET ATTRIBUTE POINTER
TYPE,2         ! SETS THE ELEMENT TYPE ATTRIBUTE POINTER TO THE SPRING
               ! ELEMENTS ACTING IN THE X DIRECTION
EN,9,20,23,    ! DEFINES A FINITE ELEMENT BY ITS NUMBER AND NODE
               ! CONNECTIVITY
TYPE,3         ! SETS THE ELEMENT TYPE ATTRIBUTE POINTER TO THE SPRING
               ! ELEMENTS ACTING IN THE Y DIRECTION
EN,10,26,29,   ! DEFINES A FINITE ELEMENT BY ITS NUMBER AND NODE
               ! CONNECTIVITY
TYPE,4         ! SETS THE ELEMENT TYPE ATTRIBUTE POINTER TO THE SPRING
               ! ELEMENTS ACTING IN THE Z  DIRECTION
EN,11,32,35,   ! DEFINES A FINITE ELEMENT BY ITS NUMBER AND NODE
               ! CONNECTIVITY

CE,13,0,17,UX,-1,23,UX,1,17,ROTY,6.25
CE,14,0,17,UY,-1,29,UY,1,17,ROTx,-6.25
CE,15,0,17,UZ,-1,35,UZ,1,,,

CE,16,0,9,UX,-1,21,UX,1,9,ROTY,-6.25
CE,17,0,9,UY,-1,27,UY,1,9,ROTx,6.25
CE,18,0,9,UZ,-1,33,UZ,1,,,

! FOR A FIXED X-Y POSITION ALONG THE DELAMINATION FRONT
! DEFINE THE CONNECTIONS BETWEEN NODES
! BY MEANS OF SPRING ELEMENTS AND CONSTRAINT EQUATIONS
ESYS, 11       ! SETS THE ELEMENT COORDINATE SYSTEM ATTRIBUTE POINTER
REAL,3         ! SETS THE ELEMENT REAL CONSTANTS SET ATTRIBUTE POINTER
TYPE,2         ! SETS THE ELEMENT TYPE ATTRIBUTE POINTER TO THE SPRING
               ! ELEMENTS ACTING IN THE X DIRECTION
EN,12,21,24,   ! DEFINES A FINITE ELEMENT BY ITS NUMBER AND NODE
               ! CONNECTIVITY
TYPE,3         ! SETS THE ELEMENT TYPE ATTRIBUTE POINTER TO THE SPRING
               ! ELEMENTS ACTING IN THE Y DIRECTION
EN,13,27,30,   ! DEFINES A FINITE ELEMENT BY ITS NUMBER AND NODE
               ! CONNECTIVITY
TYPE,4         ! SETS THE ELEMENT TYPE ATTRIBUTE POINTER TO THE SPRING
               ! ELEMENTS ACTING IN THE Z  DIRECTION
EN,14,33,36,   ! DEFINES A FINITE ELEMENT BY ITS NUMBER AND NODE
               ! CONNECTIVITY

CE,19,0,18,UX,-1,24,UX,1,18,ROTY,6.25
CE,20,0,18,UY,-1,30,UY,1,18,ROTx,-6.25
CE,21,0,18,UZ,-1,36,UZ,1,,,

CE,22,0,5,UX,-1,19,UX,1,5,ROTY,-6.25
CE,23,0,5,UY,-1,25,UY,1,5,ROTx,6.25
CE,24,0,5,UZ,-1,31,UZ,1,,,

! FOR A FIXED X-Y POSITION ALONG THE DELAMINATION FRONT DEFINE
! THE CONNECTIONS BETWEEN NODES
```

```
! BY MEANS OF SPRING ELEMENTS AND CONSTRAINT EQUATIONS
ESYS, 11     ! SETS THE ELEMENT COORDINATE SYSTEM ATTRIBUTE POINTER
REAL,4       ! SETS THE ELEMENT REAL CONSTANTS SET ATTRIBUTE POINTER
TYPE,2       ! SETS THE ELEMENT TYPE ATTRIBUTE POINTER TO THE SPRING
             ! ELEMENTS ACTING IN THE X DIRECTION
EN,15,19,22,! DEFINES A FINITE ELEMENT BY ITS NUMBER AND NODE
             ! CONNECTIVITY
TYPE,3       ! SETS THE ELEMENT TYPE ATTRIBUTE POINTER TO THE SPRING
             ! ELEMENTS ACTING IN THE Y DIRECTION
EN,16,25,28,! DEFINES A FINITE ELEMENT BY ITS NUMBER AND NODE
             ! CONNECTIVITY
TYPE,4       ! SETS THE ELEMENT TYPE ATTRIBUTE POINTER TO THE SPRING
             ! ELEMENTS ACTING IN THE Z  DIRECTION
EN,17,31,34,! DEFINES A FINITE ELEMENT BY ITS NUMBER AND NODE
             ! CONNECTIVITY

CE,25,0,14,UX,-1,22,UX,1,14,ROTY,6.25
CE,26,0,14,UY,-1,28,UY,1,14,ROTx,-6.25
CE,27,0,14,UZ,-1,34,UZ,1,,,

NSEL,S,LOC,X,250.09,251.01,,0   ! SELECTS A SUBSET OF NODES
NSEL,R,LOC,Z,0.09,1.01,,0
D,ALL,UZ,0,,,,,,,,,             ! DEFINES DOF CONSTRAINTS AT
! PREVIOUSLY DEFINED NODES CONSTRAINING THE Z TRANSLATION
NSEL,ALL,,,,,,                  ! REACTIVATES SUPPRESSED NODES
NSEL,S,LOC,X,0.99,1.01,,0       ! SELECTS A SUBSET OF NODES
NSEL,R,LOC,Z,0.99,1.01,,0
D,ALL,UZ,0,,,,UX,UY,,,, ! DEFINES DOF CONSTRAINTS AT PREVIOUSLY
! DEFINED NODES CONSTRAINING THE Z, X AND Y TRANSLATIONS
NSEL,ALL,,,,,,                  ! REACTIVATES SUPPRESSED NODES
```

10.2.3 Unilateral Contact and Damaging Interface

To prevent interpenetration between delaminated sublaminates in the delaminated region Ω_D, a unilateral frictionless contact interface can be introduced, characterized by a zero stiffness for opening relative displacements ($\Delta w \geq 0$) and a positive stiffness for closing relative displacements ($\Delta w < 0$)

$$\sigma_{zz} = \frac{1}{2} \left(1 - sign\left(\triangle w\right)\right) k_z \, \triangle w \qquad (10.19)$$

where σ_{zz} is the contact stress, k_z is the penalty number imposing contact constraint and *sign* is the signum function. A very large value for k_z restricts sublaminate overlapping and simulates the contact condition [77]. Unilateral contact conditions may be implemented in ANSYS using COMBIN39, which is a unidirectional element with nonlinear constitutive relationships, by appropriate specialization of the nonlinear constitutive law according to (10.19).

Introducing a scalar damage variable D (see Chapter 8), taking the value of *1* for no adhesion and the value *0* for perfect adhesion, leads to a single extended interface model, whose constitutive law is valid both for undelaminated $\Omega - \Omega_D$ and delaminated Ω_D areas. Its constitutive law, therefore, can be expressed as

$$\{\sigma\} = (1 - D)\,[K]\,\{\triangle\} \tag{10.20}$$

Although the governing equations in terms of plate variables are available in [64, 60, 61], in this textbook we emphasize the formulation via finite elements, and thus make use directly of commercial finite element implementations for all aspects related to plate elements, interface elements and Lagrange multipliers. It is worth noting that in commercial FEA packages the Lagrange multipliers are represented by either CE or rigid links, whereas interface elements are implemented by the analyst using a combination of spring elements (COMBIN14 for ANSYS) and CE.

10.2.4 ERR-Interface Model

In the present section the problem of determining the total ERR and its individual mode components is considered from the theoretical point of view. Classical plate theory does not allow for accurate prediction of interlaminar stresses, thus excluding the possibility of a local fracture mechanics approach. On the contrary, these stresses are captured by means of interface elements, being able to model delamination-front stress singularities as the interface stiffness approaches infinity. The accuracy of these stresses depends on how well the plate model simulates the three-dimensional behavior. The problem of ERR computation can be solved locally by using interface variables (interlaminar stresses and relative displacements). The connection between the interface approach and fracture mechanics approach will be established, pointing out that the interface approach corresponds to the limit physical situation resulting when the thickness of a thin adhesive layer tends to zero. Using the interface constitutive equation (10.17) to compute interlaminar stresses (for additional details see [25], [48]), leads to the following total ERR expression

$$G(s) = \frac{1}{2} \lim_{k_z,k_{xy} \to \infty} \left[k_z \,\triangle w^2(s) + k_{xy} \,\triangle u^2(s) + k_{xy} \triangle v^2(s) \right], \quad \triangle w(s) \geq 0 \tag{10.21}$$

where s is a curvilinear coordinate along the delamination front. In (10.21) $G(s)$ is the local ERR function along the delamination front $\partial\Omega_D$, defined by

$$\dot{\Pi}_r(\Omega_D(\tau),\lambda) = -\int_{\partial\Omega_D} G(s)\dot{\Omega}_D(s)ds \tag{10.22}$$

where Π_r is the total potential energy of the system at equilibrium, a dot denotes total differentiation with respect to a time-like parameter τ governing a virtual monotonic delamination growth $\Omega_D(\tau)$, and $\dot{\Omega}_D(s) \geq 0$ ($\forall s \in \partial\Omega_D$) is the rate of normal extension of the delamination front $\partial\Omega_D$ (Figure 10.8) which describes the rate of variation of $\Omega_D(\tau)$ [12, 69]. For the limit (10.21) to be finite, interlaminar stresses must approach infinity at the delamination front whereas interlaminar displacements must tend to zero.

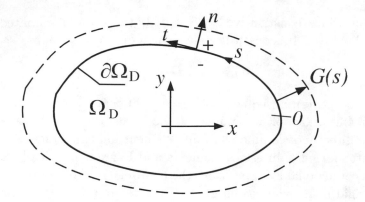

Fig. 10.8: Propagation of the delamination front.

In order to determine the individual ERR, the relative interface displacements must be expressed in the local coordinate system attached to the delamination front shown in Figure 10.8. Denoting the unit vectors in the normal and tangential directions to the delamination front as n and t, respectively, the relative interface displacement in the global x-y-z system $\{\Delta\} = \{\Delta u, \Delta v, \Delta w\}^T$, is related to that in the n-t-z system $\{\Delta'\} = \{\Delta u_n, \Delta u_t, \Delta w\}^T$ by the following transformation

$$\{\Delta\} = [A]^T\{\Delta'\} \tag{10.23}$$

where $[A]$ is the matrix of direction cosines

$$[A] = \begin{bmatrix} n_x & n_y & 0 \\ -n_y & n_x & 0 \\ 0 & 0 & 1 \end{bmatrix} \tag{10.24}$$

Similarly for the interface traction vector we have

$$\{\sigma\} = [A]^T\{\sigma'\} \tag{10.25}$$

Therefore, after introduction of (10.23)-(10.25) and use of (10.17) in (10.21) we have

$$\begin{aligned} G(s) &= \frac{1}{2} \lim_{k_z,k_{xy}\to\infty} (\{\Delta'\}^T[K']\{\Delta'\}) \\ &= \frac{1}{2} \lim_{k_z,k_{xy}\to\infty} \left[k_z\,\Delta w^2(s) + k_{xy}\Delta u_n^2(s) + k_{xy}\Delta u_t^2(s)\right] \end{aligned} \tag{10.26}$$

where $\Delta w(s) \geq 0$ and

$$[K'] = [A][K][A]^T = [K] \tag{10.27}$$

Consequently, the point-wise G and its mode I, II, III components G_I, G_{II}, G_{III}, take the following form

$$
\begin{aligned}
G(s) &= G_I(s) + G_{II}(s) + G_{III}(s) \\
G_I(s) &= \left\{ \begin{array}{ll} \lim\limits_{k_z, k_{xy} \longrightarrow \infty} \tfrac{1}{2} k_z \triangle w^2(s) & \text{if} \quad \triangle w(s) \geq 0 \\ 0 & \text{if} \quad \triangle w(s) < 0 \end{array} \right\} \\
G_{II}(s) &= \lim\limits_{k_z, k_{xy} \longrightarrow \infty} \frac{1}{2} k_{xy} \triangle u_n^2(s) \\
G_{III}(s) &= \lim\limits_{k_z, k_{xy} \longrightarrow \infty} \frac{1}{2} k_{xy} \triangle u_t^2(s)
\end{aligned} \tag{10.28}
$$

The relation between the interface approach (10.21) and the fracture mechanics approach can be elucidated by applying the definition (10.22) and taking the limit of the total ERR derivative of the modified total potential energy as interface stiffness parameters approach infinity

$$
\int_{\partial \Omega_D} G(s) \dot{\Omega}_D(s) ds = \lim_{k \to \infty} -\frac{d}{dt} \Pi_k \tag{10.29}
$$

Since Π_k does not involve stress resultant discontinuities until the penalty parameters approach infinity [60, 61], (10.29) furnishes

$$
-\int_{\partial \Omega_D} G(s) \dot{\Omega}_D(s) ds = \lim_{k \to \infty} \left\{ \left[W' + I' + L' - P'(\lambda) \right] - \int_{\partial \Omega_D} \Lambda \dot{\Omega}_D(s) ds \right\} \tag{10.30}
$$

where Λ is the interface strain energy density per unit area defined as

$$
\Lambda = \frac{1}{2} \left[k_z \triangle w^2 (x, y) + k_{xy} \triangle u^2 (x, y) + k_{xy} \triangle u^2 (x, y) \right] \tag{10.31}
$$

Furthermore, a prime denotes partial differentiation in the time-like parameter τ, W is the strain energy of the whole plate, I the strain energy of the interface representing the penalty functional, L the Lagrange functional imposing the adhesion constraint between undelaminated layers and P is the work of the external loads.

The last term in (10.30) represents the flux of interface strain energy through the delamination front $\partial \Omega_D$ and it arises as a consequence of the growth of Ω_D. Taking into account equilibrium requirements, boundary, and continuity conditions, the term in square brackets in (10.30) vanishes, leading to

$$
G(s) = \lim_{k \to \infty} \Lambda(s) = \lim_{k_z, k_{xy} \to \infty} \frac{1}{2} (k_z \triangle w^2 + k_{xy} \triangle u^2 + k_{xy} \triangle v^2) \tag{10.32}
$$

An alternative approach to that provided by the interface model for the calculation of the ERR is to use plate variables [60] in which case mode partition can be obtained by applying the virtual crack closure technique (VCC) or the Jacobian derivative method (JDM) [70] as a post-computation.

10.2.5 Mixed Mode Analysis

In order to predict crack propagation in laminates for general loading conditions, energy release rates distributions along the delamination front are needed. Fracture mechanics assumes that delamination propagation is controlled by the critical energy release rate. The growth condition can be defined by means of a power law criterion, similar to (10.10). Delamination grows on the region of the delamination front where the following condition is satisfied

$$\left(\frac{G_I(s)}{G_I^c}\right)^a + \left(\frac{G_{II}(s)}{G_{II}^c}\right)^\beta + \left(\frac{G_{III}(s)}{G_{III}^c}\right)^\gamma = 1 \qquad (10.33)$$

where α, β and γ are mixed mode fracture parameters determined by fitting to experimental test results.

The critical ERR ($G_I^c, G_{II}^c, G_{III}^c$) are assumed to be material properties, independent of their location along the delamination front, and evaluated from experimental procedures. The closed-form expressions for the ERR, (10.21) and (10.28), obtained by means of the interface model, can be used in conjunction with a finite element code to check whether propagation occurs. Therefore, the calculation of $G(s)$ along the delamination front reduces to a simple post-computation, once a global FEA of the laminate is carried out. The extent of the propagation of the delamination area may be established by releasing the node in which the relation (10.33) is first satisfied, leading to a modification of the delamination front, which in turn requires another equilibrium solution. It follows that the delamination growth analysis must be accomplished iteratively.

For simplicity, only the computation of ERR is described here. The study of the propagation for a three-dimensional planar delamination requires the use of nonlinear incremental numerical computation, which often must be accompanied by simplified assumptions (for instance, the use of the total ERR to detect growth, neglecting of mode III effects) in order to save computational time [15, 44, 45]. These methods are outside the scope of this chapter. However, it is worth noting that a good prediction of delamination propagation relies on an accurate ERR distribution computation, which for a generic delaminated composite under generic loading is itself a complex task due to the three-dimensional nature of the mode-decomposition problem.

The delaminated laminate is represented by using two sublaminates, one for the portion above the delamination plane and the other for the part below (Figure 10.6). When only the total ERR is to be computed, sufficient accuracy may be achieved by representing each sublaminate with a single plate element. In this case, the model is called a *two-layer plate model*. When the mode components are needed, an assembly of plate elements in each sublaminate is necessary to achieve sufficient accuracy. In this case the model is called a *multilayer plate model*.

Sublaminates can be modeled by using standard shear deformable elements (AN-SYS SHELL181), whereas interface elements can be used for the interface model. Since available interface elements (ANSYS INTER204) are only compatible with

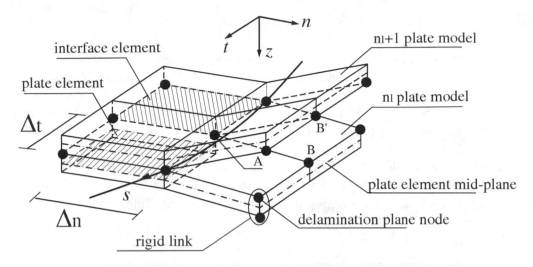

Fig. 10.9: Plate assembly in the neighborhood of the delamination front for the calculation of ERR. Nodes on the plane of the delamination and the location of rigid links are shown.

solid elements, interface elements are simulated here by coupling CE with spring elements (ANSYS COMBIN14). Plate and interface models must be described by the same in-plane mesh.

The FE model of the plates adjacent to the delamination plane in proximity of the delamination front is illustrated in Figure 10.9. Interface elements model the undelaminated region $\Omega - \Omega_D$ up to the delamination front.

In the FE model, ERR are computed by using (10.34), which is a modified version of (10.28) to avoid excessive mesh refining at the delamination front. The mesh of interface and plate elements must be sufficiently refined in order to capture the high interface stress gradient in the neighborhood of the delamination front, which occurs because high values for interface stiffness must be used to simulate perfect adhesion. To this end, the individual ERR at the generic node A of the delamination front are calculated by using the reactions obtained from spring elements and the relative displacements between the nodes already delaminated and located along the normal direction. This leads to the following expressions

$$G_I(A) = \frac{1}{2}\frac{R_A^z \triangle w_{B-B'}}{\triangle_n \triangle_t}, G_{II}(A) = \frac{1}{2}\frac{R_A^n \triangle u_{nB-B'}}{\triangle_n \triangle_t}, G_{III}(A) = \frac{1}{2}\frac{R_A^t \triangle u_{tB-B'}}{\triangle_n \triangle_t}$$
$$(10.34)$$

where $R_A{}^z$ is the reaction in the spring element connecting node A in the z-direction, whereas $\triangle w_{B-B'}$ is the relative z-displacement between the nodes B and B', located immediately ahead of the delamination front along its normal direction passing through A. Similar definitions apply for reactions and relative displacement related to modes II and III. The characteristic mesh sizes in the normal and tangential directions of the delamination front are denoted by $\triangle_t$ and $\triangle_n$. In (10.34), the same element size is assumed for elements ahead of and behind the delamination front. When the node is placed at a free edge, $\Delta_t/2$ must be used in (10.34) instead of Δ_t.

In order to simplify the FE modeling procedure, it is possible to introduce spring elements only along the delamination front instead of the entire undelaminated region. In this case, perfect adhesion along the remaining portion of the undelaminated region can be imposed by CE. However, when the delamination propagation must be simulated, it is necessary to introduce interface elements in the whole undelaminated region $\Omega - \Omega_D$.

In the next examples, the delamination modeling techniques presented so far are applied to analyze typical 3D delamination problems in laminated plates. The ERR distribution along the delamination front are computed for different laminates and loading conditions.

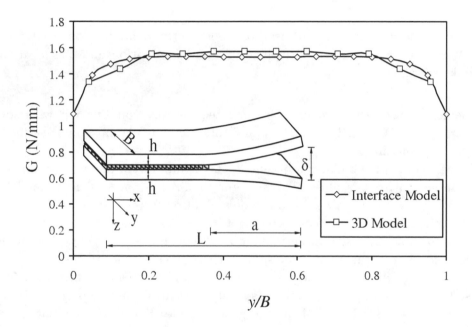

Fig. 10.10: ERR distribution along the delamination front.

Example 10.3 *A double cantilever beam (DCB) specimen of width $B = 30$ mm, thickness $2h = 3$ mm, length $L = 135$ mm, with a delamination of length $a = 29$ mm (Figure 10.10) is subjected to an opening displacement of 2.5 mm. Unidirectional carbon/epoxy layers are used with the material properties given in Table 10.1.*

Solution to Example 10.3 *Two plates have been used to model the delaminated plate in the thickness direction, one for each sublaminate, i.e., $n_u = n_l = 1$. Each is modeled with ANSYS SHELL181 elements.*

The projection on the $x - y$ plane of the finite element mesh, assumed equal for the plate and the interface models, is shown in Figure 10.11. The mesh is refined in a zone of 5×30 mm^2 centered with respect to the delamination front. In this zone, the length of the plate and interface elements in the x-direction is 0.125 mm, whereas along the y-direction

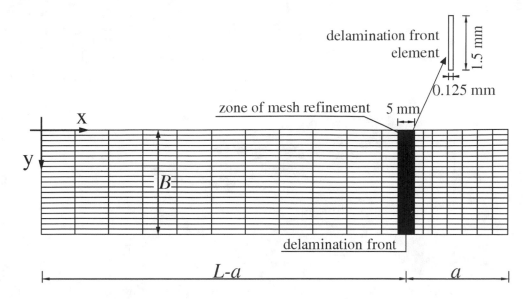

Fig. 10.11: x-y view of the FE mesh used for the mode I example, illustrating the region chosen for mesh refinement and convergence studies.

it is 1.5 mm. Due to symmetry, it is sufficient to use only z-translational springs. A displacement of 2.5 mm is imposed in the z-direction at the edge of the delaminated plate. Two plates have been used to model the delaminated plate in the thickness direction.

Interface elements, necessary to connect the two sublaminates in the region $\Omega - \Omega_D$, are implemented by using a combination of rigid links (ANSYS MPC184, defined by two nodes and three degrees of freedom at each node) and spring elements (ANSYS COMBIN14). Spring elements are placed among the offset nodes. The offset nodes can be generated by rigid links but in this example we choose to use CE, which are easier to work with than rigid links in ANSYS. For simplicity the linear elastic interface has been modeled only along the delamination front, whereas the undelaminated portion of the interface has been modeled only by means of rigid links or constraint equations.

When CE are used in place of rigid links, for each mid-plane node of the upper plate model, three coincident nodes located on the lower surface of the plate are created, and constrained to deform as embedded in a plate segment normal to the mid-plane by means of constraint equations. Similarly, for each mid-plane node of the lower plate model, three coincident nodes located on the upper surface are created. Three COMBIN14 elements connected to the three pairs of coincident nodes placed at the delamination plane are then introduced, each one acting in a different translational direction. It is worth noting that for the pure mode I example it suffices to model only z-translational springs due to symmetry. However, in order to provide a general code to be used for general mixed-mode loading conditions, a complete modeling of interface elements has been implemented. The deformed mesh is shown in Figure 10.12.

The ERR distribution along the straight delamination front is evaluated by using the first of (10.34) and it is illustrated in Figure 10.10. Since the ERR assumes its maximum at the center of the specimen, it is expected that the delamination proceeds from the middle. Comparison with results using the modified virtual crack closure technique and 20-node anisotropic three-dimensional finite element models are also shown in Figure 10.10 [44].

It can be seen that the present model, referred to as the "interface model," provides sat-

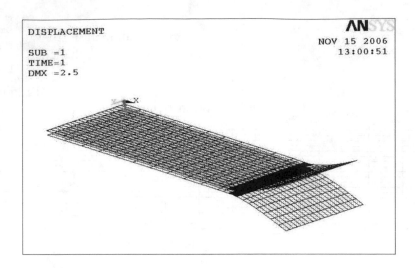

Fig. 10.12: FEA model for a symmetric delaminated plate under mode I opening.

isfactory accuracy in comparison with the three-dimensional solution. The 3D methodology, clearly, may involve a larger computational cost with respect to the present model, in which each sublaminate is represented by only one shear deformable plate. Moreover, Figure 10.10 points out that for symmetric loading and geometry, a double plate model suffices to determine accurately the ERR.

Investigations performed to choose values for interface stiffness have shown that a reasonable choice is 0.5 10^8 N/mm^3 [60]. These analyses were conducted on the basis of mesh refinement and convergence studies, taking care that interface stiffness must assume a relatively high value to reflect the actual interface bonding, but not too high to cause numerical instabilities. The numerical results are shown in Table 10.2. The expression Mesh denotes the number of plate and interface elements (x-direction-y-direction) used in one quarter of the mesh refinement region (2.5 × 15 mm^2) illustrated in Figure 10.11. The delamination front center values of ERR for different interface stiffness and meshes are shown in Table 10.2. It can be noticed that for k_z = 0.5 10^8 N/mm^3 a mesh grid of 20 × 10 elements suffices to obtain an accurate value of G. This corresponds to the mesh used in Figure 10.11. Moreover, refining the mesh in the y-direction does not lead to a sensible variation of ERR accuracy, as shown in Table 10.2.

The ANSYS^{TM} command file containing this example is provided in [76]. To run the example, go to the File menu and select "read input from".

Example 10.4 *For Example 10.3 (mode I), enforce the classical delamination model, which assumes that the undelaminated portion of the laminate behaves as a single plate model, by using four-node thin plate elements that exclude shear deformability.*

Solution to Example 10.4 *Using the first of (10.34) to compute ERR, Figure 10.13 is obtained, where both thin and thick four-node plate elements have been used in conjunction with interface elements. It can be noted that neglecting shear deformability causes a notable underestimation of ERR results, especially near the edges of the delamination front.*

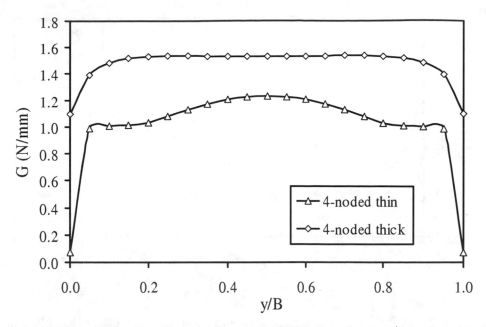

Fig. 10.13: Effect of the plate kinematics on the ERR distribution (interface model).

Example 10.5 *Mode II loading condition is now considered for a three-dimensional DCB geometry. The properties of the plates are as follows:*

a = 256 mm, h=16 mm, B = 400 mm, L = 512 mm, E = 80.00 GPa, ν = 0.3, N = 6.25N/mm

Use ANSYS to model the delaminated plate under mode II loading condition with results shown in Figures 10.14 and 10.15. Use a $(n_u - n_l) : (2 - 2)$ plate assembly, which gives reasonable accuracy. Calculate the mode II and III ERR distributions along the delamination front. Expression (10.35), used to normalize results, takes the value $6.943 \ 10^{-6} N/mm$.

Solution to Example 10.5 *Because of edge effects, mode II loading produces both mode II and mode III energy release rates. However, mode II ERR predominates under mode II loading. Four-node shear deformable plate elements are used. The size of the elements*

Table 10.2: Mesh refinement and convergence studies for the DCB example in terms of G at the delamination center-point

Mesh	Interface Stiffness kz			
	0.50E+06	0.50E+07	0.50E+08	0.50E+09
20×10	1.5130	1.5278	1.5282	1.5282
40×10	1.4331	1.5250	1.5281	1.5281
60×10	1.3316	1.5158	1.5280	1.5282
80×10	1.2446	1.4984	1.5276	1.5282
100×10	1.1762	1.4738	1.5269	1.5282
100×20	1.1763	1.4738	1.5267	1.5293

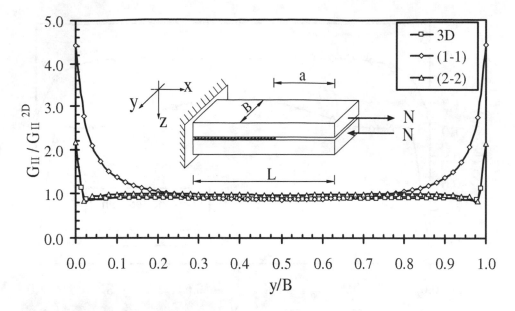

Fig. 10.14: G_{II} distribution along the delamination front for mode II loading shows convergence to 3D results with increasing number of plate elements through-the-thickness in each sublaminate (u_u, u_l): (1-1) and (2-2).

at the delamination front is 1 mm in the x-direction and 8 mm in the y-direction. The maximum ratio between the x and y dimensions is 8 for all the elements. Mesh refinement and convergence studies on the behavior of ERR as the interface stiffness parameters approach infinity have been conducted showing that choosing 10^8 N/mm^3 for the normal and tangential interface stiffness parameters is appropriate.

The analysis shows that when mixed mode conditions are involved, a double plate model is able to capture accurately the mode decomposition in the region near the midpoint of the delamination front. On the other hand, an accurate mode decomposition near the free edges of the delamination front, where 3D effects are more complex, necessitates more than one plate element in each sublaminate along the thickness direction, due to the large gradients in ERR. However, the solution converges quickly because a small number of plates is needed to obtain a reasonable approximation. On the contrary, a large number of solid finite elements may be needed along the thickness to obtain an accurate 3D modeling.

Results from the proposed methodology have been compared to those obtained in [55] by using solid finite elements while using, in the x-y plane, the same mesh used for plate models. The use of the same mesh for the plate elements and the solid elements allows a direct comparison in term of computational cost. Results show that if each sublaminate is divided uniformly into two plate elements, the model captures with satisfactory accuracy the ERR distribution along the whole delamination front and gives better predictions than the double plate model, especially near the free edges. This is illustrated in Figures 10.14 and 10.15.

In these figures, the ERR have been normalized with respect to the approximate values G_{II}^{2D}. Using a classical delamination model with a single plate element in the undelaminated region and assuming a straight delamination propagation, yields

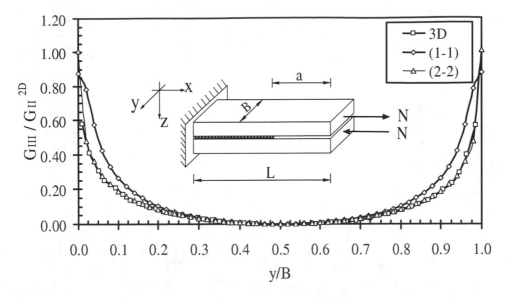

Fig. 10.15: G_{III} distribution along the delamination front for mode II loading shows convergence to 3D results with increasing number of plate elements through-the-thickness in each sublaminate (u_u, u_l): (1-1) and (2-2).

$$G_{II}^{2D} = \frac{1}{2}N^2 \left[\alpha_{11}^1 + \alpha_{11}^2 - \frac{(h_1 + h_2)^2}{4} \delta_{11} \right] \qquad (10.35)$$

where, following the conventional notation of classical laminated plate theory ([75, (6.18-6.19)]), $[\alpha^1], [\alpha^2]$ are the extensional compliance matrix of the upper (lower) sublaminate and $[\delta]$ is the bending compliance of the whole laminate.

It can be concluded that the (2-2) model is able to capture satisfactorily the actual in-plane displacements distribution both along the thickness and the delamination front. For mode II loading, the (2-2) model gives a maximum error within 10% with respect to the 3D model for G_{II}. This implies that along the delamination front the model gives an accuracy much better than 10% (see Table 10.3 at the in-plane loading column). On the other hand, the (1-1) model does not reflect accurately the 3D strain state near the edges, with unacceptable errors away from the delamination midpoint.

The ANSYS command file containing the command sequence for this example is provided in [76]. To run the example, go to the File menu and select "Read input from". The SHELL181 element type is used to mesh the plate models, whereas the COMBIN14 element type is used for the interface model. The same methodology of Example 10.3 is used to connect COMBIN14 elements to the mid-plane plate nodes. Displacement continuity at the interfaces between sublaminates is imposed by means of constraint equations.

Example 10.6 Use ANSYS to model the delaminated plate under mode III loading condition whose results are shown in Figures 10.16 and 10.17. Calculate the corresponding ERR distributions along the delamination front. Use the (2-2) plate assembly which gives reasonable accuracy. Expression (10.36) used to normalize results takes the value $2.89 \ 10^{-4} \ N/mm$. Use the same methodology of Example 10.5.

Table 10.3: Comparisons between interface-model- and 3D-FEM-results with an increasing number of plates in both sublaminates

y/B	In-Plane Loading		Out-of-Plane Loading	
	$G_{II(1-1)}/G_{II3D}$	$G_{II(2-2)}/G_{II3D}$	$G_{III(1-1)}/G_{III3D}$	$G_{III(2-2)}/G_{III3D}$
0.5	0.985	1.075	0.400	0.960
0.6	0.994	1.059	0.392	0.950
0.7	1.042	1.029	0.377	0.961
0.8	1.165	1.096	0.354	0.938
0.9	1.516	1.093	0.339	0.910
1.0	2.210	1.085	2.021	1.025

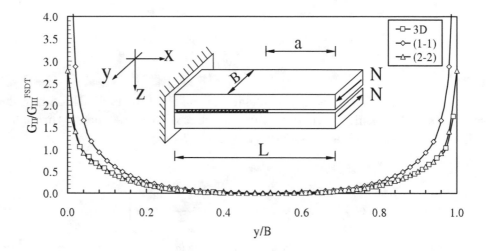

Fig. 10.16: G_{II} distribution along the delamination front for mode III loading shows convergence to 3D results with increasing number of plate elements through-the-thickness in each sublaminate (u_u, u_l): (1-1) and (2-2).

Solution to Example 10.6 *The solution of this example is similar to that of Example 10.5. The ANSYS command file containing the command sequence for this example is provided in [76].*

Results from the proposed methodology are compared to those obtained in [78] by using solid finite elements while using, in the x-y plane, the same mesh used for plate models as illustrated in Figures 10.16 and 10.17.

In these figures, the ERR have been normalized with respect to the approximate values G_{III}^{FSDT}. Using a delamination model based on the first-order shear deformable plate theory and taking the derivative with respect to the crack area of the sum of the strain energies of the three regions comprising the delaminated plate.

$$G_{III}^{FSDT} = \frac{1}{2}N^2 \left[a_{66}^1 + a_{66}^2 - \frac{(h_1+h_2)^2}{4}\delta_{66} + \frac{12a^2}{B^2}\left(\frac{1}{E_{11}^1 h_1} + \frac{1}{E_{11}^2 h_2}\right) \right.$$
$$\left. + \frac{3}{2}\left(\frac{1}{G_{12}^1 h_1} + \frac{1}{G_{12}^2 h_2}\right)\right]$$
(10.36)

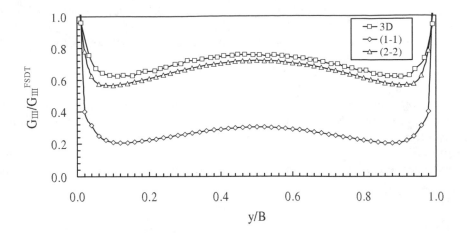

Fig. 10.17: G_{III} distribution along the delamination front for mode III loading shows convergence to 3D results with increasing number of plate elements through-the-thickness in each sublaminate (u_u, u_l): (1-1) and (2-2).

Two first-order shear deformable plates were used for the delaminated members whereas one first-order shear deformable plate has been introduced for the undelaminated region.

It can be concluded that the (2-2) model is able to capture satisfactorily the actual in-plane displacements distribution both along the thickness and the delamination front. For mode III loading, the (2-2) model gives a maximum error within 10% with respect to the 3D model for G_{III}. This implies that along the delamination front the model gives an accuracy much better than 10% (see Table 10.3 at the out-of-plane loading column). On the other hand, the (1-1) model does not reflect accurately the 3D strain state near the edges, with unacceptable errors away from the delamination midpoint.

Suggested Problems

Problem 10.1 *Using a three-dimensional finite element model and the modified virtual crack closure technique (MVCCT, [79]) compute the ERR distribution along the straight delamination front for the DCB problem in Figures 10.14 and 10.15. Compare these results with those from the interface model (details can be found in [60]). Use the same mesh used for results of Figure 10.15 in the x-y plane and one row of solid finite element in the thickness direction. The results from the interface model must show, away from the edges, a satisfactory accuracy in comparison with those obtained by a three-dimensional formulation, although each sublaminate is assumed as composed by simply one shear deformable plate. This confirms the results obtained in the 2D delamination problem, where it was proved that for special loading and geometrical schemes a double plate model suffices to determine accurately the mode mix. It is expected that for more general loading and geometry conditions (mixed-mode loadings, offset delaminations, for instance) more than one plate is needed to model each sublaminate to capture the actual 3D character of the problem near the delamination front.*

Problem 10.2 *Use ANSYS to analyze a double plate assembly comprising two orthotropic layers with different thicknesses, and subjected to a pair of opposed opening forces as shown in Figure 10.18.*

The upper sublaminate has the following properties: $E_x^1 = 35,000$ *MPa,* $E_y^1 = E_z^1 = 10,500$, $G_{yz}^1 = 10,500$ *MPa,* $G_{xy}^1 = G_{xz}^1 = 1,167$, $\nu_{xy}^1 = \nu_{xz}^1 = \nu_{yz}^1 = 0.3$.

The lower sublaminate has the following properties: $E_x^2 = 70,000$ *MPa,* $E_y^2 = E_z^2 = 21,000$, $G_{yz}^2 = 2,100$ *MPa,* $G_{xy}^2 = G_{xz}^2 = 2,333$, $\nu_{xy}^2 = \nu_{xz}^2 = \nu_{yz}^2 = 0.3$.

The opening force T *is equal to 1 N/mm and the dimensions of the laminate are:* $a = 10$ *mm,* $B = 20$ *mm,* $L = 20$ *mm,* $h_1 = 0.5$ *mm,* $h_2 = 1$ *mm.*

Use mesh refinement at the delamination front similar to that used in Example 10.5, in which the maximum ratio between the x *and* y *dimensions is 8. Use interface stiffness* k_z *and* k_{xy} *equal to* 10^8 *N/mm³ and the (2-4) model, which is sufficiently accurate since this subdivision reflects plate geometry by using more plate elements for the thicker sublaminate. Calculate the mode I, II and III components of ERR along the delamination front. Results may be found in [61].*

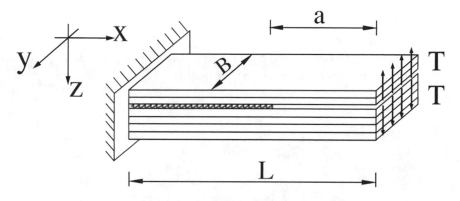

Fig. 10.18: Bi-material delaminated plate in problems 10.2-10.3.

Problem 10.3 *Solve the example of Problem 10.2 by using solid finite elements and perform mesh refinement analysis to evaluate the behavior of the total ERR and its mode components when the size of the delamination front element in the x-direction decreases.*

Consistently with the (2-4) multilayer modeling, two and four layers of solid finite elements can be placed in the thickness direction for the upper and lower sublaminates, respectively. The laminate geometry and the 3D FE model are illustrated in Figure 10.19b. The ERR mode components for the 3D model must be calculated by using the MVCCT [79] which leads to an expression similar to (10.34), by evaluating the node forces as Lagrange multipliers related to the adhesion constraints along the undelaminated region of the delamination plane.

The analysis should show that both the individual components and the total ERR converge as the delamination front elements are smaller. On the contrary, results obtained by using solid finite elements should show a non-convergence behavior for the individual ERR. Results may be found in [61]. This is a consequence of the mismatch of material properties across the interface which leads to an oscillatory singularity behavior of stresses and displacements near the delamination front [41], in place of the inverse square-root singularity which occurs when delamination is placed between two equal orthotropic or isotropic layers, as is the case of the examples analyzed in the previous sections. Therefore, the proposed method can be used as a computationally efficient method to eliminate the oscillatory singularity that causes non-convergent behavior when solid finite elements are used.

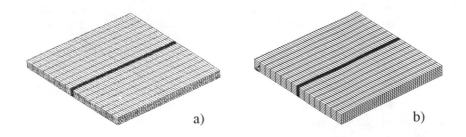

Fig. 10.19: Bi-material delaminated plate in problems 10.2-10.3: a) Multi-layer FEM; b) 3D FEM.

References

[1] H. Chai, C. D. Babcock, and W. G. Knauss, One Dimensional Modeling of Failure in Laminated Plates by Delamination Buckling, International Journal of Solids and Structures, 17(11) (1981) 1069-1083.

[2] M.-K. Yeh and L.-B. Fang, Contact Analysis and Experiment of Delaminated Cantilever Composite Beam, Composites: Part B, 30(4) (1999) 407-414.

[3] W. L. Yin, S. N. Sallam, and G. J. Simitses, Ultimate Axial Load Capacity of a Delaminated Beam-Plate, AIAA Journal, 24(1) (1986) 123-128.

[4] D. Bruno, Delamination Buckling in Composite Laminates with Interlaminar Defects, Theoretical and Applied Fracture Mechanics, 9(2) (1988) 145-159.

[5] D. Bruno and A. Grimaldi, Delamination Failure of Layered Composite Plates Loaded in Compression, International Journal of Solids and Structures, 26(3) (1990) 313-330.

[6] G. A. Kardomateas, The Initial Post-Buckling and Growth Behaviour of Internal Delaminations in Composite Plates, Journal of Applied Mechanics, 60(4) (1993) 903-910.

[7] G. A. Kardomateas and A. A. Pelegri, The Stability of Delamination Growth in Compressively Loaded Composite Plates, International Journal of Fracture, 65(3) (1994) 261-276.

[8] I. Sheinman, G. A. Kardomateas, and A. A. Pelegri, Delamination Growth During Pre- and Post-Buckling Phases of Delaminated Composite Laminates, International Journal of Solids and Structures, 35(1-2) (1998) 19-31.

[9] D. Bruno and F. Greco, An Asymptotic Analysis of Delamination Buckling and Growth in Layered Plates, International Journal of Solids and Structures, 37(43) (2000) 6239-6276.

[10] H.-J. Kim, Postbuckling Analysis of Composite Laminates with a Delamination, Computer and Structures, 62(6) (1997) 975-983.

[11] W. G. Bottega and A. Maewal, Delamination Buckling and Growth in Laminates – Closure, Journal of Applied Mechanics, 50(14) (1983) 184-189.

[12] B. Cochelin and M. Potier-Ferry, A Numerical Model for Buckling and Growth of Delaminations in Composite Laminates, Computer Methods in Applied Mechanics, 89(1-3) (1991) 361-380.

[13] P.-L. Larsson, On Delamination Buckling and Growth in Circular and Annular Orthotropic Plates, International Journal of Solids and Structures, 27(1) (1991) 15-28.

[14] W.-L. Yin, Axisymmetric Buckling and Growth of a Circular Delamination in a Compressed Laminate, International Journal of Solids and Structures, 21(5) (1985) 503-514.

[15] K.-F. Nilsson, L. E. Asp, J. E. Alpman, and L. Nystedt, Delamination Buckling and Growth for Delamiantions at Different Depths in a Slender Composite Panel, International Journal of Solids and Structures, 38(17) (2001) 3039-3071.

[16] H. Chai and C. D. Babcock, Two-Dimensional Modeling of Compressive Failure in Delaminated Laminates, Journal of Composite Materials, 19(1) (1985) 67-98.

[17] J. D. Withcomb and K. N. Shivakumar, Strain-Energy Release Rate Analysis of Plates with Postbuckled Delaminations, Journal of Composite Materials, 23(7) (1989) 714-734.

[18] W. J. Bottega, A Growth Law for Propagation of Arbitrary Shaped Delaminations in Layered Plates, International Journal of Solids and Structures, 19(11) (1983) 1009-1017.

[19] B. Storåkers and B. Andersson, Nonlinear Plate Theory Applied to Delamination in Composites, Journal of the Mechanics and Physics of Solids, 36(6) (1988) 689-718.

[20] P.-L. Larsson, On Multiple Delamination Buckling and Growth in Composite Plates, International Journal of Solids and Structures, 27(13) (1991) 1623-1637.

[21] M. A. Kouchakzadeh and H. Sekine, Compressive Buckling Analysis of Rectangular Composite Laminates Containing Multiple Delaminations, Composite Structures, 50(3) (2000) 249-255.

[22] J. M. Comiez, A. M. Waas, and K. W. Shahwan, Delamination Buckling: Experiment and Analysis, International Journal of Solids and Structures, 32(6/7) (1995) 767-782.

[23] G. A. Kardomateas, Postbuckling Characteristics in Delaminated Kevlar/Epoxy Laminates: An Experimental Study, Journal of Composites Technology and Research, 12(2) (1990) 85-90.

[24] O. Allix and A. Corigliano, Modeling and Simulation of Crack Propagation in Mixed-Modes Interlaminar Fracture Specimens, International Journal of Fracture, 77(2) (1996) 111-140.

[25] D. Bruno and F. Greco, Delamination in Composite Plates: Influence of Shear Deformability on Interfacial Debonding, Cement and Concrete Composites, 23(1) (2001) 33-45.

[26] J. W. Hutchinson and Z. Suo, Mixed Mode Cracking in Layered Materials, Advances in Applied Mechanics, 29 (1992) 63-191.

[27] N. Point and E. Sacco, Delamination of Beams: An Application to the DCB Specimen, International Journal of Fracture, 79(3) (1996) 225-247.

[28] J. G. Williams, On the Calculation of Energy Release Rate for Cracked Laminates, International Journal of Fracture, 36(2) (1988) 101-119.

[29] A. Tylikowsky, Effects of Piezoactuator Delamination on the Transfer Functions of Vibration Control Systems, International Journal of Solids and Structures, 38(10-13) (2001) 2189-2202.

[30] C. E. Seeley and A. Chattopadhyay, Modeling of Adaptive Composites Including Debonding, International Journal Solids and Structures, 36(12) (1999) 1823-1843.

[31] W. J. Bottega and A. Maewal, Dynamics of Delamination Buckling, International Journal of Non-Linear Mechanics, 18(6) (1983) 449-463.

[32] G. Alfano and M. A. Crisfield, Finite Element Interface Models for the Delamination Analysis of Laminated Composites: Mechanical and Computational Issues, International Journal for Numerical Methods in Engineering, 50(7) (2001) 1701-1736.

[33] P. P. Camanho, C. G. Dávila, and M. F. Moura, Numerical Simulation of Mixed-Mode Progressive Delamination in Composite Materials, Journal of Composite Materials, 37(16) (2003) 1415-1438.

[34] Z. Zou, S. R. Reid, and S. Li, A Continuum Damage Model for Delaminations in Laminated Composites, Journal of the Mechanics and Physics of Solids, 51(2) (2003) 333-356.

[35] J. R. Reeder, A Bilinear Failure Criterion for Mixed-Mode Delamination in Composite Materials: Testing and Design, American Society for Testing and Materials, ASTM STP 1206 (1993) 303-322.

[36] H. Chai, Three-Dimensional Fracture Analysis of Thin-Film Debonding, International Journal of Fracture, 46(4) (1990) 237-256.

[37] J. D. Withcomb, Three Dimensional Analysis of a Post-Buckled Embedded Delamination, Journal of Composite Materials, 23(9) (1989) 862-889.

[38] J. G. Williams, The Fracture Mechanics of Delamination Tests, Journal of Strain Analysis, 24(4) (1989) 207-214.

[39] W.-L. Yin and J. T. S. Wang, The Energy-Release Rate in the Growth of a One-Dimensional Delamination, Journal of Applied Mechanics, 51(4) (1984) 939-941.

[40] J. L. Beuth, Separation of Crack Extension Modes in Orthotropic Delamination Models, International Journal of Fracture, 77(4) (1996) 305-321.

[41] I. S. Raju, J. H. Crews Jr., and M. A. Aminpour, Convergence of Strain Energy Release Rate Components for Edge-Delaminated Composite Laminates, Engineering Fracture Mechanics, 30(3) (1988) 383-396.

[42] C. T. Sun and W. Qian, The Use of Finite Extension Strain Energy Release Rates in Fracture of Interface Cracks, International Journal of Solids and Structures, 34(20) (1997) 2595-2609.

[43] B. D. Davidson, R. Kruger, and M. Konig, Effect of Stacking Sequence on Energy Release Rate Distributions in Multidirectional DCB and ENF specimens, Engineering Fracture Mechanics, 55(4) (1996) 557-569.

[44] D. Hitchings, P. Robinson, and F. Javidrad, A Finite Element Model for Delamination Propagation in Composites, Computers and Structures, 60(6) (1996) 1093-1104.

[45] A. Riccio, F. Scaramuzzino, and P. Perugini, Embedded Delamination Growth in Composite Panels Under Compressive Load, Composites: Part B, 32(3) (2001) 209-218.

[46] M. F. Kanninen, An Augmented Double Cantilever Beam Model for Studying Crack Propagation and Arrest, International Journal of Fracture, 9(1) (1973) 83-91.

[47] D. J. Chang, R. Muki, and R. A. Westmann, Double Cantilever Beam Models in Adhesive Mechanics, International Journal of Solids and Structures, 12(1) (1976) 13-26.

[48] D. Bruno and F. Greco, Mixed Mode Delamination in Plates: A Refined Approach. International Journal of Solids and Structures. 38(50-51) (2001) 9149-9177.

[49] Z. Suo and J. W. Hutchinson, Interface Crack Between Two Elastic Layers, International Journal of Fracture, 43(1) (1990) 1-18.

[50] Z. Suo, Delamination Specimen for Othotropic Materials, Journal of Applied Mechanics, 57(3) (1990) 627-634.

[51] R. A. Schapery and B. D. Davidson, Prediction of Energy Release Rate for Mixed-Mode Delamination Using Classical Plate Theory, Applied Mechanics Reviews, 43 (1990) S281-S287.

[52] B. D. Davidson, Hurang Hu, and R. A. Schapery, An Analytical Crack-Tip Element for Layered Elastic Structures, Journal of Applied Mechanics, 62(2) (1995) 294-305.

[53] V. Sundararaman and B. D. Davidson, An Unsymmetric Double Cantilever Beam Test for Interfacial Fracture Toughness Determination, International Journal of Solids and Structures, 34(7) (1997) 799-817.

[54] B. D. Davidson, H. Hu, and H. Yan, An Efficient Procedure for Determining Mixed-Mode Energy Release Rate in Practical Problems of Delamination, Finite Elements in Analysis and Design, 23(2-4) (1996) 193-210.

[55] B. D. Davidson, L. J. Yu, and H. Hu, Determination of Energy Release Rate and Mode Mix in Three-Dimensional Layered Structures Using Plate Theory, International Journal of Fracture, 105(1) (2000) 81-104.

[56] J. Wang and P. Qiao, Interface Crack Between Two Shear Deformable Elastic Layers, Journal of the Mechanics and Physics of Solids, 52(4) (2004) 891-905.

[57] Z. Zou, S. R. Reid, P. D. Soden, and S. Li, Mode Separation of Energy Release Rate for Delamination in Composite Laminates Using Sublaminates, Journal of Solids and Structures, 38(15) (2001) 2597-2613.

[58] V. Q. Bui, E. Marechal, and H. Nguyen-Dang, Imperfect Interlaminar Interfaces in Laminated Composites: Interlaminar Stresses and Strain-Energy Release Rates, Composites Science and Technology, 60(1) (2000) 131-143.

[59] O. Allix, P. Ladevèze, and A. Corigliano, Damage Analysis of Interlaminar Fracture Specimens, Composite Structures, 31(1) (1995) 61-74.

[60] D. Bruno, F. Greco, and P. Lonetti, Computation of Energy Release Rate and Mode Separation in Delaminated Composite Plates by Using Plate and Interface Variables, Mechanics of Advanced Materials and Structures, 12(4) (2005) 285–304.

[61] D. Bruno, F. Greco, and P. Lonetti, A 3D Delamination Modeling Technique Based on Plate and Interface Theories for Laminated Structures, European Journal of Mechanics–A/Solids, 24(1) (2005) 127-149.

[62] F. Greco, P. Lonetti and R. Zinno, An Analytical Delamination Model for Laminated Plates Including Bridging Effects, International Journal of Solids and Structures, 39(9) (2002) 2435-2463.

[63] D. Bruno and F. Greco, An Efficient Model of Mixed-Mode Delamination in Laminated Composites Including Bridging Mechanisms, Simulation Modelling Practice and Theory, 11(5-6) (2003) 465-481.

[64] D. Bruno, F. Greco, and P. Lonetti, A Coupled Interface-Multilayer Approach for Mixed Mode Delamination and Contact Analysis in Laminated Composites, International Journal of Solids and Structures, 40(26) (2003) 7245-7268.

[65] E. J. Barbero and J. N. Reddy, Modeling of Delamination in Composites Laminates Using a Layer-Wise Plate Theory, International Journal of Solid and Structures, 28(3) (1991) 373-388.

[66] J. R. Rice, A Path Independent Integral and the Approximate Analysis of Strain Concentrations by Notches and Cracks, Journal of Applied Mechanics, 35 (1968) 379-386.

[67] E. J. Barbero, F. Greco, and P. Lonetti, Continuum Damage-Healing Mechanics with Application to Self-Healing Composites, International Journal of Damage Mechanics, 14(1) (2005) 51-81.

[68] A. C. Eringen, Mechanics of Continua, John Wiley & Sons, Inc., New York, 1967.

[69] Q. S. Nguyen, Stability and Nonlinear Solid Mechanics, John Wiley & Sons, Chichester, 2000.

[70] E. J. Barbero and J. N. Reddy, The Jacobian Derivative Method for Three-Dimensional Fracture Mechanics, Communications in Applied Numerical Methods, 6 (1990) 507-518.

[71] J. N. Reddy, E. J. Barbero, and J. L. Teply, A Plate Bending Element Based on a Generalized Laminate Plate Theory, International Journal for Numerical Methods in Engineering, 28(10) (1989) 2275-2292.

[72] E. J. Barbero, A 3-D Finite Element for Laminated Composites With 2-D Kinematic Constraints, Computers and Structures, 45(2) (1992) 263-271.

[73] R. Zinno and E. J. Barbero, Total Lagrangian Formulation for Laminated Composite Plates Analysed by Three-Dimensional Finite Elements With Two-Dimensional Kinematic Constraints, Computers and Structures, 57(3) (1995) 455-466.

[74] R. Zinno and E. J. Barbero, Three-Dimensional Layer-Wise Constant Shear Element for General Anisotropic Shell-Type Structures, International Journal for Numerical Methods in Engineering, 37(14) (1994) 2445-2470.

[75] E. J. Barbero, Introduction to Composite Materials Design, Taylor & Francis, Philadelphia, 1999.

[76] E. J. Barbero, Web resource: http://www.mae.wvu.edu/barbero/feacm/

[77] E. J. Barbero, R. Luciano, and E. Sacco, Three-Dimensional Plate and Contact/Friction Elements for Laminated Composite Joints, Computers and Structures, 54(4) (1995) 689-703.

[78] P. Robinson and D. Q. Song, A Modified DCB Specimen for Mode I Testing of Multidirectional Laminates, Journal of Composite Materials, 26(11) (1992) 1554-1577.

[79] E. F. Rybicki and M. F. Kanninen, A Finite Element Calculation of Stress Intensity Factors by a Modified Crack Closure Integral, Engineering Fracture Mechanics, 9 (1977) 931-938.

Appendix A

Tensor Algebra

Tensor operations are needed for the derivation of some of the equations in this textbook. Since most of these operations are not easily found in textbooks, they are presented here for reference [1].

A.1 Principal Directions of Stress and Strain

Since stress and strain tensors are symmetric and of second order, they have three real principal values and three orthogonal principal directions. The principal values λ^q and directions n_i^q of the stress tensor σ_{ij} satisfy the following

$$[\sigma_{ij} - \lambda^q \delta_{ij}]n_i^q = 0 \tag{A.1}$$
$$n_i^q n_j^q = 1 \tag{A.2}$$

where δ_{ij} is the Kronecker delta ($\delta_{ij} = 1$ if $i = j$, zero otherwise). Each of the principal directions is described by its direction cosines with respect to the original coordinate system.

The principal directions are arranged by rows into a matrix $[A]$. Then, the diagonal matrix $[A^*]$ of the principal values is

$$[A^*] = [a][A][a]^T \tag{A.3}$$

It can be shown that $[a]^{-1} = [a]^T$, where $[a]$ is the transformation matrix given by (1.21)

A.2 Tensor Symmetry

Minor symmetry provides justification for using contracted notation (Section 1.5). Minor symmetry refers to identical values of tensor components when adjacent subscripts are swapped. For example, minor symmetry of the stiffness tensor $\mathbf{C}$ means

$$A_{ijkl} = A_{jilk} = A_{\alpha\beta} \tag{A.4}$$

299

Major symmetry refers to identical values when adjacent pairs of subscripts are swapped, or when contracted subscripts are swapped. For example,

$$A_{ijkl} = A_{klij}$$
$$A_{\alpha\beta} = A_{\beta\alpha} \tag{A.5}$$

A.3   Matrix Representation of a Tensor

A tensor A_{ijkl} with a minor symmetry has only 36 independent constants. Therefore it can be represented in contracted notation by a 6×6 matrix. Let $[a]$ be the contracted form of the tensor A. Each element of $[a]$ corresponds to an element in the tensor A according to the following transformation

$$a_{\alpha\beta} = A_{ijkl} \tag{A.6}$$

with

$$\alpha = i \text{ when } i = j$$
$$\alpha = 9 - (i + j) \text{ when } i \neq j \tag{A.7}$$

The same transformations apply between β and k and l, or in matrix representation, as

$$[a] = \begin{bmatrix} A_{1111} & A_{1122} & A_{1133} & A_{1123} & A_{1113} & A_{1112} \\ A_{2211} & A_{2222} & A_{2233} & A_{2223} & A_{2213} & A_{2212} \\ A_{3311} & A_{3322} & A_{3333} & A_{3323} & A_{3313} & A_{3312} \\ A_{2311} & A_{2322} & A_{2333} & A_{2323} & A_{2313} & A_{2312} \\ A_{1311} & A_{1322} & A_{1333} & A_{1323} & A_{1313} & A_{1312} \\ A_{1211} & A_{1222} & A_{1233} & A_{1223} & A_{1213} & A_{1212} \end{bmatrix} \tag{A.8}$$

It is convenient to perform tensor operations using contracted form, especially if the result can be represented also in contracted form. This saves memory and time since it is faster to operate on 36 elements than on 81 elements. Examples of these operations are the inner product of two fourth-order tensors and the inverse of a fourth-order tensor. However, tensor operations in index notation do not translate directly into matrix operations in contracted form. For example, the double contraction of two fourth-order tensors is

$$\mathbf{C} = \mathbf{A} : \mathbf{B}$$
$$C_{ijkl} = A_{ijmn} B_{mnkl} \tag{A.9}$$

Let $[a]$, $[b]$, and $[c]$ the 6×6 matrix representations of the above tensors. Then, it can be shown that

$$[a][b] \neq [c] \text{ or}$$
$$a_{\alpha\beta} b_{\beta\gamma} \neq c_{\alpha\gamma} \text{ (matrix multiplication)} \tag{A.10}$$

The rest of this appendix presents formulas for adequate representation of tensor operations in their contracted form.

A.4 Double Contraction

In (A.9), an element like C_{1211} can be expanded as

$$
\begin{aligned}
C_{1211} \;=\; & A_{1211}B_{1111} + A_{1222}B_{2211} + A_{1233}B_{3311} \\
& + 2A_{1212}B_{1211} + 2A_{1213}B_{1311} + 2A_{1223}B_{2311}
\end{aligned}
\tag{A.11}
$$

In order to achieve the same result by matrix multiplication, multiply the last three columns of the matrix $[a]$ by 2, and then perform the multiplication

$$
[c] \;=\;
\begin{bmatrix}
A_{1111} & A_{1122} & A_{1133} & 2A_{1123} & 2A_{1113} & 2A_{1112} \\
A_{2211} & A_{2222} & A_{2233} & 2A_{2223} & 2A_{2213} & 2A_{2212} \\
A_{3311} & A_{3322} & A_{3333} & 2A_{3323} & 2A_{3313} & 2A_{3312} \\
A_{2311} & A_{2322} & A_{2333} & 2A_{2323} & 2A_{2313} & 2A_{2312} \\
A_{1311} & A_{1322} & A_{1333} & 2A_{1323} & 2A_{1313} & 2A_{1312} \\
A_{1211} & A_{1222} & A_{1233} & 2A_{1223} & 2A_{1213} & 2A_{1212}
\end{bmatrix}
$$
$$
\begin{bmatrix}
B_{1111} & B_{1122} & B_{1133} & B_{1123} & B_{1113} & B_{1112} \\
B_{2211} & B_{2222} & B_{2233} & B_{2223} & B_{2213} & B_{2212} \\
B_{3311} & B_{3322} & B_{3333} & B_{3323} & B_{3313} & B_{3312} \\
B_{2311} & B_{2322} & B_{2333} & B_{2323} & B_{2313} & B_{2312} \\
B_{1311} & B_{1322} & B_{1333} & B_{1323} & B_{1313} & B_{1312} \\
B_{1211} & B_{1222} & B_{1233} & B_{1223} & B_{1213} & B_{1212}
\end{bmatrix}
\tag{A.12}
$$

This transformation can be produced by using the Reuter matrix $[R]$

$$
[R] =
\begin{bmatrix}
1 & 0 & 0 & 0 & 0 & 0 \\
0 & 1 & 0 & 0 & 0 & 0 \\
0 & 0 & 1 & 0 & 0 & 0 \\
0 & 0 & 0 & 2 & 0 & 0 \\
0 & 0 & 0 & 0 & 2 & 0 \\
0 & 0 & 0 & 0 & 0 & 2
\end{bmatrix}
\tag{A.13}
$$

Substituting in (A.12) we have

$$
[c] = [a]\,[R]\,[b]
\tag{A.14}
$$

A.5 Tensor Inversion

First, it is convenient to define the fourth-order identity tensor I_{ijkl} as a tensor that multiplied innerly by another fourth-order tensor yields this same tensor, or

$$
I_{ijmn}A_{mnkl} = A_{ijkl}
\tag{A.15}
$$

If A_{ijkl} has a minor symmetry, the following tensor achieves (A.15)

$$I_{ijkl} = \frac{1}{2}\left(\delta_{ik}\delta_{jl} + \delta_{il}\delta_{jk}\right) \tag{A.16}$$

where δ_{ij} is the Kronecker delta, defined as

$$\begin{aligned} \delta_{ij} &= 1 \quad \text{if } i = j \\ \delta_{ij} &= 0 \quad \text{if } i \neq j \end{aligned} \tag{A.17}$$

In Voigt contracted notation, the fourth-order identity tensor is denoted as $[i]$, which is equal to the inverse of the Reuter matrix

$$[i] = \begin{bmatrix} 1 & 0 & 0 & 0 & 0 & 0 \\ 0 & 1 & 0 & 0 & 0 & 0 \\ 0 & 0 & 1 & 0 & 0 & 0 \\ 0 & 0 & 0 & 1/2 & 0 & 0 \\ 0 & 0 & 0 & 0 & 1/2 & 0 \\ 0 & 0 & 0 & 0 & 0 & 1/2 \end{bmatrix} = [R]^{-1} \tag{A.18}$$

Now, the inverse of a tensor is a tensor that multiplied by the original tensor yields the identity tensor, as follows:

$$A_{ijmn}A^{-1}_{mnkl} = I_{ijkl} \tag{A.19}$$

Let us introduce the following notation:

$[a]^{-1}$ = inverse of the reduced form of A_{ijkl}

$[a^{-1}]$ = reduced form of the inverse of A_{ijkl}

If A_{ijkl} has a minor symmetry, the components of $a^{-1}_{\alpha\beta}$ are:

1. Multiply the last three columns of $[a]$ by 2 by using the matrix $[R]$

2. Invert the obtained matrix.

3. Multiply the matrix by $[i]$

In order words, the matrix $[a]^{-1}$ is computed as

$$[a^{-1}] = [[a]\,[R]]^{-1}\,[i] = [i]\,[a]^{-1}\,[i] \tag{A.20}$$

A.6 Tensor Differentiation

A.6.1 Derivative of a Tensor With Respect to Itself

Any symmetric second-order tensor Φ_{ij} satisfies the following:

$$d\Phi_{ij} = d\Phi_{ji} \tag{A.21}$$

Therefore, differentiating a second-order symmetric tensor with respect to itself is accomplished as follows

$$\frac{\partial \Phi_{ij}}{\partial \Phi_{kl}} = J_{ijkl} \tag{A.22}$$

where J_{ijkl} is a fourth-order tensor defined as

$$\begin{aligned} J_{ijkl} &= 1 && \text{if } i = k, \text{ and } j = l \\ J_{ijkl} &= 1 && \text{if } i = l, \text{ and } j = k \\ J_{ijkl} &= 0 && \text{otherwise} \end{aligned} \tag{A.23}$$

In contracted notation, the tensor J_{ijkl} is represented by

$$[j] = \begin{bmatrix} 1 & 0 & 0 & 0 & 0 & 0 \\ 0 & 1 & 0 & 0 & 0 & 0 \\ 0 & 0 & 1 & 0 & 0 & 0 \\ 0 & 0 & 0 & 1 & 0 & 0 \\ 0 & 0 & 0 & 0 & 1 & 0 \\ 0 & 0 & 0 & 0 & 0 & 1 \end{bmatrix} \tag{A.24}$$

A.6.2 Derivative of the Inverse of a Tensor With Respect to the Tensor

A second-order tensor contracted with its inverse yields the second-order identity tensor, or Kronecker delta

$$A_{ij} A_{jk}^{-1} = \delta_{ik} \tag{A.25}$$

Differentiating (A.25) with respect to A_{mn} and rearranging terms yields

$$A_{ij} \frac{\partial A_{jk}^{-1}}{\partial A_{mn}} = -\frac{\partial A_{ij}}{\partial A_{mn}} A_{jk}^{-1} \tag{A.26}$$

Pre-multiplying both sides by A_{li}^{-1} and rearranging yields

$$\frac{\partial A_{ij}^{-1}}{\partial A_{mn}} = -A_{ik}^{-1} \frac{\partial A_{kl}}{\partial A_{mn}} A_{lj}^{-1} \tag{A.27}$$

Finally, using (A.22) yields

$$\frac{\partial A_{ij}^{-1}}{\partial A_{mn}} = -A_{ik}^{-1} J_{klmn} A_{lj}^{-1} \tag{A.28}$$

Appendix B

Strain Concentration Tensors

The following closed form formulation for the elastic strain concentration tensor is based on [2]. For a unidirectional composite with long, circular cylindrical fibers embedded in isotropic matrix, the strain concentration tensor is derived as follows. The composite stiffness tensor $\mathbf{C}$ can be calculated [2, page 25], using contracted notation (6×6 matrix), as follows

$$\mathbf{C} = \mathbf{C}^m - V_f \left[\left(\mathbf{C}^m - \mathbf{C}^f \right)^{-1} - \mathbf{P} \right]^{-1} \tag{B.1}$$

where the tensor $\mathbf{P}$, which accounts for the geometry of the inclusions and microstructure, is calculated by (B.7). Here $V_i, \mathbf{C}^i$, with $i = f, m$ are the volume fraction and stiffness tensors of the fiber (f) and matrix (m), respectively.

Alternatively, the composite stiffness tensor in contracted notation (6×6 matrix) can be calculated in terms of the strain concentration tensors $\mathbf{A}^i$ [3, (2.9)], as follows

$$\mathbf{C} = V_f \mathbf{C}^f \mathbf{A}^f + (1 - V_f) \mathbf{C}^m \mathbf{A}^m \tag{B.2}$$

Furthermore, the strain concentration tensors obey the following [3]

$$V_f \mathbf{A}^f + (1 - V_f) \mathbf{A}^m = \mathbf{I} \tag{B.3}$$

where $\mathbf{I}$ is the 6×6 identity matrix. Therefore,

$$\mathbf{A}^m = \frac{\mathbf{I} - V_f \mathbf{A}^f}{(1 - V_f)} \tag{B.4}$$

Introducing (B.4) into (B.2) yields

$$\mathbf{C} = \mathbf{C}^m - V_f \left(\mathbf{C}^m - \mathbf{C}^f \right) \mathbf{A}^f \tag{B.5}$$

Comparing (B.5) to (B.1), yields an expression for the strain concentration tensor in terms of symmetric tensors

$$\mathbf{A}^f = \left(\mathbf{C}^m - \mathbf{C}^f \right)^{-1} \left[\left(\mathbf{C}^m - \mathbf{C}^f \right)^{-1} - \mathbf{P} \right]^{-1} \tag{B.6}$$

For composites reinforced by long fibers aligned with the x_1 axis, all the coefficients in the first row and column of P are equal to zero. Other expressions of $\mathbf{A}^f$ (e.g., [3, (2.86)] or [1, 4]) become intractable for this case, but (B.6) remains amenable to simple matrix computation in contracted notation.

The tensor P is defined as

$$P_{ijkl} = \frac{1}{\mu_m} \sum_{\xi}^{\pm\infty} t(\xi) \left(sym \left(\xi \otimes I^{(2)} \otimes \xi \right) - \frac{1}{2(1 - \nu_m)} \right) (\xi \otimes \xi \otimes \xi \otimes \xi) \right) \qquad \text{(B.7)}$$

The MATLAB symbolic code `Ptensor.m`, available in [5], can be used to expand B.7, but it must be noted that all the odd powers of ξ, such as $\sum \xi_1 \xi_3$ vanish due to orthogonality of the Fourier base functions. In (B.7), sym indicates minor symmetry; that is, a symmetrization with respect to the minor indices is enforced. Therefore, the P tensor can be written in contracted notation, as shown in (B.10), where it can be observed that P also has major symmetry.

The following series are now defined

$$
\begin{aligned}
S_1 &= \sum_{\xi}^{\pm\infty} t(\xi) \bar{\xi}_1^2 \quad ; \quad S_2 = \sum_{\xi}^{\pm\infty} t(\xi) \bar{\xi}_2^2 \\
S_3 &= \sum_{\xi}^{\pm\infty} t(\xi) \bar{\xi}_3^2 \quad ; \quad S_4 = \sum_{\xi}^{\pm\infty} t(\xi) \bar{\xi}_1^4 \\
S_5 &= \sum_{\xi}^{\pm\infty} t(\xi) \bar{\xi}_2^4 \quad ; \quad S_6 = \sum_{\xi}^{\pm\infty} t(\xi) \bar{\xi}_3^4 \\
S_7 &= \sum_{\xi}^{\pm\infty} t(\xi) \bar{\xi}_2^2 \bar{\xi}_3^2 \quad ; \quad S_8 = \sum_{\xi}^{\pm\infty} t(\xi) \bar{\xi}_1^2 \bar{\xi}_3^2 \\
S_9 &= \sum_{\xi}^{\pm\infty} t(\xi) \bar{\xi}_1^2 \bar{\xi}_2^2
\end{aligned}
\qquad \text{(B.8)}
$$

with

$$\bar{\xi}_i = \frac{\xi_i}{\xi} \qquad \text{(B.9)}$$

In terms of the series S_i the tensor P can be expressed in contracted notation as the following 6×6 matrix

$$[p] = \begin{bmatrix}
\frac{S_1}{\mu_m} - \frac{S_4}{2\mu_m(1-\nu_m)} & -\frac{S_9}{2\mu_m(1-\nu_m)} & -\frac{S_8}{2\mu_m(1-\nu_m)} \\
\text{symm} & \frac{S_2}{\mu_m} - \frac{S_5}{2\mu_m(1-\nu_m)} & -\frac{S_7}{2\mu_m(1-\nu_m)} \\
\text{symm} & \text{symm} & \frac{S_3}{\mu_m} - \frac{S_6}{2\mu_m(1-\nu_m)} \\
\text{symm} & \text{symm} & \text{symm} \\
\text{symm} & \text{symm} & \text{symm} \\
\text{symm} & \text{symm} & \text{symm} \\
& & \\
0 & 0 & 0 \\
0 & 0 & 0 \\
0 & 0 & 0 \\
\frac{S_3+S_2}{4\mu_m} - \frac{S_7}{2\mu_m(1-\nu_m)} & 0 & 0 \\
\text{symm} & \frac{S_1+S_3}{4\mu_m} - \frac{S_8}{2\mu_m(1-\nu_m)} & 0 \\
\text{symm} & \text{symm} & \frac{S_1+S_2}{4\mu_m} - \frac{S_9}{2\mu_m(1-\nu_m)}
\end{bmatrix}$$

$$(B.10)$$

For unidirectional composites with long circular cylindrical fibers the following expressions apply

$$\begin{aligned}
S_1 &= S_4 = S_8 = S_9 = 0 \\
S_2 &= S_3 \\
S_5 &= S_6
\end{aligned}$$

$$(B.11)$$

The series S_3, S_6, and S_7 can be expressed with these parabolic expressions

$$\begin{aligned}
S_3 &= 0.49247 - 0.47603\,V_f - 0.02748\,V_f^2 \\
S_6 &= 0.36844 - 0.14944\,V_f - 0.27152\,V_f^2 \\
S_7 &= 0.12346 - 0.32035\,V_f + 0.23517\,V_f^2
\end{aligned}$$

$$(B.12)$$

where V_f is the fiber volume fraction. The MATLAB code `FEAcomp_ExC1.m` implements these equations. It performs the calculations using material properties from `materialC1.dat` and writes the results to `materialC1_results.dat`.

The average strain in each phase $(i) = (f, m)$ can be found in terms of the far field strain ε^0 as

$$\bar{\varepsilon}^{(i)} = \mathbf{A}^i \varepsilon^0$$

$$(B.13)$$

from which the average stress in each phase can be found as

$$\bar{\sigma}^{(i)} = \mathbf{C}^i \mathbf{A}^i \varepsilon^0$$

$$(B.14)$$

with $\mathbf{A}^i$ given by (B.4, B.6).

Appendix C

Second-Order Diagonal Damage Models

Explicit expressions associated to second-order diagonal damage models are presented here for completeness.

C.1 Effective and Damaged Spaces

A *second-order damage tensor* can be represented as a diagonal tensor (see (8.61))

$$D_{ij} = d_i \, \delta_{ij} \quad ; \quad \text{non sum on } i \tag{C.1}$$

in a coordinate system coinciding with the principal directions of $\mathbf{D}$, which may coincide with the fiber, transverse, and thickness directions, and d_i are the eigenvalues of the damage tensor, which represent the damage ratio along these directions. The dual variable of the damage tensor is the *integrity tensor*, $\mathbf{\Omega} = \sqrt{\mathbf{I} - \mathbf{D}}$, which represents the undamaged ratio.

The *second-order damage tensor* $\mathbf{D}$ and the *integrity tensor* $\mathbf{\Omega}$ are diagonal and have the following explicit forms

$$D_{ij} = \begin{bmatrix} d_1 & 0 & 0 \\ 0 & d_2 & 0 \\ 0 & 0 & d_3 \end{bmatrix} \tag{C.2}$$

$$\Omega_{ij} = \begin{bmatrix} \sqrt{1-d_1} & 0 & 0 \\ 0 & \sqrt{1-d_2} & 0 \\ 0 & 0 & \sqrt{1-d_3} \end{bmatrix} = \begin{bmatrix} \Omega_1 & 0 & 0 \\ 0 & \Omega_2 & 0 \\ 0 & 0 & \Omega_3 \end{bmatrix} \tag{C.3}$$

A symmetric fourth-order tensor, $\mathbf{M}$, called the *damage effect tensor*, is defined (see (8.63)) as

$$M_{ijkl} = \frac{1}{2} \left(\Omega_{ik}\Omega_{jl} + \Omega_{il}\Omega_{jk} \right) \tag{C.4}$$

The *damage effect tensor* in contracted form multiplied by the Reuter matrix, takes the form of a 6×6 array as follows

$$\mathbf{M} = \mathbf{M}_{\alpha\beta} = \begin{bmatrix} \Omega_1^2 & 0 & 0 & 0 & 0 & 0 \\ 0 & \Omega_2^2 & 0 & 0 & 0 & 0 \\ 0 & 0 & \Omega_3^2 & 0 & 0 & 0 \\ 0 & 0 & 0 & \Omega_2\Omega_3 & 0 & 0 \\ 0 & 0 & 0 & 0 & \Omega_1\Omega_3 & 0 \\ 0 & 0 & 0 & 0 & 0 & \Omega_1\Omega_2 \end{bmatrix} \quad (C.5)$$

The damaged stiffness tensor $\mathbf{C}$ multiplied by the Reuter matrix can be written in explicit contracted notation for an orthotropic material by a 6×6 array as a function of the undamaged stiffness tensor $\overline{\mathbf{C}}$ as follows

$$\mathbf{C}_{\alpha\beta} = \begin{bmatrix} \overline{C}_{11}\Omega_1^4 & \overline{C}_{12}\Omega_1^2\Omega_2^2 & \overline{C}_{13}\Omega_1^2\Omega_3^2 & 0 & 0 & 0 \\ \overline{C}_{12}\Omega_1^2\Omega_2^2 & \overline{C}_{22}\Omega_2^4 & \overline{C}_{23}\Omega_2^2\Omega_3^2 & 0 & 0 & 0 \\ \overline{C}_{13}\Omega_1^2\Omega_3^2 & \overline{C}_{23}\Omega_2^2\Omega_3^2 & \overline{C}_{33}\Omega_3^4 & 0 & 0 & 0 \\ 0 & 0 & 0 & 2\overline{C}_{44}\Omega_2^2\Omega_3^2 & 0 & 0 \\ 0 & 0 & 0 & 0 & 2\overline{C}_{55}\Omega_1^2\Omega_3^2 & 0 \\ 0 & 0 & 0 & 0 & 0 & 2\overline{C}_{66}\Omega_1^2\Omega_2^2 \end{bmatrix}$$
$$(C.6)$$

where $\overline{C}_{44} = \overline{G}_{23}$, $\overline{C}_{55} = \overline{G}_{13}$ and $\overline{C}_{66} = \overline{G}_{12}$. The Voigt contracted notation for fourth-order elasticity tensors is used here: $C_{\alpha\beta}$ replaces C_{ijkl} where α, β take the values 1, 2, 3, 4, 5, 6, corresponding to the index pairs 11, 22, 33, 23, 13 and 12, respectively.

The relations between the effective and actual stress components assume the following expressions

$$\begin{array}{ll} \overline{\sigma}_1 = \sigma_1 \, \Omega_1^{-2}; & \overline{\sigma}_4 = \sigma_4 \, \Omega_2^{-1}\Omega_3^{-1}; \\ \overline{\sigma}_2 = \sigma_2 \, \Omega_2^{-2}; & \overline{\sigma}_5 = \sigma_5 \, \Omega_1^{-1}\Omega_3^{-1}; \\ \overline{\sigma}_3 = \sigma_3 \, \Omega_3^{-2}; & \overline{\sigma}_6 = \sigma_6 \, \Omega_1^{-1}\Omega_2^{-1}; \end{array} \quad (C.7)$$

and the strain components

$$\begin{array}{ll} \overline{\varepsilon}_1 = \varepsilon_1 \, \Omega_1^2; & \overline{\varepsilon}_4 = \varepsilon_4 \, \Omega_2\Omega_3; \\ \overline{\varepsilon}_2 = \varepsilon_2 \, \Omega_2^2; & \overline{\varepsilon}_5 = \varepsilon_5 \, \Omega_1\Omega_3; \\ \overline{\varepsilon}_3 = \varepsilon_3 \, \Omega_3^2; & \overline{\varepsilon}_6 = \varepsilon_6 \, \Omega_1\Omega_2; \end{array} \quad (C.8)$$

where the over-line indicates an effective property.

C.2   Thermodynamic Force Y

By satisfying the Clausius-Duhem inequality, thus assuring non-negative dissipation, the following thermodynamic forces (see (8.127)) are defined

$$Y_{ij} = -\frac{\partial\psi}{\partial D_{ij}} = -\frac{1}{2}\left(\varepsilon_{kl} - \varepsilon_{kl}^p\right)\frac{\partial C_{klpq}}{\partial D_{ij}}\left(\varepsilon_{pq} - \varepsilon_{pq}^p\right) = -\frac{1}{2}\varepsilon_{kl}^e\frac{\partial C_{klpq}}{\partial D_{ij}}\varepsilon_{pq}^e \quad (C.9)$$

The second-order tensor of the conjugate thermodynamic forces associated to the damage variables takes the following form

$$\mathbf{Y} = Y_{ij} = \begin{bmatrix} Y_{11} & 0 & 0 \\ 0 & Y_{22} & 0 \\ 0 & 0 & Y_{33} \end{bmatrix} \tag{C.10}$$

or in Voigt contracted notation as

$$\mathbf{Y} = Y_\alpha = \{Y_{11}, Y_{22}, Y_{33}, 0, 0, 0\}^T \tag{C.11}$$

Using (9.8), the explicit expressions for the thermodynamic forces written in terms of effective strain are found as

$$
\begin{aligned}
Y_{11} &= \frac{1}{\Omega_1^2} \left(\overline{C}_{11} \bar{\varepsilon}_1{}^2 + \overline{C}_{12} \bar{\varepsilon}_2 \bar{\varepsilon}_1 + \overline{C}_{13} \bar{\varepsilon}_3 \bar{\varepsilon}_1 + 2\overline{C}_{55} \bar{\varepsilon}_5{}^2 + 2\overline{C}_{66} \bar{\varepsilon}_6{}^2 \right) \\
Y_{22} &= \frac{1}{\Omega_2^2} \left(\overline{C}_{22} \bar{\varepsilon}_2{}^2 + \overline{C}_{12} \bar{\varepsilon}_2 \bar{\varepsilon}_1 + \overline{C}_{23} \bar{\varepsilon}_3 \bar{\varepsilon}_2 + 2\overline{C}_{44} \bar{\varepsilon}_4{}^2 + 2\overline{C}_{66} \bar{\varepsilon}_6{}^2 \right) \\
Y_{33} &= \frac{1}{\Omega_3^2} \left(\overline{C}_{33} \bar{\varepsilon}_3{}^2 + \overline{C}_{13} \bar{\varepsilon}_3 \bar{\varepsilon}_1 + \overline{C}_{23} \bar{\varepsilon}_3 \bar{\varepsilon}_2 + 2\overline{C}_{44} \bar{\varepsilon}_4{}^2 + 2\overline{C}_{55} \bar{\varepsilon}_5{}^2 \right)
\end{aligned} \tag{C.12}
$$

The thermodynamic forces written in terms of actual stress are

$$
\begin{aligned}
Y_{11} &= \frac{1}{\Omega_1^2} \left(\frac{\overline{S}_{11}}{\Omega_1^4} \sigma_1{}^2 + \frac{\overline{S}_{12}}{\Omega_1^2 \Omega_2^2} \sigma_2 \sigma_1 + \frac{\overline{S}_{13}}{\Omega_1^2 \Omega_3^2} \sigma_3 \sigma_1 + \frac{2\overline{S}_{55}}{\Omega_1^2 \Omega_3^2} \sigma_5{}^2 + \frac{2\overline{S}_{66}}{\Omega_1^2 \Omega_2^2} \sigma_6{}^2 \right) \\
Y_{22} &= \frac{1}{\Omega_2^2} \left(\frac{\overline{S}_{22}}{\Omega_2^4} \sigma_2{}^2 + \frac{\overline{S}_{12}}{\Omega_2^2 \Omega_1^2} \sigma_2 \sigma_1 + \frac{\overline{S}_{23}}{\Omega_2^2 \Omega_3^2} \sigma_3 \sigma_2 + \frac{2\overline{S}_{44}}{\Omega_2^2 \Omega_3^2} \sigma_4{}^2 + \frac{2\overline{S}_{66}}{\Omega_2^2 \Omega_1^2} \sigma_6{}^2 \right) \\
Y_{33} &= \frac{1}{\Omega_3^2} \left(\frac{\overline{S}_{33}}{\Omega_3^4} \sigma_3{}^2 + \frac{\overline{S}_{13}}{\Omega_3^2 \Omega_1^2} \sigma_3 \sigma_1 + \frac{\overline{S}_{23}}{\Omega_3^2 \Omega_2^2} \sigma_3 \sigma_2 + \frac{2\overline{S}_{44}}{\Omega_3^2 \Omega_2^2} \sigma_4{}^2 + \frac{2\overline{S}_{55}}{\Omega_3^2 \Omega_1^2} \sigma_5{}^2 \right)
\end{aligned} \tag{C.13}
$$

The derivative of the thermodynamic forces with respect to the damage $(\partial \mathbf{Y}/\partial \mathbf{D})$ is given by

$$
\frac{\partial \mathbf{Y}}{\partial \mathbf{D}} = \begin{bmatrix}
\dfrac{Y_{11}}{\Omega_1^4} & 0 & 0 & 0 & 0 & 0 \\
0 & \dfrac{Y_{22}}{\Omega_2^4} & 0 & 0 & 0 & 0 \\
0 & 0 & \dfrac{Y_{33}}{\Omega_3^4} & 0 & 0 & 0 \\
0 & 0 & 0 & 0 & 0 & 0 \\
0 & 0 & 0 & 0 & 0 & 0 \\
0 & 0 & 0 & 0 & 0 & 0
\end{bmatrix} \tag{C.14}
$$

The derivative of the thermodynamic forces with respect to the actual strain is given by

$$
\frac{\partial \mathbf{Y}}{\partial \boldsymbol{\varepsilon}} =
\begin{bmatrix}
-\dfrac{P_{11}}{\Omega_1^2} & -\dfrac{\overline{C}_{12}\,\overline{\varepsilon}_1}{\Omega_1^2} & -\dfrac{\overline{C}_{13}\,\overline{\varepsilon}_1}{\Omega_1^2} & 0 & -2\dfrac{\overline{C}_{55}\,\overline{\varepsilon}_5}{\Omega_1^2} & -2\dfrac{\overline{C}_{66}\,\overline{\varepsilon}_6}{\Omega_1^2} \\[2mm]
-\dfrac{\overline{C}_{12}\,\overline{\varepsilon}_2}{\Omega_2^2} & -\dfrac{P_{22}}{\Omega_2^2} & -\dfrac{\overline{C}_{23}\,\overline{\varepsilon}_2}{\Omega_2^2} & -2\dfrac{\overline{C}_{44}\,\overline{\varepsilon}_4}{\Omega_2^2} & 0 & -2\dfrac{\overline{C}_{66}\,\overline{\varepsilon}_6}{\Omega_2^2} \\[2mm]
-\dfrac{\overline{C}_{13}\,\overline{\varepsilon}_3}{\Omega_3^2} & -\dfrac{\overline{C}_{23}\,\overline{\varepsilon}_3}{\Omega_3^2} & -\dfrac{P_{33}}{\Omega_3^2} & -2\dfrac{\overline{C}_{44}\,\overline{\varepsilon}_4}{\Omega_3^2} & -2\dfrac{\overline{C}_{55}\,\overline{\varepsilon}_5}{\Omega_3^2} & 0 \\[2mm]
0 & 0 & 0 & 0 & 0 & 0 \\[2mm]
0 & 0 & 0 & 0 & 0 & 0 \\[2mm]
0 & 0 & 0 & 0 & 0 & 0
\end{bmatrix}
\tag{C.15}
$$

where

$$
\begin{aligned}
P_{11} &= 2\overline{C}_{11}\,\overline{\varepsilon}_1 + \overline{C}_{12}\,\overline{\varepsilon}_2 + \overline{C}_{13}\,\overline{\varepsilon}_3 \\
P_{22} &= \overline{C}_{12}\,\overline{\varepsilon}_1 + 2\overline{C}_{22}\,\overline{\varepsilon}_2 + \overline{C}_{23}\,\overline{\varepsilon}_3 \\
P_{33} &= \overline{C}_{13}\,\overline{\varepsilon}_1 + \overline{C}_{23}\,\overline{\varepsilon}_2 + 2\overline{C}_{33}\,\overline{\varepsilon}_3
\end{aligned}
\tag{C.16}
$$

The derivative of the thermodynamic forces with respect to the actual unrecoverable strain is given by

$$
\frac{\partial \mathbf{Y}}{\partial \boldsymbol{\varepsilon}^p} = -\frac{\partial \mathbf{Y}}{\partial \boldsymbol{\varepsilon}}
\tag{C.17}
$$

The derivative of the actual stress with respect to damage is given by

$$
\frac{\partial \boldsymbol{\sigma}}{\partial \mathbf{D}} =
\begin{bmatrix}
P'_{11} & 0 & 0 & 0 & 0 & 0 \\[2mm]
0 & P'_{22} & 0 & 0 & 0 & 0 \\[2mm]
0 & 0 & P'_{33} & 0 & 0 & 0 \\[2mm]
0 & -\dfrac{1}{2}\dfrac{\Omega_3\overline{C}_{44}\,\overline{\varepsilon}_4}{\Omega_2} & -\dfrac{1}{2}\dfrac{\Omega_2\overline{C}_{44}\,\overline{\varepsilon}_4}{\Omega_3} & 0 & 0 & 0 \\[2mm]
-\dfrac{1}{2}\dfrac{\Omega_3\overline{C}_{55}\,\overline{\varepsilon}_5}{\Omega_1} & 0 & -\dfrac{1}{2}\dfrac{\Omega_1\overline{C}_{55}\,\overline{\varepsilon}_5}{\Omega_3} & 0 & 0 & 0 \\[2mm]
-\dfrac{1}{2}\dfrac{\Omega_2\overline{C}_{66}\,\overline{\varepsilon}_6}{\Omega_1} & -\dfrac{1}{2}\dfrac{\Omega_1\overline{C}_{66}\,\overline{\varepsilon}_6}{\Omega_2} & 0 & 0 & 0 & 0
\end{bmatrix}
\tag{C.18}
$$

where

$$
\begin{aligned}
P'_{11} &= -\overline{C}_{11}\,\overline{\varepsilon}_1 - \overline{C}_{12}\,\overline{\varepsilon}_2 - \overline{C}_{13}\,\overline{\varepsilon}_3 \\
P'_{22} &= -\overline{C}_{12}\,\overline{\varepsilon}_1 - \overline{C}_{22}\,\overline{\varepsilon}_2 - \overline{C}_{23}\,\overline{\varepsilon}_3 \\
P'_{33} &= -\overline{C}_{13}\,\overline{\varepsilon}_1 - \overline{C}_{23}\,\overline{\varepsilon}_2 - \overline{C}_{33}\,\overline{\varepsilon}_3
\end{aligned}
\tag{C.19}
$$

C.3   Damage Surface

An anisotropic damage criterion expressed in tensorial form, introducing two fourth-order tensors, $\mathbf{B}$ and $\mathbf{J}$ defines a multiaxial limit surface in the thermodynamic force space, $\mathbf{Y}$, that bounds the damage domain. The damage evolution is defined by a damage potential associate to the damage surface and by an isotropic hardening function. The proposed damage surface g^d is given (see (9.13)) by

$$
g^d = \left(\hat{Y}_{ij}^N J_{ijhk} \hat{Y}_{hk}^N\right)^{1/2} + \left(Y_{ij}^S B_{ijhk} Y_{hk}^S\right)^{1/2} - \left(\gamma(\delta) + \gamma_0\right)
\tag{C.20}
$$

where γ_0 is the initial damage threshold value and $\gamma(\delta)$ defines the hardening.

The derivative of the damage surface with respect to thermodynamic forces is given by

$$\frac{\partial g^d}{\partial \mathbf{Y}} = \begin{bmatrix} \dfrac{J_{11}\, Y_{11}^N}{\Phi^N} + \dfrac{B_{11}\, Y_{11}^S}{\Phi^S} \\[2mm] \dfrac{J_{22}\, Y_{22}^N}{\Phi^N} + \dfrac{B_{22}\, Y_{22}^S}{\Phi^S} \\[2mm] \dfrac{J_{33}\, Y_{33}^N}{\Phi^N} + \dfrac{B_{33}\, Y_{33}^S}{\Phi^S} \\[2mm] 0 \\ 0 \\ 0 \end{bmatrix} \qquad (C.21)$$

where

$$\begin{aligned} \Phi^N &= \sqrt{J_{11}\,(Y_{11}^N)^2 + J_{22}\,(Y_{22}^N)^2 + J_{33}\,(Y_{33}^N)^2} \\ \Phi^S &= \sqrt{B_{11}\,(Y_{11}^S)^2 + B_{22}\,(Y_{22}^S)^2 + B_{33}\,(Y_{33}^S)^2} \end{aligned} \qquad (C.22)$$

The derivative of the damage surface with respect to damage hardening is

$$\frac{\partial g^d}{\partial \gamma} = -1 \qquad (C.23)$$

C.4 Unrecoverable-Strain Surface

The unrecoverable-strain (yield) surface g^p is a function of the thermodynamic forces in the effective configuration $(\overline{\sigma}, R)$. Therefore, the unrecoverable-strain surface (see (9.19)) is

$$g^p = \sqrt{f_{ij}\overline{\sigma}_i\overline{\sigma}_j + f_i\overline{\sigma}_i} - (R\,(p) + R_0) \qquad (C.24)$$

where $(i = 1, 2...6)$, R_0 is the initial unrecoverable-strain threshold and R is the hardening function.

The derivative of the unrecoverable-strain surface with respect to effective stress is given by

$$\frac{\partial \mathbf{g^p}}{\partial \overline{\sigma}} = \begin{bmatrix} \dfrac{1}{2}\,\dfrac{f_1 + 2\,f_{11}\,\overline{\sigma}_2 + 2\,f_{12}\,\overline{\sigma}_2 + 2\,f_{13}\,\overline{\sigma}_3}{\Phi^p} \\[3mm] \dfrac{1}{2}\,\dfrac{f_2 + 2\,f_{22}\,\overline{\sigma}_1 + 2\,f_{12}\,\overline{\sigma}_1 + 2\,f_{23}\,\overline{\sigma}_3}{\Phi^p} \\[3mm] \dfrac{1}{2}\,\dfrac{f_3 + 2\,f_{33}\,\overline{\sigma}_3 + 2\,f_{13}\,\overline{\sigma}_1 + 2\,f_{23}\,\overline{\sigma}_1}{\Phi^p} \\[3mm] \dfrac{f_4\,\overline{\sigma}_4}{\Phi^p} \\[3mm] \dfrac{f_5\,\overline{\sigma}_5}{\Phi^p} \\[3mm] \dfrac{f_6\,\overline{\sigma}_6}{\Phi^p} \end{bmatrix} \qquad (C.25)$$

where
$$\begin{aligned}
\Phi^p \;=\; \big(&f_1\,\overline{\sigma}_1 + f_2\,\overline{\sigma}_2 + f_3\,\overline{\sigma}_3 + \\
&+ f_{11}\,\overline{\sigma}_1{}^2 + f_{22}\,\overline{\sigma}_2{}^2 + f_{33}\,\overline{\sigma}_3{}^2 + \\
&+ 2\,f_{12}\,\overline{\sigma}_1\,\overline{\sigma}_2 + 2\,f_{13}\,\overline{\sigma}_1\,\overline{\sigma}_3 + 2\,f_{23}\,\overline{\sigma}_2\,\overline{\sigma}_3 + \\
&+ f_6\,\overline{\sigma}_6{}^2 + f_5\,\overline{\sigma}_5{}^2 + f_4\,\overline{\sigma}_4{}^2 \big)^{1/2}
\end{aligned} \tag{C.26}$$

The derivative of the yield surface with respect to unrecoverable-strain hardening is
$$\frac{\partial g^p}{\partial R} = -1 \tag{C.27}$$

Appendix D

Numerical Inverse Laplace Transform

The following is a description of a numerical method to obtain discrete values in the time domain for the time domain function $f(t)$ from the Laplace function $f(s)$, or $f(t) = L^{-1}[f(s)]$. The method falls in the category of collocation methods [7]. The solution is approximated in terms of Legendre polynomials P_N of order N, that are orthogonal in the interval $(-1, 1)$, that is

$$\int_{-1}^{1} P_i P_j dx = 0 \quad if \quad i \neq j \tag{D.1}$$

For example, the Legendre polynomial of order $N = 5$ is [8]

$$P_5 = \frac{63}{8}x^5 - \frac{35}{4}x^3 + \frac{15}{8}x \tag{D.2}$$

(A) The shape functions used are defined as

$$w_i = \int_{-1}^{1} \left\{ P_N / \left((x - x_i) \left[\frac{dP_N}{dx} \right]_{x=x_i} \right) \right\} dx \tag{D.3}$$

with x_i being the roots of P_N.

(B) For convenience, the roots and shape functions are shifted to the interval $(0, 1)$, so that

$$x_i' = (x_i - 1)/2 \tag{D.4}$$
$$w_i' = w_i/2$$

The roots, expressed in terms of time are

$$t_i = -\ln(x_i') \tag{D.5}$$

(C) The time must be scaled in the interval $(0, t_{\max})$ with $t_{\max} = \max(t_i)$.

If the maximum time for which a solution is sought is t_N, then the roots are adjusted as

$$t'_i = ct_i \tag{D.6}$$

with the time-scale constant

$$c = t_N/t_{max} \tag{D.7}$$

The collocation method [7] results in a system of algebraic equations

$$K_{ij}f_j = F_i \tag{D.8}$$

with the coefficients of the matrix $[K]$ given by

$$K_{ij} = \left(x'_j\right)^{(i-1)} w'_j \tag{D.9}$$

Denoting by $f(s)$ the function to be back-transformed from the Laplace domain to the time domain, the components of the vector $\{F\}$ are given by

$$F_i = \frac{1}{c}f\left(\frac{i}{c}\right) \tag{D.10}$$

that is replacing s by i/c. The next step is to solve to get the values f_i of $f(t)$ at the times given by t'_i. There are N points of the solution of $f(t) = L^{-1}[f(s)]$. One more point can be obtained at $t = 0$ using the initial value theorem (7.44). Now, fit the numerical solution with a power law of the form (7.26). For $t = 0$, $D_o = f(0)$, then

$$f(t) - D_o = D_1 t^m \tag{D.11}$$

Taking the natural log

$$y = a + mx \tag{D.12}$$

with $y = \ln(f(t) - D_o)$ and $x = \ln(t)$. A linear regression is then used to obtain the parameters a and m in terms of the discrete values $y_i = \ln(f(t_i) - D_o)$ and $x_i = \ln(t_i)$, as

$$\bar{x} = \frac{1}{n}\sum x_i$$
$$\bar{y} = \frac{1}{n}\sum y_i \tag{D.13}$$
$$m = \frac{\sum y_i(x_i - \bar{x})}{\sum(x_i - \bar{x})^2}$$
$$a = \bar{y} - \eta\bar{x}$$

and

$$D_1 = \exp(a) \tag{D.14}$$

A computer program to compute an approximate solution in the time domain from the Maxwell model in the Laplace domain is available as *collocation.for* in [5]. Since the Maxwell model can be inverted to the time domain exactly, a comparison

Table D.1: K-matrix for $N = 5$

time[s]	1	2	3	4	5
0.693147	0.284444	0.118463	0.118463	0.239314	0.239314
0.048046	0.142222	0.112906	0.005557	0.184088	0.055225
3.059522	0.071111	0.107609	0.00026	0.141607	0.012744
0.262359	0.035555	0.102561	1.22288E-05	0.108929	0.00294
1.466353	0.017777	0.09775	5.73653E-07	0.083792	0.000678

between the approximate solution and the exact one is performed. Also, the program computes the linear regression for the parameters in the power law, and evaluates the power law at the same time values used in the numerical method to compare its accuracy. Note the use of *function* definition at the beginning of the Fortran code for the power law (power), the Maxwell model in time (Jt), and the same in Laplace domain (Js). The shifted time values, not scaled for a particular problem, and the coefficients matrix $[K]$ are listed in Table D.1 for $N = 5$ (collocation.dat, [5]). The solution of the system of equations is performed with the code for Gauss elimination (gauss.for, [5]).

Appendix E

Introduction to the Software Interface

Only four software applications are used throughout this textbook. ANSYS is by far the most used. BMI3 is used only in Chapter 4. MATLAB is used for symbolic as well as numerical computations. Finally, Intel Fortran must be available to compile and link ANSYS with user programmed material subroutines, but its usage is transparent to the user because it is called by a batch file requiring no user intervention. Of course, some knowledge of Fortran is required to program new user material subroutines, but programming is made easier by several example subroutines, which are provided and used in the examples.

The aim of this section is to present an introduction to the software used in this textbook, namely ANSYS and BMI3, as well as how to use Intel Fortran to compile and link user subroutines with ANSYS. It is assumed that the reader can use MATLAB without help besides that provided by the self-explanatory MATLAB code included with the examples, either printed in this textbook or downloadable from the Web site [5].

Operation of the software is illustrated for a Windows XP platform but operation in a Linux environment is very similar. For the sake of space, this section is very brief. The vendors of these applications have a wealth of information, training sessions, user groups, and so on, that the reader can use to get familiar with the software interface. One such source of information is the Web site for this textbook at http://www.mae.wvu.edu/barbero/feacm/. Another source of information is the book's user group at http://tech.groups.yahoo.com/group/feacomposites/.

E.1 ANSYS

ANSYS is a commercial finite element analysis (FEA) application. It has a friendly graphical user interface (GUI) and an extensive help system. Once started, the user should have no difficulty navigating menus and so on. Since all the mouse clicks in the GUI generate ANSYS command lines, which are saved in a .log file, it is easy to use the GUI to learn what the various commands do. The ANSYS help can then be used to enhance the user's knowledge of the ANSYS command structure. Ultimately, a `.log` file that automatically performs all the tasks of model creation, execution, and post-processing is desired because it can be debugged, refined, adapted to similar situations, recalled later on, and it provides excellent documentation for project reporting to the client and even auditing by third parties. The `.log` files provided with the examples in this textbook demonstrate their usefulness, even if the GUI was used to help generate most of them.

The examples in this textbook were produced using ANSYS version 10. They work identically in Windows XP and Linux platforms. On Windows, ANSYS is accessible from the START menu, through two icons: "ANSYS" and "ANSYS Product Launcher". It is best to use at least once the "ANSYS Product Launcher", as it allows us to set the default location for the model files.

In this Appendix, it is assumed that the user has created a folder `c:\ansys\`, where all the model files reside. Therefore, in "ANSYS Product Launcher", under "File Management", the "Working Directory" should be set to `c:\ansys\`. Clicking on `Run` invokes the GUI.

The ANSYS' GUI has a command bar at the top and a menu list on the left. Below the command bar, there is single-line command window. ANSYS commands typed in this window are executed immediately and have the same effect as equivalent GUI operations.

Although the GUI is user friendly, it is very challenging to describe (in a textbook) all the mouse clicks one has to do in order to set up and solve a problem. It is also challenging to remember what one did during a previous session using the GUI. And there is no use trying to write down the myriad mouse clicks needed to accomplish a task. Fortunately, all GUI operations (mouse clicks, menu selections, data entry, and so on) are saved by ANSYS into a `.log` file in the current directory (`c:\ansys`). The `.log` file is a text file that can be edited and cleaned up of the many commands that represent dead ends that one reached during a session. Cleaned up `.log` files can be recalled into ANSYS and executed to reproduce a prior session.

The `.log` files can be recalled in three ways. First, each line in the `.log` file can be typed in the command window and executed one at a time (by pressing `enter`, of course). This is very useful in order to learn the effect that each command line has on the model generation, execution, and so on. Second, a portion or the whole `.log` file can be pasted into the command window and executed. Finally, once a `.log` file is polished, the most computationally efficient way to enter a model is to type the following command `/input,file,log`, in the command window. This will retrieve `file.log` and execute it. The equivalent GUI operation is: `File, Read Input From, OK`.

As was mentioned before, this section is very brief. It has been my experience that students successfully teach themselves ANSYS by figuring out the commands used in the examples in this textbook, which are available on the Web site [5], along with the help system and the documentation included with ANSYS. Video recordings illustrating the execution of the examples by using the ANSYS GUI will be published on the Web site [5] as they become available.

E.1.1 ANSYS USERMAT, Compilation and Execution

Compilation and execution of ANSYS programmable features can be accomplished following the procedure described in this section or as explained in [9]. In this section, it is assumed that ANSYS 10.0, Microsoft Visual Studio .NET 2003 or newer, and FORTRAN Intel 9.1.3 or newer, are all available on a Windows XP system. Note that the path to software components will change with time as new versions and different platforms are released. Therefore, the paths given in this section may have to be adjusted. For example `v100` refers to version 10.0 and `intel` refers to the Intel^{TM} processor, which may have to be adjusted to reflect the software and hardware configuration available to you. This is the procedure that we recommend:

- Create a directory in a local disk, such as `c:\ansys` for the model files (.log and other such files)

- Create a directory in a local disk, such as `c:\ansyscustom` for the Fortran code files (.F and other such files)

- Go to `C:\Program Files\Ansys Inc\v100\ANSYS\custom\user\intel` and copy the following files to `c:\ansyscustom`

  - `ansyslarge.def`
  - `ansyssmall.def`
  - `MAKEFILE`

- From the Web site [5], copy the following files into `c:\ansyscustom`

  - `ANSCUSTWVU.BAT` (this is a modified version of `ANSCUST.BAT`)
  - the user subroutine (e.g., `usermat1d.F` to solve Example 3.13)

- From the Web site [5], copy the following files into `c:\ansys`

  - the `.log` file (e.g., `FEAcomp_Ex313.log`)

- On an explorer window showing `c:\ansyscustom`, double click `ANSCUSTWVU.BAT`. You should respond `N` to the prompt in order to build a large version of the executable file.

  This will compile all .F files in the current directory and link them with ANSYS, thus creating a custom version of ansys.exe that will be moved to the appropriate directory, that is to `C:\Program Files\Ansys Inc\v100\ANSYS\custom\user\intel\ANSYS.exe`. The process takes a long time. Be patient.

- At least the first time, run your custom version of ANSYS as follows:

  - START, ANSYS 10, ANSYS Product Launcher
  - On the File Management tab, set "Working Directory" to c:\ansys
  - On the Customization Preferences tab, set "Custom ANSYS Exe" to C:\Program Files\Ansys Inc\v100\ANSYS\custom\user\intel\ANSYS.exe
  - Run

- This will set the custom version of ANSYS and the working directory until you change them with the ANSYS Product Launcher. Until such time, you need only to run START, ANSYS 10, ANSYS

- In the input field at the top of the screen, the input file (say FEAcomp_Ex313. log) can be read by entering /INPUT,FEAcomp_Ex313,log, which should execute within your customized version of ANSYS. Alternatively, the input file can be read from the GUI left menu panel, as follows: File, Read Input From

Additional information can be obtained from the ANSYS documentation [9] and online sources such as [6].

E.2   BMI3

Most users will run BMI3 within ANSYS as explained in Section E.2.2 but for troubleshooting it is useful to know how to operate it outside ANSYS, as explained in Section E.2.1 next.

E.2.1   Stand Alone BMI3

Native BMI3 code accepts an input file in ABAQUS format, as long as the input file is filtered by the program I2B [5]. Not all of the ABAQUS commands are accepted by I2B. For example it only accepts models with concentrated forces on nodes.
    Most commercial CAD packages such as I-DEAS^{TM} and FEMAP^{TM} can output an ABAQUS file. Then, it is easy to modify the file to make it comply with the restrictions of I2B. Run I2B to generate DEMO.inp, ABAQUS.inp, and DEMO.dat. If ABAQUS.inp were to be executed within ABAQUS, it would give the bifurcation loads $\Lambda^{(cr)}$.
    The material properties and perturbation parameters are in DEMO.dat. The last line contains *modenum, nodenum, component*. This is the mode, node, and component used as perturbation parameter. If all are zeros, BMI3 picks the lowest mode and the node-component combination that yields the largest mode amplitude. The results are printed in DEMO.out and the mode shapes saved in MODES file.

E.2.2   BMI3 within ANSYS

It is possible to use the program BMI3 directly from ANSYS, with some restrictions:

- Use only element type SHELL99.

- Introduce the laminate properties using $ABDH$ matrices, with KEYOPT(2)=2.

- Only apply loads on nodes or keypoints using concentrated forces (do not use moments). If the model has distributed loads, calculate the equivalent nodal forces and apply them at the nodes.

- Use only one real constant set for all the model.

The procedure to compute the post-critical path parameters using BMI3 within ANSYS is described next.

- In the working directory (c:\ansys), copy the APDL macro ans2i.mac, and the programs bmi3.exe and i2b_ans.exe from [5].

- Define the model in ANSYS and solve it using the "Eigenvalue buckling analysis" procedure for obtaining the bifurcation loads $\Lambda^{(cr)}$ (e.g., Example 4.2).

- Run the APDL macro ans2i simply by entering ans2i in the ANSYS command line [5] to calculate parameters of the post-critical path.

- Look for the c:\ANSYS 10.0 Output Window, which is minimized in the Windows task bar, and bring it to the foreground.

- In the c:\ANSYS 10.0 Output Window, respond to the two prompts: (i) activate or not sorting of the nodes in order to minimize the bandwidth of the system of equations (sorting along the longest dimension is recommended), (ii) introduce the *mode*, *node*, and *component* used as perturbation parameter s or let BMI3 choose the default. By default, the lowest *mode* and the *node*, *component*, with the largest mode amplitude is used. If an error message of "INSUFFICIENT STORAGE" appears, try sorting along another direction. If that fails, BMI3 needs to be recompiled with larger arrays.

- In addition to the critical load $\Lambda^{(cr)}$, BMI3 computes the slope $L(1) = \Lambda^{(1)}$ and the curvature $L(2) = \Lambda^{(2)}$ of the bifurcation mode selected. These results are shown in the c:\ANSYS 10.0 Output Window and they are printed in DEMO.out

- Do not close the c:\ANSYS 10.0 Output Window, just minimize it. Otherwise, it will abort ANSYS. ANSYS should be closed from the GUI.

Note that the results (bifurcation loads, slopes, and curvatures) appear with negative sign. This is usual in stability analysis. If a model is constructed with tensile loads (instead of the usual compression), one can type REVERS=-1 in the ANSYS command line before executing the APDL macro ANS2I. Another peculiarity of the BMI3 software is that transverse deflections w (perpendicular to the plate) have opposite sign to that used by ANSYS. Since transverse deflections w are often used as perturbation parameters, the change in sign must be taken into account during interpretation of results.

References

[1] A. Caceres, Local Damage Analysis of Fiber Reinforced Polymer Matrix Composites, Ph.D. dissertation, West Virginia University, Morgantown, WV, 1998.

[2] R. Luciano and E. J. Barbero, Formulas for the Stiffness of Composites with Periodic Microstructure, Int. J. of Solids Structures, 31(21) (1995) 2933-2944.

[3] J. Aboudi, Mechanics of Composite Materials: A Unified Micromechanical Approach, volume 29 of Studies in Applied Mechanics. Elsevier, Amsterdam, 1991.

[4] G. F. Abdelal, A Three-Phase Constitutive Model for Macrobrittle Fatigue Damage of Composites, Ph.D. dissertation, West Virginia University, Morgantown, WV, 2000.

[5] E. J. Barbero, Web resource: http://www.mae.wvu.edu/barbero/feacm/

[6] E. J. Barbero, Web resource: http://tech.groups.yahoo.com/group/feacomposites/

[7] J. N. Reddy, Energy and Variational Methods in Applied Mechanics, Wiley, New York, 1984.

[8] Wikipedia, Legendre Polynomials
URL: http://en.wikipedia.org/wiki/Legendre_polynomial

[9] Guide to ANSYS Programmable Features, ANSYS Inc., Canonsburg, PA, August 2005.

Index

FINITE ELEMENT ANALYSIS OF COMPOSITE MATERIALS

FINITE ELEMENT ANALYSIS
OF COMPOSITE MATERIALS

Ever J. Barbero

Department of Mechanical and Aerospace Engineering
West Virginia University
USA

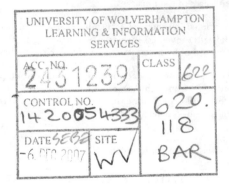

CRC Press
Taylor & Francis Group
Boca Raton   London   New York

CRC Press is an imprint of the
Taylor & Francis Group, an **informa** business

CRC Press
Taylor & Francis Group
6000 Broken Sound Parkway NW, Suite 300
Boca Raton, FL 33487-2742

© 2008 by Taylor & Francis Group, LLC
CRC Press is an imprint of Taylor & Francis Group, an Informa business

---

Library of Congress Cataloging-in-Publication Data

Barbero, Ever J.
   Finite element analysis of composite materials / Ever J. Barbero.
     p. cm.
   ISBN-13: 978-1-4200-5433-0 (alk. paper)
   ISBN-10: 1-4200-5434-1
   1. Composite materials--Mathematical models. 2. Finite element method. I. Title.

TA418.9.C6B368 2007
620.1'18--dc22
                                                    2007007536

Visit the Taylor & Francis Web site at
http://www.taylorandfrancis.com

and the CRC Press Web site at
http://www.crcpress.com